BUSINESS MATHEMATICS IN A CHANGING WORLD

BUSINESS MATHEMATICS IN A CHANGING WORLD

Herbert Berry
Morehead State University

Lila S. Berry

Wm. C. Brown Publishers

Book Team

Editor *Earl McPeek*
Developmental Editor *Janette S. Scotchmer*
Production Editor *Diane Beausoleil*
Visuals/Design Freelance Specialist *Barbara J. Hodgson*
Permissions Editor *Gail Wheatley*
Visuals Processor *Andréa Lopez-Meyer*
Visuals/Design Consultant *Marilyn Phelps*

Cover photo © Stephen F. Grohe/The Picture Cube

Cover and interior design Deborah Schneck

Illustrations Schneck/DePippo Graphics

Copyeditor Deborah Reinbold

Library of Congress Catalog Card Number: 91–70301

ISBN 0–697–11259–4

Printed in the United States of America by Wm. C. Brown Publishers, 2460 Kerper Boulevard, Dubuque, IA 52001

10 9 8 7 6 5 4 3 2 1

To our children with love

Jack J. Berry
Scott F. Berry
Alice J. Berry

Contents

Preface

Business Mathematics in a Changing World will be used by students in community colleges, technical colleges, and business colleges offering an arithmetic-based business mathematics course. The text can also be successfully utilized in four-year institutions that offer the business mathematics course.

A strong formal mathematics background is not a prerequisite. Although algebra is not emphasized, it is introduced early with some equation solving. The evaluation of algebraic expressions by electronic tools, as well as more equation solving, is presented in part 8—Supplementary Topics.

Approach

As the title suggests, we have attempted to reflect real-life business practices as currently as possible, discussing computers, financial packages, NOW accounts, money market accounts, and interest-sensitive instruments. In the past two decades, interest rates have ranged from 6 to 16 percent; banking and credit have evolved more drastically than anyone anticipated. How can students cope in the face of rapid change? Fortunately, the fundamentals of business mathematics remain steady buoys in a turbulent sea. *Business Mathematics in a Changing World* covers the basics in light of historical and current business issues.

Organization

The textbook is organized into parts, chapters, and sections. Organization by parts enhances learning by grouping similar topics together. Parts 1 through 3 deal with commercial arithmetic reasoning based on the four arithmetic operations, the percentage formula, the two percentage variations, the simple interest formula, and the three simple interest variations.

Parts 3 through 5 relate to consumer mathematics. Parts 6 and 7 pertain to financial mathematics. Some formulas must be evaluated on a calculator with exponential key, but tables are provided for students lacking such equipment.

Each part contains three to six chapters. Each chapter deals with an essential subject that is related to a particular area. For example, part 4, Consumer Applications, contains chapters on revolving credit and insurance. Each section introduces a specific subtopic of the subject; a section on revolving credit explains charge accounts. Subdividing in this manner enables the student to assimilate the material more readily.

Features

- Although calculators have revolutionized computing and have been in use for many years, few texts at this level genuinely incorporate them. All students should be able to perform the four arithmetic operations with paper and pencil, but we assume that most students will use a calculator for all but simple values. *Business Mathematics in a Changing World* thoroughly integrates calculator instruction throughout the text, augmenting the traditional formulas with the corresponding calculator sequences.

  Unfortunately, a standard educational calculator does not exist. Many desk calculators employ arithmetic logic for addition and subtraction, while most hand-held calculators employ algebraic logic for these operations. The calculator sequences in this text utilize a common core of operations that can be performed on any inexpensive desk or hand-held machine featuring M-memory. Both arithmetic and algebraic keystroke sequences are given for addition and subtraction.
- Ratio and Proportion are given full chapter treatment so that students can successfully work ratio-related problems.
- Units of Measurement: The Metric System features a new approach. Instead of emphasizing conversion, the chapter provides parallel explanations and exercises in order for students to "think" metric, not merely convert to metric.
- Percentage is covered early, in the four chapters of part 2. Once students have mastered percent, they can successfully complete the remainder of the course.
- Common Statistical Measures and Presentation Graphics shows students how to interpret graphs rather than construct them manually. (Directions are given for computerized construction through Lotus.)
- Each chapter begins with a brief introduction and a list of objectives.
- Most sections begin with a list of key terms. Each section contains practice assignments that allow students to work problems before attempting the tear-out end-of-chapter assignment. Assignment contain both word problems and numeric problems.
- Most practice assignments are preceded by a plan summarizing strategy and requirements for solving the problems.
- In chapters 1 through 29, an optional *Challenger* problem follows the final practice assignment for students seeking further mental stimulation.
- As an aid in referencing chapter material, end-of-chapter summaries in table format display examples, procedures, and formulas. The formulas are shown

in algebra form, as calculator sequences, and as computer expressions. Color and screens distinguish the different forms.

Optional Electronic Spreadsheets

Twenty-two chapters feature a computer spreadsheet exercise related to a chapter topic. In these exercises we emphasize Lotus 1-2-3®, a popular spreadsheet software. To prevent students from being hindered by a host of esoteric commands, a diskette containing partially completed spreadsheets (called templates) is available. The templates are designed for DOS users of Lotus 1-2-3, release 2.01, 2.01S, 2.2, and 2.3. (Release 2.3 requires a hard disk.) These exercises do not merely substitute the computer for the calculator nor do they require trivial data entry. Instead, they enable students to adapt some powerful spreadsheet features—for example, the copying of formulas—to mathematical applications that would be tedious to accomplish even with a calculator.

The use of calculators and spreadsheets is briefly introduced in chapter 5. If computer equipment is not available, the spreadsheet exercises may be omitted without loss of continuity.

Suggested Lesson Plans

Business Mathematics in a Changing World can be used in lecture classes, modified lecture classes in which students complete some exercises during class time, or self-study nonlecture classes where most of the class time is spent on student difficulties.

Subject to time restraints, this text may be adapted to the following levels of courses:

1. Commercial Arithmetic Reasoning (sequence 1)
2. Commercial Arithmetic Reasoning and Consumer Mathematics (sequences 1 and 2)
3. Commercial Arithmetic Reasoning and Mathematics of Finance (sequences 1 and 3)
4. Commercial Arithmetic Reasoning, Consumer Mathematics, and Mathematics of Finance (sequences 1, 2, and 3)

Sequence 1 Commercial Arithmetic Reasoning

Computing with Decimal Numbers	Chapter 1
Fractions: Addition and Subtraction	Chapter 2
Fractions: Multiplication and Division	Chapter 3
Ratio and Proportion: Finding the Unknown	Chapter 4
Electronic Tools	Chapter 5
Units of Measurement: The Metric System	Chapter 6
Percentage	Chapter 7
Trade Discounts	Chapter 8
Cash Discounts and the Invoice	Chapter 9
Markup	Chapter 10
Simple Interest	Chapter 11
Interest Variables	Chapter 12
Payroll	Chapter 20

Sequence 2 Consumer Mathematics

Promissory Notes, Present Value	Chapter 13
Bank Discount	Chapter 14
Installment Plans and Truth-in-Lending	Chapter 15
Revolving Credit	Chapter 16
Checking Accounts	Chapter 17
Stocks and Bonds	Chapter 18
Insurance: Life and Auto	Chapter 19
Sales and Property Taxes	Chapter 21
Property Insurance	Chapter 22

Sequence 3 Mathematics of Finance

Inventory	Chapter 23
Depreciation	Chapter 24
Financial Statements	Chapter 25
Compound Interest, Present and Future Values	Chapter 26
Savings Accounts	Chapter 27
Annuities, Present and Future Values	Chapter 28
Amortization and Sinking Funds	Chapter 29

Material from the supplementary topics in chapters 30 through 33 may be inserted into any of the sequences at the convenience of the instructor and students.

Supplements

The spreadsheet diskette provides students with valuable experience in electronic spreadsheet techniques including the creation and use of formulas.

A practice set (*Electric City*) applies much of the text material to a fictional retail chain and offers students the opportunity for more intensive practice. Students work math problems encountered by this model company.

The TestPak/QuizPak offers about 1,500 additional problems organized according to level of difficulty, objective, and section. A Test Item File contains a printout of all questions found in the TestPak/QuizPak.

Acknowledgments

To insure accuracy, we have consulted individuals and organizations in many areas of expertise, and we gratefully acknowledge their contributions.

Accounting	Professor John W. Osborne Morehead State University Morehead, Kentucky
Bank accounts	Peoples First Bank Morehead, Kentucky
Interest-sensitive insurance policies	C. Roger Lewis Agency Morehead, Kentucky
Discover Financial Services Card, stocks and bonds	Dean Witter Financial Services Group Dean Witter Reynolds, Inc. New York, New York
MasterCard	Exchange Bank of Kentucky Mount Sterling, Kentucky
Stocks and bonds	Merrill Lynch and Company, Inc.
APR tables	Board of Governors Federal Reserve System

With the help of Wm. C. Brown Publishers, we acquired the ideas for chapter 6, Units of Measurement, from the Metric Conversion Committee of the U.S. Department of Commerce.

We would like to express our appreciation to the following reviewers:

Donnie N. Byers
Johnson County Community College

Harold L. Caddell, Ph.D.
California Baptist College

Mark Ciampa
Volunteer State Community College

Don D. Coates
Neosho County Community College

Beth DeWees
Indiana Vocational Technical College

John Gould, M.A.
Garrett Community College

Ann G. Grubbs
Darton College

Jo Ann Hoiles
Raritan Valley Community College

Louise Holcomb
Gainesville College

Marlys Johansen
Blackhawk College

Estelle Kochis
Suffolk Community College

Jan Paprskar
El Camino College

Patrick J. Quinn
University of Wisconsin–Superior

Richard VanBeek
Northern State University

Emma J. Watts
Westark Community College

Mary Etta Williams
Central State University

Randy L. Joyner of East Carolina University was particularly helpful in examining the spreadsheet chapter and exercises.

We would also like to thank Ann Grubbs of Darton College for the accuracy check.

PART 1 FUNDAMENTAL PROCESSES AND COMPUTATIONAL DEVICES

Chapter 1 Computing with Decimal Numbers

OBJECTIVES

- Understand the decimal number system
- Operate on decimal numbers with paper and pencil
- Round whole and mixed numbers

SECTIONS

INTRODUCTION

Good math skills are essential in the daily operations of a successful business. While machines automatically perform many mathematical functions, both simple and complex, a keen grasp of the fundamentals makes it possible to read and understand mathematical data; carry out basic operations; and plan, program, and monitor automated processes. In this chapter, set aside your calculator and computer, and work only with paper and pencil. Working out fundamental arithmetic operations by hand reinforces your knowledge and strengthens your grasp of all numerical concepts, and also helps prevent panic in the absence of calculators or other mathematical tools. You will also find yourself better prepared to take employment tests in which the use of calculators is generally not permitted.

Section 1.1 Decimal Numbers

KEY TERMS **Whole number** **Decimal digit** **Mixed number** **Decimal point** **Decimal fraction** **Place value**

In the decimal system, all numbers are expressed in multiples of ten. (Since ancient times people have counted to ten using their fingers.) Computations in this system can be carried out with paper and pencil, calculator, or computer. (While computers operate on binary (base two) numbers, data is displayed in decimal form.)

0,1,2,3,4,5,6,7,8,9,10,11,12,13, . . . are **whole numbers.**

The first ten whole numbers are the **decimal digits.**

Most calculations produce **mixed numbers** containing a whole number and a fractional part.

The **decimal point** separates a whole number from a fractional part. Digits to the left of the decimal point (if any) comprise a whole number; digits to the right of the decimal point (if any) comprise a fraction.

> 658 is a whole number.
>
> 0,1,2,3,4,5,6,7,8,9 are the ten decimal digits.
>
> 0.875 is a **decimal fraction.**
>
> 564.23 is a mixed number. 564 is the whole part. 23 is the fractional part; it represents $\frac{23}{100}$.
>
> 5,873,975 is large; 0.0000421 is small. Any number, whether small or large, can be written as an arrangement of decimal digits.

Place Values in the Decimal Number System

Hundred thousands	Ten thousands	Thousands	Hundreds	Tens	Ones	Decimal point	Tenths	Hundredths	Thousandths	Ten thousandths	Hundred thousandths
100000	10000	1000	100	10	1	.	1	01	001	0001	00001
100000	10000	1000	100	10	1	.	$\frac{1}{10}$	$\frac{1}{100}$	$\frac{1}{1000}$	$\frac{1}{10000}$	$\frac{1}{100000}$

Place values for 5839421	5 represents	5 times	1000000	which equals	5000000
	8 represents	8 times	100000	which equals	800000
	3 represents	3 times	10000	which equals	30000
	9 represents	9 times	1000	which equals	9000
	4 represents	4 times	100	which equals	400
	2 represents	2 times	10	which equals	20
	1 represents	1 times	1	which equals	1
					5839421
Place values for 784.3612	7 represents	7 times	100	which equals	700
	8 represents	8 times	10	which equals	80
	4 represents	4 times	1	which equals	4
	3 represents	3 times	0.1	which equals	0.3
	6 represents	6 times	0.01	which equals	0.06
	1 represents	1 times	0.001	which equals	0.001
	2 represents	2 times	0.0001	which equals	0.0002
					784.3612

Any number can be written as an arrangement of digits.

Place value assigns a value to each digit.

Commas are often edited into a whole number, separating it into three-digit groups. While commas make a number easier to read, they are neither necessary nor desirable for computation. Some calculators display commas; some do not. Computers may display commas, but ignore them in computation. For numbers containing a decimal point, commas may be inserted into the whole number, to the left of the decimal point but not into the fractional part, to the right of the decimal point. A decimal point has strong mathematical meaning, while commas merely improve readability.

Computational Form	Edited Form
2387	2,387
73509	73,509
5839421	5,839,421
5572.8297	5,572.8297
68492.07	68,492.07

Numbers can also be written in word form. When writing out the numbers from 21 to 99 insert a hyphen. In numbers above 999, separate the thousands, millions, etc., with commas. Use the word "and" to represent the decimal point in **mixed numbers.**

Computational Digital Form	Edited Digital Form	Word Form
8	8	eight
17	17	seventeen
489	489	four hundred eighty-nine
7365	7,365	seven thousand, three hundred sixty-five
73509	73,509	seventy-three thousand, five hundred nine
5688743	5,688,743	five million, six hundred eighty-eight thousand, seven hundred forty-three
65387.3914	65,387.3914	sixty-five thousand, three hundred eighty-seven and three thousand, nine hundred fourteen ten thousandths
435572.829	435,572.829	four hundred thirty-five thousand, five hundred seventy-two and eight hundred twenty-nine thousandths

PRACTICE ASSIGNMENT

1. In the number 6045.99178:

a. the digit in the thousands place is:
b. the digit in the thousandths place is:
c. the digit in the hundreds place is:
d. the digit in the hundredths place is:
e. the digit in the ones place is:
f. the digit in the tenths place is:

2. In the number 7389.25816

a. the digit in the ____________ place is 5.
b. the digit in the ________________ place is 1.
c. the digit 8 is in the _______ place and in the ____________ place.
d. the digit in the thousands place is _____ .
e. the digit in the _________ place is 2.
f. the digit in the ____________________ place is 6.

3. Write the following numbers in edited digital form. (Insert appropriate commas.)

a. 4423.66219:
b. 18603.1170:
c. 136410.6795:
d. 9132.03970:
e. 37291.575014:

4. Write the following numbers in word form.

a. 52.7:

b. 7738.15:

c. 33,820.2252:

d. 0.74621:

5. Write the following numbers in edited digital form.

a. four thousand, seven hundred sixty-two:
b. five hundred eighty-six thousand, seventy-three:

c. three hundred nineteen million, six thousand, five hundred two:
d. sixty-two thousand, one hundred forty and three thousand eighty-one ten thousandths:

6. Write

$$7 \times 1000000 + 3 \times 100000 + 4 \times 10000 + 9 \times 1000 + 0 \times 100 + 8 \times 10 + 1 \times 1$$

a. in computational digital form:
b. in edited digital form:
c. in word form:

Section 1.2 Addition

KEY TERMS **Sum** **Carry digits** **Total**

Remember, in this chapter you have been asked to set aside your calculators and computers and rely only on paper-and-pencil computations. This practice will strengthen your grasp of all numerical concepts and form a strong base for computer competence.

To find the **sum** of two or more numbers, use addition.

Consider the addition problem:

753 + 2682 + 1961.

Align the numbers vertically from the right into four columns. Add the digits in each column, beginning with the first column from the right.

21	**carry digits**	
753	3 + 2 + 1 = 6	write 6 under the ones place
2682	5 + 8 + 6 = 19	write 9 under the tens place, carry 1
1961	1 + 7 + 6 + 9 = 23	write 3 under the hundreds place, carry 2
	2 + 0 + 2 + 1 = 5	write 5 under the thousands place
5396	sum or **total**	

When adding mixed numbers, align the digits by the decimal point, because digits possessing a particular place value must be added to digits possessing the same place value. When you vertically align the decimal point, you align the place values. Notice in the box to the right that the digits 0, 2, 3, and 5 are in the hundredths place and are added together.

15.0004
0.82
38.9
208.333
1.75
264.8034

It is helpful to expand the fractional part of each mixed number to the same number of digits by appending (attaching) zeros to the right. Now the column is right justified, and the decimal points remain aligned.

112 1	**carry digits**	
83.2000	0 + 8 + 0 + 0 = 8	write 8
142.4588	0 + 8 + 0 + 1 = 9	write 9
0.5400	0 + 5 + 4 + 5 = 14	write 4, carry 1
28.9510	1 + 2 + 4 + 5 + 9 = 21	write 1, carry 2
	2 + 3 + 2 + 0 + 8 = 15	write 5, carry 1
255.1498	1 + 8 + 4 + 0 + 2 = 15	write 5, carry 1
	1 + 0 + 1 + 0 + 0 = 2	write 2

PRACTICE ASSIGNMENT

Plan: Compute with paper and pencil only.

1. Write each set of numbers in a vertically aligned column and compute the sum.

a. 7642 + 948 + 7250 + 76313

b. 108300 + 42762 + 205775 + 3445

c. 23.907 + 7.858 + 210.44 + 877 + 9452.9

d. 528.38 + 995.202 + 2005 + 250.777 + 1000.02

2. Crazy Clyde's, a used car dealership, recorded the following sales for a four-year period:

1st year	2d year	3d year	4th year
$82,077	$259,453	$238,979	$431,670

What were the total sales for this period?

3. The Tenth Street Warehouse has 485,715 cubic feet of storage space on the first floor, 372,983 on the second floor, 369,054 on the third floor, 277,481 on the fourth floor, and 317,552 in the basement. What is the total number of cubic feet in the warehouse?

4. Rosie's Restaurant pays the following weekly gross salaries to its employees:

$350.35 to the head cook
$315.15 to another cook
$118.25 to a table server
$121.13 to another table server
$156.00 to a dishwasher
$ 98.65 to a busperson
$101.25 to a part-time cashier

What is the total weekly payroll?

Section 1.3 Subtraction

KEY TERMS **Difference** **Borrowing**

To find the **difference** between two numbers, use subtraction. Vertically align two whole numbers into a right-justified column, so that ones are beneath ones, tens are beneath tens, and so forth.

Consider the subtraction problem: 5793 − 3452.

```
         3 − 2 = 1   write 1 under the ones place
         9 − 5 = 4   write 4 under the tens place
  5793   7 − 4 = 3   write 3 under the hundreds place
− 3452   5 − 3 = 2   write 2 under the thousands place
  2341   difference
```

When a digit of the lower number is greater than the corresponding digit of the upper number, take ten from the next higher place (to the left). This process is called **borrowing.**

Now consider the subtraction problem: 5793 − 3956.

```
                3 − 6 cannot be done; borrow 10 from the next place to the left (the tens place).
  4 17 8 13    13 − 6 = 7
  5  7 9  3     8 − 5 = 3
− 3  9 5  6     7 − 9 cannot be done; borrow 10 from the place to the left (the thousands place).
  1  8 3  7    17 − 9 = 8
                4 − 3 = 1
```

When subtracting one mixed number from another, digits possessing a particular place value must be subtracted from digits possessing the same place value. As in addition, vertically aligning the decimal points aligns the place values. Notice that 7 and 2 are in the tenths place; subtract 2 from 7.

```
               9 − 5 = 4
  4 8 . 7 9    7 − 2 = 5
−   3 . 2 5    8 − 3 = 5
  4 5 . 5 4    4 − 0 = 4
```

Remember to expand the fractional parts of each mixed number by appending zeros so that the two fractional parts have the same number of digits. This right justifies the two mixed numbers, and the decimal points remain aligned.

```
               34.6400
             −  1.7538
Difference     32.8862
```

235.89 − 83.725

```
 1 13        8 10
 2  3 5 . 8 9  0
−   8 3 . 7 2  5
 1  5 2 . 1 6  5
```

0 − 5 cannot be done; borrow 10 from the place to the left.
10 − 5 = 5
8 − 2 = 6
8 − 7 = 1
5 − 3 = 2
3 − 8 cannot be done; borrow 10 from the place to the left.
13 − 8 = 5
1 − 0 = 1

PRACTICE ASSIGNMENT

Plan: Compute with paper and pencil only.

1. Write each set of numbers in a vertically aligned column and compute the difference.

   a. 78000 − 8982

   b. 17.7207 − 14.64

   c. 532.49 − 87.50

   d. 0.0295 − 0.0059

   e. 135 − 118.25

2. Jean Shin invested $3,015.42 in 200 shares of MCI stock. Two months later the value of the 200 shares rose to $3,405.95, and Jean sold them. How much did she gain on her investment?

3. Jack Berryman received a scholarship worth $2,000 for each of his last two undergraduate years. In each of those two years he earned $3,640.15 working weekends at a local radio station. His yearly expenses were $1,050.25 for the meal plan, $1,500 for the dormitory, $300 for books and supplies, and $650 for automobile insurance and maintenance. Did Jack have any money left after expenses for entertainment and other incidentals? If so, how much annually?

4. The Electromat Company grossed $181,995.50 in profit on vacuum cleaners sold by door-to-door salespersons. The company paid the following salaries and commissions:

Salesperson	Salary	Commission	Total
R. Anderson	$6,339.25	$3,226.83	______
J. Cioffi	6,574.50	4,074.64	______
F. McGrath	7,183.00	4,604.51	______
L. Silberstein	6,953.75	4,396.08	______
M. Torres	6,395.25	4,319.55	______
Total	______	______	______

What was Electromat's profit after sales salaries and commissions were paid? $127,928.14

Section 1.4 Multiplication

KEY TERMS **Multiplicand** **Multiplier** **Product**

Multiplication is a convenient substitute for repeated addition. The number that is multiplied by another number is called the **multiplicand.** You multiply the multiplicand by the **multiplier.** The result of multiplication is called the **product.**

multiplicand × multiplier = product

When multiplying mixed numbers remember:

Number of decimal places* in product
equals
Number of decimal places in multiplicand
plus
Number of decimal places in multiplier

*In common usage, "decimal places" refers to the places to the right of the decimal point (fractional decimal places).

When a zero is in the multiplier, instead of writing an all-zero partial product, shift the next partial product two places to the left rather than one place.

Consider the multiplication problem: 715 × 83.

```
                                                715
                                             ×   83
First partial product 3 × 715                  2145
Second partial product 8 × 715 shifted left once 5720
Sum of partial products = product =           59345
                                    Edited   59,345
```

Shifting the digits to the left multiplies by 10. In this case, 715 × 83 = 715 × 3 + 715 × 80. You create the second partial product by multiplying by 8.

First partial product:

3 × 5 = 15	write 5, carry 1
3 × 1 + 1 = 4	write 4
3 × 7 = 21	write 1, carry 2
	write 2

Second partial product:

8 × 5 = 40	write 0, carry 4
8 × 1 + 4 = 12	write 2, carry 1
8 × 7 + 1 = 57	write 7, carry 5
	write 5

```
                                                  352.34
                                                ×    4.8
First partial product 8 × 35234                   281872
Second partial product 4 × 35234 shifted once    140936
Sum of partial products                          1691232
2 places + 1 place = 3 places          Product = 1691.232
```

First partial product:

8 × 4 = 32	write 2, carry 3
8 × 3 + 3 = 27	write 7, carry 2
8 × 2 + 2 = 18	write 8, carry 1
8 × 5 + 1 = 41	write 1, carry 4
8 × 3 + 4 = 28	write 8, carry 2
	write 2

Second partial product:

4 × 4 = 16	write 6, carry 1
4 × 3 + 1 = 13	write 3, carry 1
4 × 2 + 1 = 9	write 9
4 × 5 = 20	write 0, carry 2
4 × 3 + 2 = 14	write 4, carry 1
	write 1

```
                                                  4.16
                                                × 2.08
First partial product 8 × 416                     3328
Second partial product 2 × 416 shifted twice     832
Sum of partial products                          86528
2 places + 2 places = 4 places       Product = 8.6528
```

PRACTICE ASSIGNMENT

Plan: Compute with paper and pencil only.

1. Perform the following multiplications.

   a. 176 × 325 =

   b. 3971 × 620 =

   c. 127 × 6.2 =

   d. 23.23 × 67.01 =

   e. 3.446 × 82.5 =

2. The star forward of a basketball team averages 23.5 points per game. During a 56-game season, how many points is he expected to score, assuming he plays in each game?

3. One liter equals 0.264 gallons. If a car's fuel tank capacity is 65 liters, how many gallons will it hold?

4. A vending machine contains 107 quarters, 83 dimes, and 154 nickels. How much money does it contain?

Section 1.5 Rounding Numbers

KEY TERM **Estimate**

A manufacturer produced 3,458,916 units during the year. However, the annual report showed 3,460,000 units. Businesses report rounded figures because people who use the reports find rounded figures easier to read and remember. The precise production figure of 3,458,916 was rounded to the nearest ten thousand.

When rounding a whole number:

Replace all digits to the right of the last retained digit with zeros.

If the digit following the last one to be retained is 5 or greater, the last digit to be retained is increased by 1.

If the digit following the last one to be retained is less than 5, the last digit to be retained remains unchanged.

Example 1 Round 2,869,467 to the nearest hundred thousand.

The hundred thousand digit is 8 and the digit to the right of the hundred thousand position is a 6, which is greater than 5.

Add 1 to the hundred thousand position and replace all digits to the right of it with zeros.

The rounded number is 2,900,000.

Example 2 Round 283,943 to the nearest ten thousand.

The ten thousand digit is 8 and the digit to the right of the ten thousand position is a 3, which is less than 5.

Replace all digits to the right of the ten thousand position with zeros.

The rounded number is 280,000.

Whether you compute with paper and pencil, calculator, or computer, **estimate** the answers after (or before) the calculation; calculated and estimated answers ought to be close. For example, if your estimate is 600, the precise result might be 623.72. You would recognize an answer of 6,237.20 as an error.

Obtain an estimate by first rounding each number to its largest place value. This simplifies the computation so that in many cases it can be performed mentally.

Precise	Rounded
894	900
+ 3235	+ 3000
+ 18	+ 20
+ 154	+ 200
4301	4120
94153	90000
− 7276	− 7000
86877	83000

Multiplication or division may produce more fractional decimal places than are useful. For example, a student worked $8\frac{1}{2}$ hours at $4.35 per hour. Gross pay amounts to $8.5 \times 4.35 = 36.975$ or 36 dollars and $97\frac{1}{2}$ cents. The product must be stated as 36.97 or 36.98.

Money is usually rounded to the nearest hundredth (two decimal places). However, quantities may be rounded to any place value. While there are cases in which intermediate money amounts are rounded, usually only the final result is rounded in order to prevent inaccuracies.

When rounding whole numbers, replace the digits to the right of a particular place with zeros; when rounding decimal fractions, drop the digits to the right of a particular place.

Example 3 Round 283.9467 to the nearest hundredth.

Keep two places to the right of the decimal point.

The digit to the right of the hundredth position is a 6, which is greater than 5.

Add 1 to the hundredth place and drop all digits to the right.

The rounded number is 283.95.

Example 4 Round 283.9437 to the nearest hundredth.

Keep two places to the right of the decimal point.

The digit to the right of the hundredth position is a 3, which is less than 5.

Drop all digits to the right of the hundredth position.

The rounded number is 283.94.

PRACTICE ASSIGNMENT

1. Round each number to the indicated place.
   - **a.** 736 to the nearest ten:
   - **b.** 83,357 to the nearest thousand:
   - **c.** 52,888,870 to the nearest hundred thousand:
   - **d.** 951,202,000 to the nearest million:
   - **e.** 1.258 to the nearest tenth:
   - **f.** 2,894.666666 to the nearest thousandth:
   - **g.** $4,855.001 to the nearest cent (hundredth):
   - **h.** $87.99552 to the nearest cent:

2. Estimate the results of the following arithmetic problems without performing the computations on the precise numbers.

a.
```
   734
 + 339
 + 771
 + 384
```

b.
```
  9041
 + 635
 + 3570
 + 7729
```

c.
```
   2128
 + 8017
 +  992
 + 19441
```

d.
```
   827
 - 609
```

e.
```
  7509
 - 2597
```

f.
```
 38468
 - 9058
```

Section 1.6 Division

KEY TERMS **Dividend** **Divisor** **Quotient** **Remainder**

To find how many times one number is contained in another, use division. The number being divided is called the **dividend.** The number you divide by is called the **divisor.** (The divisor is never zero.) The result is called the **quotient.** When the divisor does not divide the dividend evenly, the number that is left over is called the **remainder.**

```
        quotient
divisor)dividend
```

Example 1

```
   quotient
25)3025
```

Both dividend and divisor are whole numbers. This division leaves a remainder of zero.

```
    121
25)3025
   25↓ | 25 into 30 = 1   1 × 25 = 25
    52 | 30 − 25    = 5   first remainder; bring down 2
    50↓  25 into 52 = 2   2 × 25 = 50
     25  52 − 50    = 2   second remainder; bring down 5
     25  25 into 25 = 1   1 × 25 = 25
      0  25 − 25    = 0   third remainder
```

When all digits of a whole number dividend have been used up and the remainder is zero, the division process is finished and the quotient is exact.

quotient × divisor = dividend
121 × 25 = 3025

Example 2

```
  quotient
4)23
```

This division problem leaves a nonzero remainder. Since the quotient will not be sufficiently accurate in this form, expand the dividend into a mixed number by appending a decimal point and zeros.

```
  quotient
4)23.00
```

If the dividend is a mixed number, make sure that the decimal point in the quotient is directly above the decimal point in the dividend.

```
   5.75
4)23.00
  20↓ |  4 into 23 = 5   5 × 4 = 20
   30 |  23 − 20   = 3   first remainder; bring down 0
   28↓   4 into 30 = 7   7 × 4 = 28
    20   30 − 28   = 2   second remainder; bring down 0
    20   4 into 20 = 5   5 × 4 = 20
     0   20 − 20   = 0   third remainder
```

Once the remainder is zero, the division process is finished, and the quotient is exact.

quotient × divisor = dividend
5.75 × 4 = 23.00

Example 3

```
     quotient
2.7)3.824
```

Both dividend and divisor are mixed numbers.

If the divisor is not a whole number, move the decimal point to the right until the divisor becomes a whole number.

Next, move the decimal point in the dividend an equal number of places to the right; append zeros if necessary.

```
    quotient
27)38.24
```

If the division process does not terminate with a remainder of zero, round the quotient to the desired accuracy. To round the quotient to the nearest hundredth, create a quotient with three decimal places.

```
      1.416
 27)38.240
    27
    11 2
    10 8
       44
       27
       170
       162
         8
```

27 into 38	= 1	1 × 27 = 27
38 − 27	= 11	first remainder; bring down 2
27 into 112	= 4	4 × 27 = 108
112 − 108	= 4	second remainder; bring down 4
27 into 44	= 1	1 × 27 = 27
44 − 27	= 17	third remainder; bring down 0
27 into 170	= 6	6 × 27 = 162
170 − 162	= 8	fourth remainder

The quotient contains three fractional digits, and the remainder is not zero.

The quotient rounded to the nearest hundredth is 1.42.

PRACTICE ASSIGNMENT

Plan: Compute with pencil and paper only.

1. Find the quotients. If a fractional part does not terminate, find the thousandth position and round the result to the nearest hundredth.

   a. 8)28376 **b.** 25)8575

   c. 17)5576 **d.** 67)57.30

   e. 2.8)578.41 **f.** 4.07)512

   g. .77)30.158 **h.** .044).0822

2. If 575 reams of paper cost $3,588, what is the cost per ream?

3. Phil Gautier's gross earnings for the week ending May 15 were $317.30. If he worked 38 hours, what was his hourly wage?

4. A traveler begins an automobile trip with a full tank of gas. After driving 152.5 miles, she tops off the tank with 7.4 gallons. Calculate the miles per gallon to the nearest tenth of a gallon.

Section 1.7 Multiplication and Division by 10, 100, 1000, . . .

To multiply a whole number by 10, 100, 1000, . . . append one, two, three, . . . zeros, respectively.

72 × 10 = 720
638 × 100 = 63800
51 × 1000 = 51000

To multiply a mixed number by 10, 100, 1000, . . . move the decimal point one, two, three, . . . places to the right, respectively. Append zeros if necessary.

3.76 × 10 = 37.6
572.98 × 100 = 57298
1.64 × 1000 = 1640

To divide a whole number by 10, 100, 1000, . . . append the decimal point and move it one, two, three, . . . places to the left, respectively. Insert zeros to the right of the decimal point if necessary.

508 ÷	10		
508. ÷	10	=	50.8
7245 ÷	100		
7245. ÷	100	=	72.45
82 ÷	1000		
82. ÷	1000	=	0.082

To divide a mixed number by 10, 100, 1000, . . . move the decimal point one, two, three, . . . places to the left, respectively. Insert zeros to the right of the decimal point if necessary.

9.82 ÷	10	=	0.982
515.334 ÷	100	=	5.15334
78.895 ÷	1000	=	0.078895

PRACTICE ASSIGNMENT

Plan: Compute problems 1–4 mentally.

1. **a.** 25 × 10 =
b. 635 × 100 =
c. 772 × 1000 =
d. 54 × 1000 =
e. 350 × 100 =

2. **a.** 3.667 × 10 =
b. 68.0909 × 100 =
c. 344.707 × 1000 =
d. 0.7889 × 1000 =
e. 0.03556 × 100 =

3. **a.** 334 ÷ 10 =
b. 8400 ÷ 100 =
c. 93873 ÷ 1000 =
d. 17 ÷ 10 =
e. 82 ÷ 1000 =

4. **a.** 17.3333 ÷ 10 =
b. 6.88989 ÷ 100 =
c. 0.42 ÷ 1000 =
d. 152.525 ÷ 100 =
e. 3.1415 ÷ 1000 =

Challenger

Last year, earnings on 31,000 outstanding shares of the Superight Corporation were $98,270. This year, earnings on 46,000 outstanding shares were $187,680. Does this represent an increase or decrease in earnings per share? What is the increase or decrease?

CHAPTER 1 SUMMARY

Concept	*Example*	*Procedure*
SECTION 1.1 Decimals		
Digits	0,1,2,3,4,5,6,7,8,9	Any number can be written as an arrangement of digits.
Place value	. . . thousands, hundreds, tens, ones, tenths, hundredths, thousandths . . .	Place value assigns a value to each digit.
Whole number	53782	Digits to the left of the decimal point comprise a whole number.
Decimal point	324.	The decimal point separates a whole number from a fractional part.
Decimal fraction	0.125	Digits to the right of the point comprise a fraction.
Mixed number	324.125	
Edited form	560342 = 560,342	Commas separating three-digit groups improve readability.
Word form	3478.52 = three thousand, four hundred seventy-eight and fifty-two hundredths	The word form of a number is required on some documents. The word "and" separates a whole number from a fractional part.
SECTION 1.2 Addition		
Carry	11 1 4318.984 +2507.062	Add digits with a particular place value to digits with the same place value. Partial sums over ten carry into the next place to the left.
Sum	6826.046	

Concept	*Example*	*Procedure*
SECTION 1.3 Subtraction		
Difference	397.534 −124.503 273.031	Subtract digits with a particular place value from digits with the same place value.
Borrowing	4 14 6 15 ~~5~~ ~~4~~~~7~~ . ~~5~~3 4 −1 6 5 . 6 2 1 3 8 1 . 9 1 3	When a partial subtraction cannot be done, borrow ten from the next place to the left.
SECTION 1.4 Multiplication		
Multiplicand Multiplier Partial products Product	48.52 × 3.4 19408 14556 164968 164.968	Product = Sum of partial products. Fractional places in product = Fractional places in multiplicand + Fractional places in multiplier.
	6.18 × 3.04 2472 1854 187872 18.7872	When a digit in the multiplier is zero, shift two places to the left in the partial product.
SECTION 1.5 Rounding		
Rounding in order to approximate	54876 55000 54376 54000	Round up on digits of 5 or greater. Round down on digits less than 5. Replace digits to the right with zeros.
Estimate	17534 20000 + 6488 + 6000 24022 26000	Estimate by operating on rounded numbers.
Rounding fractional parts to a particular accuracy	537.3584 537.36 537.3524 537.35	Round up on digits of 5 or greater. Round down on digits less than 5. Drop digits to the right.
SECTION 1.6 Division		
Quotient Divisor, Dividend	16 15)240 15 90 90 0	When remainder is zero division is finished. Quotient is exact.
	3.125 8)25.000 24 1 0 8 20 16 40 40 0	Dividend and quotient must be expanded with decimal places. When remainder is zero division is finished. Quotient is exact.
	37.5)189.34 5.049 375)1893.400 1875 18 4 0 18 40 15 00 3 400 3 375 25	If divisor is a fraction or mixed number, dividend and divisor must be adjusted so that divisor is a whole number. If a zero remainder does not appear, round quotient to desired accuracy.
SECTION 1.7 Multiplication and Division by 10, 100, 1000, . . .		
Multiplication by 10, 100, 1000, . . .	37 × 100 = 3700 38.56 × 100 = 3856	Append one, two, three, . . . zeros. Move decimal point one, two, three, . . . places to the right.
Division by 10, 100, 1000, . . .	34900 ÷ 10 = 3490 3968.2 ÷ 10 = 396.82	Remove one, two, three, . . . zeros. Move decimal point one, two, three, . . . places to the left.

NAME ______________________

CLASS SECTION ____________ DATE ____________

SCORE ____________ MAXIMUM 150 PERCENT ____________

CHAPTER 1 ASSIGNMENT

(Use paper and pencil for this assignment.)

1. In the number 73,582.7668

a. the digit in the tens place is:

b. the digit in the tenths place is:

c. the digit in the thousandths place is:

d. the digit in the ten thousandths place is:

e. the digit in the hundreds place is:

f. the digit in the hundredths place is: *(6 points)*

2. Write 305.04346 in word form: *(1 point)*

3. Write thirty-three million, two hundred ninety-eight thousand, four hundred seventeen in edited digital form: *(1 point)*

4. Write:

$$2 \times 1000 + 0 \times 100 + 1 \times 10 + 4 \times 1 + 3 \times \frac{1}{10} + 3 \times \frac{1}{100} + 6 \times \frac{1}{1000} + 6 \times \frac{1}{10000}$$

a. In decimal notation:

b. In word form:

(2 points)

5. Match the numbers in the left-hand column with the numbers in the right-hand column.

i. thirty-one thousand, sixty-two _____

ii. three hundred thousand, one hundred sixty-two _____

iii. three thousand, one hundred sixty-two _____

iv. three hundred sixteen thousand, two hundred _____

v. thirty thousand, one hundred sixty-two _____

a. 3,162

b. 30,162

c. 31,062

d. 300,162

e. 316,200

(5 points)

6. Round 622,629,581 to the nearest ten million: *(1 point)*

7. Round 5,679.040404 to the nearest thousandth: *(1 point)*

8. Suppose a national magazine reports that the United States population is 274,632,850, that the dollar value of its imports is 372,344,720,200, and that its food exports total 4,582,230 tons. Use approximations to make these figures easier to understand.

a. Population size:

b. Imports in dollars:

c. Food exports in tons: *(3 points)*

9. Estimate the results of the following arithmetic problems without performing the computations on the precise numbers.

a.
```
   733
+  188
+ 4121
+ 3570
+ 7729
```

b.
```
  84525
- 39006
```

(4 points)

10. Compute the following:

a. $0.006 + 15.0 + 17.02 + 75.00533$

b. $25.7 - 0.523$

c. $3.409 \times 0.87 =$

d. $286\overline{)54.238}$ (Round to the nearest thousandth.)

(12 points)

11. Terry Sato, a bookkeeper at Centre Clinic, must bill patient Wilma Novacek for laboratory tests done at the clinic.

Family Physicians

Date 3/15/199X

Patient Name Wilma Novacek

Patient ID # 999648712

LAB TESTS	CHARGE	
blood glucose	$ 22	50
hematocrit	15	75
throat culture	45	00
urinalysis	37	65
Pap smear	17	85
X-ray	185	00
PAID		

PLEASE KEEP THIS COPY FOR YOUR RECORDS

Find the total laboratory charges to be billed to this patient: *(5 points)*

12. Operating expenses of the We Sellum Advertising Agency for the month of April were: $78.49 for office supplies, $370.54 for telephone, $296.23 for electricity, $1,622.87 for automobile expenses, $57.50 for magazine subscriptions, and $4,857.88 for entertainment expenses. What were the total operating expenses for the month? *(5 points)*

13. Lara Lykins, a part-time black belt instructor in the local karate school receives a monthly paycheck in the amount of $363.48. She needs $272.50 to cover medical expenses and plans to deposit the remainder in her savings account. How much will she deposit into her savings account? *(5 points)*

14. A hotel has 512 rooms; each room contains 238 square feet. What is the total number of square feet? *(5 points)*

15. Flight 309 takes 2.25 hours to fly from New York to the Caribbean, a distance of 1247 miles. How many miles per hour does the plane average? Round to nearest mile: *(5 points)*

16. The Heathcliff Furniture Company pays its assembly line workers $9.75 per hour. Tracy Krauss worked on the assembly line 36.5 hours this week. How much did she earn this week? *(5 points)*

17. A fleet of five trucks carries the following loads:
Truck A: 32,851.31 tons
Truck B: 9,305.4985 tons
Truck C: 15,048.227 tons
Truck D: 26,918.07 tons
Truck E: 19,840.883 tons
What is the total weight in tons carried by the entire fleet? *(5 points)*

18. The True Grit Hardware Company has patiently billed Harold Gunnarsen, a "slow pay," for the past six months for items purchased in the amount of $498.98. Mr. Gunnarsen has made payments of $235.00, $117.50, $53.75, $37.33, and $29.44. How much remains to be paid (assuming no interest charges)? *(5 points)*

19. Complete the following overtime record: ***(12 points)***

Overtime Pay for Experimental Project A						
Employee	Mon	Tue	Wed	Thur	Fri	Employee Totals
Borski	56.84	60.76	53.90	64.68	68.60	
Cousteau	50.60	74.80	61.60	68.20	57.20	
Hebert	56.35	60.95	48.30	66.70	54.05	
Risama	64.75	59.50	75.25	50.75	61.25	
Woods	53.96	62.48	65.32	55.38	56.80	
Total						

20. Nineteen employees attended a staff meeting. Each employee received a twenty-three page copy of the auditor's report. The cost to photocopy each page was $.07. How much did it cost to provide the report to everyone attending the meeting? ***(5 points)***

21. Mary Miller borrows $9,000 to purchase a new car. Assume that the interest amounts to $623.34; what will the monthly payments be if the loan is paid off in 24 equal payments? ***(5 points)***

22. A farm family sold 73.5 acres of land for $1,070.25 per acre and the house for $42,750. What was the total sale price for all of this real estate? ***(5 points)***

23. Fill in the blanks on the following sales chart:

Month	Dept. A	Dept. B	Dept. C	Dept. D	Total
Jan	$8,478.37	$7,912.44	$6,921.05	______	$29,814.21
Feb	______	5,850.71	5,078.92	5,003.38	23,772.90
Mar	8,909.63	6,710.18	7,104.06	5,985.66	______
Apr	9,245.78	______	7,133.29	6,469.03	30,589.99

(12 points)

24. The Gradys are trying to cut electricity expenses for their home. The December electric meter reading was 103,979.4 kilowatt hours; the January reading was 105,861.1. In February they lowered the thermostat and kept many of the lights out. The February reading was 107,016.5. What was the reduction in their electricity usage (in kilowatt hours) between January and February? *(10 points)*

25. The Eat-N-Run fast-food restaurant pays all cashiers the same hourly wage. During an eight-hour shift, four cashiers earned a total of $141.12. What is the hourly wage? *(5 points)*

26. Tuesday's hamburger sales at Burger Master Drive-In were as follows:

Hamburger	Quantity	Price
Plain	371	$0.79
Cheeseburger	408	1.09
Double meat (plain)	213	1.25
Double meat (cheese)	256	1.65
Superburger	252	2.27

How much money did Burger Master receive for Tuesday's hamburger sales? *(10 points)*

27. Fill in the blanks on the following payroll chart: *(10 points)*

PAYROLL DEPARTMENT *Week Ending:* 11/13

Employee Name	Hours Worked	Hourly Wage		Total Earnings	
J. Chomsky	37	$6	25		
M. Fredericks	32			190	40
C. Hernandez		6	39	255	60
L. McGuire		7	05	253	80
M. Skouras	39			268	32
		Total			

Chapter 2 Fractions: Addition and Subtraction

OBJECTIVES

- Create equivalent fractions
- Reduce fractions to lowest terms
- Convert improper fractions to mixed numbers and mixed numbers to improper fractions
- Find the least common denominator
- Add and subtract fractions

SECTIONS

INTRODUCTION

You may notice while reading the financial section of a newspaper or watching the stock market report on the evening news that Honeywell closed at $87\frac{3}{4}$ or that IBM dropped $3\frac{1}{8}$. In business, such fractions (sometimes called common fractions) are fairly simple, but since most calculators and computers cannot process them in this form, the ability to manipulate fractions is a valuable skill.

Section 2.1 Understanding and Manipulating Fractions

KEY TERMS **Denominator** **Numerator** **Terms of the fraction** **Proper fraction** **Improper fraction** **Mixed number**

A fraction represents one or more equal parts of a unit (dollar, pound, quart, inch, acre, etc.). The **denominator** shows how many equal parts the unit is divided into. The denominator is never zero. The **numerator** shows how many of these equal parts are chosen. Both numerator and denominator are called **terms of the fraction** and are separated by a horizontal line or a slash. A fraction can be expressed as

$$\frac{\text{numerator}}{\text{denominator}} \quad \text{or} \quad \text{numerator/denominator.}$$

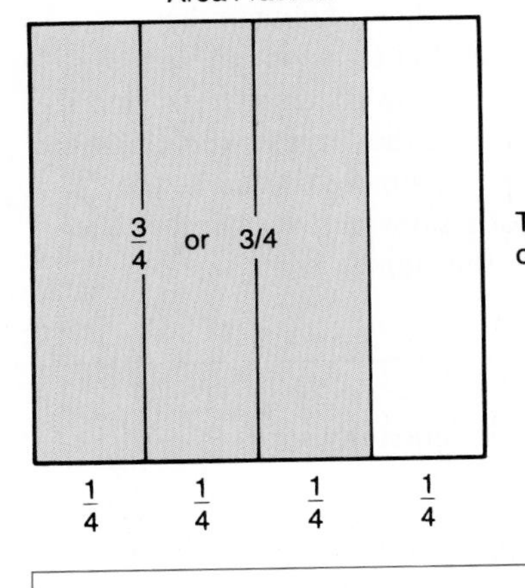

The shaded area is three-fourths of the total area.

If the numerator is smaller than the denominator, the fraction represents less than a whole unit and is called a **proper fraction.**

If the numerator is equal to or greater than the denominator, the fraction represents a whole unit or more and is called an **improper fraction.**

$\frac{3}{4}$ is proper

$\frac{4}{4}$ and $\frac{5}{4}$ are improper

$\frac{4}{4}$ represents one unit or 1

A fractional value can appear in equivalent forms with higher or lower terms.

Multiplying the numerator and denominator of a fraction by the same whole number (greater than one) raises the fraction to an equivalent fraction in higher terms.

$$\frac{3}{4} = \frac{3 \times 2}{4 \times 2} = \frac{6}{8}$$

Dividing the numerator and denominator evenly by the same whole number (greater than one) reduces the fraction to an equivalent fraction in lower terms.

$$\frac{20}{28} = \frac{20 \div 4}{28 \div 4} = \frac{5}{7}$$

If the numerator and denominator cannot be divided evenly by the same whole number, the fraction is in lowest terms.

$\frac{3}{4}$ is in lowest terms

How is a Fraction Changed to an Equivalent Fraction with a Given Higher Denominator?

You can raise a fraction to an equivalent fraction with a given higher denominator by dividing the new denominator by the old denominator. If the quotient is a whole number, multiply it by the old numerator in order to obtain the new numerator.

$$\frac{5}{6} = \frac{?}{42}$$

$$42 \div 6 = 7$$

$$\frac{5}{6} = \frac{5 \times 7}{42} = \frac{35}{42}$$

PRACTICE ASSIGNMENT 1

Plan: Raise the following fractions to equivalent fractions with the given denominator.

1. $\frac{4}{5} = \frac{\ }{20}$ **2.** $\frac{7}{20} = \frac{\ }{60}$ **3.** $\frac{5}{12} = \frac{\ }{72}$ **4.** $\frac{3}{7} = \frac{\ }{21}$

5. $\frac{8}{9} = \frac{\ }{72}$ **6.** $\frac{3}{10} = \frac{\ }{200}$ **7.** $\frac{7}{11} = \frac{\ }{77}$ **8.** $\frac{21}{40} = \frac{\ }{120}$

How is a Fraction Changed to an Equivalent Fraction in Lowest Terms?

A whole number (other than 1) that evenly divides each term of the fraction is called a common divisor. (Generally, a whole number that divides another whole number without leaving a remainder is said to be one of its divisors.) To reduce a fraction to lowest terms, first find its greatest possible common divisor, the largest whole number that divides each term evenly. In simple fractions, the greatest common divisor is often obvious. Remember the following facts in the box to the right.

A number is divisible by	If
3	The sum of the digits is divisible by 3
4	The number formed by the right-most two digits is divisible by 4
6	The number is divisible by 2 and 3
9	The sum of the digits is divisible by 9

Example 1 $\frac{27}{63}$ The greatest common divisor is 9.

$$\frac{27 \div 9}{63 \div 9} = \frac{3}{7}$$

When the greatest common divisor is not obvious, follow these steps:

1. Divide the denominator by the numerator and obtain the remainder. If the remainder is 0 or 1, skip step 2 and go to step 3.
2. Divide the divisor of the previous division by the remainder from that division step. Repeat this step until the remainder is either 0 or 1.
3. If the remainder is 0, then the last divisor becomes the greatest common divisor. If the remainder is 1, the fraction is already in lowest terms.

Example 2 Reduce $\frac{144}{672}$ to lowest terms.

Step 1	Step 2	Step 2 Repeated	Step 3
$144\overline{)672}$ quotient 4; 576; remainder 96	$96\overline{)144}$ quotient 1; 96; remainder 48	$48\overline{)96}$ quotient 2; 96; remainder 0	The remainder is 0. The greatest common divisor of 144 and 672 is 48.

$$\frac{144}{672} = \frac{144 \div 48}{672 \div 48} = \frac{3}{14}$$

Example 3 Reduce $\frac{31}{69}$ to lowest terms.

Step 1	Step 2	Step 2 Repeated	Step 3
$31\overline{)69}$ quotient 2; 62; remainder 7	$7\overline{)31}$ quotient 4; 28; remainder 3	$3\overline{)7}$ quotient 2; 6; remainder 1	The remainder is 1. The fraction is already in lowest terms.

PRACTICE ASSIGNMENT 2

Plan: Reduce each of the following fractions to lowest terms, or show that the fraction is already in lowest terms.

1. $\frac{10}{15} =$ **2.** $\frac{16}{18} =$ **3.** $\frac{28}{50} =$ **4.** $\frac{24}{36} =$ **5.** $\frac{9}{21} =$

6. $\frac{221}{513} =$ **7.** $\frac{366}{693} =$

8. $\frac{639}{1782} =$

Can a Fraction Contain a Whole Number and a Fractional Part, Similar to a Mixed Decimal Number?

Just as a mixed decimal number contains a whole number and a fractional part, a **mixed number** combines a whole number with a fraction. Because a mixed number represents more than one unit, it can be converted to an improper fraction.

Mixed decimal number: 2.75

Mixed number: $2\frac{3}{4}$

Improper fraction: $\frac{11}{4}$

To convert a mixed number to an improper fraction, multiply the whole number by the denominator of the fractional part and add the numerator. The result becomes the numerator of the improper fraction. The denominator of the fractional part remains unchanged and becomes the denominator of the improper fraction.

$$8\frac{5}{6}$$

$$8 \times 6 = 48$$

$$48 + 5 = 53$$

$$\frac{53}{6}$$

Any whole number (other than zero) can be converted to an improper fraction by making it the numerator over a denominator of one.

$$17 = \frac{17}{1}$$

To convert an improper fraction to a mixed number, divide the numerator by the denominator. The quotient is the whole number. The remainder becomes the numerator of the fractional part, and the denominator of the improper fraction becomes the denominator of the fractional part.

$$\frac{25}{7}$$

```
   3
7)25
  21
   4
```

$$3\frac{4}{7}$$

PRACTICE ASSIGNMENT 3

1. Convert the following mixed numbers to improper fractions.

a. $5\frac{2}{3} =$ **b.** $8\frac{4}{5} =$ **c.** $11\frac{5}{16} =$ **d.** $3\frac{17}{32} =$

e. $6\frac{8}{9} =$ **f.** $4\frac{11}{12} =$ **g.** $11\frac{5}{7} =$ **h.** $9\frac{9}{40} =$

2. Convert the following improper fractions to mixed numbers.

a. $\frac{19}{8} =$ **b.** $\frac{17}{4} =$ **c.** $\frac{73}{12} =$ **d.** $\frac{103}{15} =$

e. $\frac{62}{5} =$ **f.** $\frac{97}{16} =$ **g.** $\frac{118}{17} =$ **h.** $\frac{267}{32} =$

Section 2.2 Addition and Subtraction of Fractions

KEY TERMS **Least common denominator** **Prime number**

To add or subtract fractions with the same denominator, add or subtract the numerators and place the sum or difference over the denominator.

$$\frac{3}{8} + \frac{4}{8} = \frac{3+4}{8} = \frac{7}{8}$$

$$\frac{5}{7} - \frac{2}{7} = \frac{5-2}{7} = \frac{3}{7}$$

Convert the final result to standard form: as a proper fraction in lowest terms, or as a mixed number whose fractional part is in lowest terms.

$$\frac{15}{16} - \frac{9}{16} = \frac{6}{16} = \frac{3}{8}$$

To add or subtract fractions with different denominators, raise them to fractions with a common denominator. First find the lowest whole number (other than zero) into which all of the denominators evenly divide. This number is called the **least common denominator.** In the case of simple fractions, the least common denominator may be obvious.

$$\frac{2}{3} + \frac{1}{4} = \frac{8}{12} + \frac{3}{12} = \frac{11}{12}$$

Example 1 $\frac{3}{5} + \frac{7}{8}$ The least common denominator is 40.

$$\frac{3 \times 8}{5 \times 8} + \frac{7 \times 5}{8 \times 5} = \frac{24}{40} + \frac{35}{40} = \frac{59}{40} = 1\frac{19}{40}$$

Example 2 $\frac{11}{15} - \frac{7}{25}$ The least common denominator is 75.

$$\frac{11 \times 5}{15 \times 5} - \frac{7 \times 3}{25 \times 3} = \frac{55}{75} - \frac{21}{75} = \frac{34}{75}$$

Use **prime numbers** when the least common denominator is not obvious. A prime number is a whole number greater than 1 that can be evenly divided only by itself and 1.

15 has divisors 1,3,5,15 — 15 is not a prime number
22 has divisors 1,2,11,22 — 22 is not a prime number
17 has divisors 1,17 — 17 is a prime number
23 has divisors 1,23 — 23 is a prime number

The lowest prime numbers are 2,3,5,7,11,13,17,19, ...

Use the following steps to find the least common denominator:

1. List all denominators in a row.
2. Divide each value in this row by the lowest prime number that divides evenly into at least one of these values. Place the quotients into the next row. When an even division is impossible, copy the value into the next row.
3. Repeat step 2 until an even division is impossible.
4. Then repeat steps 2 and 3 with the next larger prime number that evenly divides into at least one of the values in the last row. Continue repeating steps 2 and 3 until all values in a row are 1.
5. The least common denominator is the product of all the prime divisors used.

Example 3 $\frac{3}{8} + \frac{7}{15} + \frac{11}{20}$

				Prime Numbers
8	15	20	divide by	2
4	15	10	divide by	2
2	15	5	divide by	2
1	15	5	divide by	3
1	5	5	divide by	5
1	1	1		

Least common denominator $= 2 \times 2 \times 2 \times 3 \times 5 = 120$.

$$\frac{3 \times 15}{8 \times 15} + \frac{7 \times 8}{15 \times 8} + \frac{11 \times 6}{20 \times 6} = \frac{45}{120} + \frac{56}{120} + \frac{66}{120}$$

$$= \frac{167}{120} = 1\frac{47}{120}$$

Example 4 $\frac{11}{60} - \frac{23}{135}$

			Prime Numbers
60	135	divide by	2
30	135	divide by	2
15	135	divide by	3
5	45	divide by	3
5	15	divide by	3
5	5	divide by	5
1	1		

Least common denominator $= 2 \times 2 \times 3 \times 3 \times 3 \times 5 = 540$.

$$\frac{11 \times 9}{60 \times 9} - \frac{23 \times 4}{135 \times 4} = \frac{99}{540} - \frac{92}{540} = \frac{7}{540}$$

PRACTICE ASSIGNMENT 1

Plan: Add and/or subtract the following fractions as indicated. Convert your final answers to standard form.

1. $\frac{2}{3} + \frac{5}{6} =$

2. $\frac{7}{15} + \frac{8}{15} =$

3. $\frac{9}{25} + \frac{11}{15} =$

4. $\frac{5}{8} - \frac{6}{11} =$

5. $\frac{3}{4} + \frac{2}{3} - \frac{5}{6} =$

6. $\frac{7}{8} + \frac{2}{3} - \frac{5}{12} =$

7. $\frac{29}{60} + \frac{13}{25} =$

8. $\frac{89}{120} - \frac{7}{36} =$

Can Mixed Numbers be Added or Subtracted without Converting Them to Improper Fractions?

To add mixed numbers, write them vertically. Add the whole numbers first. Then add the fractions, finding the least common denominator if necessary.

$$\begin{array}{rcr} 3\frac{11}{16} & = & 3\frac{22}{32} \\ +\ 1\frac{5}{8} & = & 1\frac{20}{32} \\ +\ 4\frac{21}{32} & = & 4\frac{21}{32} \\ \hline & & 8\frac{63}{32} = 9\frac{31}{32} \end{array}$$

When subtracting mixed numbers by the vertical method, you may find that the fractional parts cannot be subtracted without borrowing.

Example 5 $9\frac{2}{3} - 6\frac{3}{4}$

$$\begin{array}{rcr} 9\frac{2}{3} & = & 9\frac{8}{12} \\ -\ 6\frac{3}{4} & = & 6\frac{9}{12} \\ \hline \end{array}$$

(Borrow 1 from the whole number and add $\frac{12}{12}$ to the fraction.)

$$\begin{array}{r} 8\frac{20}{12} \\ -\ 6\frac{9}{12} \\ \hline 2\frac{11}{12} \end{array}$$

PRACTICE ASSIGNMENT 2

Plan: Add and/or subtract the following mixed numbers by the vertical method. Be sure your final answers are in standard form.

1. $6\frac{7}{8} + 7\frac{3}{32} =$

2. $5\frac{5}{6} + 11\frac{3}{8} =$

3. $9\frac{8}{9} - 5\frac{2}{3} =$

4. $14\frac{3}{10} - 2\frac{11}{15} =$

5. $7\frac{3}{4} - 5\frac{13}{16} =$

6. $3\frac{2}{3} + 2\frac{1}{2} + 5\frac{1}{4} =$

7. $10\frac{2}{3} - 3\frac{5}{9} - 4\frac{5}{6} =$

8. $\frac{16}{3} - 3\frac{1}{5} + \frac{5}{6} =$

Challenger

An arts and crafts project requires a decorative wooden border constructed by gluing pieces of wood together. If the border lacks ten inches for completion and there are five small wood segments available, which of these segments can be glued together to form a single piece exactly ten inches in length? The lengths are as follows:

Segment A: $2\frac{5}{12}$ inches
Segment B: $3\frac{1}{2}$ inches
Segment C: $3\frac{2}{3}$ inches
Segment D: $3\frac{3}{4}$ inches
Segment E: $3\frac{5}{6}$ inches

CHAPTER 2 SUMMARY

Concept	*Example*	*Procedure*
SECTION 2.1 Understanding and Manipulating Fractions		
Equivalent fractions	$\frac{3}{4}\ \frac{6}{8}\ \frac{9}{12}\ \frac{15}{20}$	Multiply or divide the numerator and denominator by the same whole number.
Fraction in lowest terms	$\frac{3}{4}$ is in lowest terms.	Numerator and denominator cannot be evenly divided by same whole number.
Conversion to equivalent fraction with given denominator	$\frac{2}{5} = \frac{?}{20} \quad 20 \div 5 = 4$ $\frac{2}{5} = \frac{2 \times 4}{20} = \frac{8}{20}$	
Conversion to fraction in lowest terms	$\frac{30}{75} = \frac{30 \div 15}{75 \div 15} = \frac{3}{5}$	Divide numerator and denominator by greatest common divisor.
Mixed number	$2\frac{3}{4}$	A mixed number contains a whole number and a fractional part.
Improper fraction	$2\frac{3}{4} = \frac{11}{4}$ $2 \times 4 + 3 = 8 + 3 = 11$	Any mixed number can be converted to an improper fraction.
	$\frac{11}{4}$ $\quad 4\overline{)11}$ quotient 2, -8, remainder 3 $\quad 2\frac{3}{4}$	Any improper fraction can be converted to a mixed number.
SECTION 2.2 Addition and Subtraction of Fractions		
Addition or subtraction of fractions with common denominators	$\frac{3}{7} + \frac{2}{7} = \frac{5}{7}$ $\frac{8}{9} - \frac{5}{9} = \frac{3}{9} = \frac{1}{3}$	Add or subtract numerators. Denominators remain the same. Reduce results to lowest terms.
Addition or subtraction of fractions with different denominators	$\frac{2}{3} + \frac{1}{4} = \frac{8}{12} + \frac{3}{12} = \frac{11}{12}$ $\frac{2}{3} - \frac{1}{4} = \frac{8}{12} - \frac{3}{12} = \frac{5}{12}$	Raise the fractions to those with the least common denominator and add or subtract numerators.
Addition or subtraction of mixed numbers	$2\frac{5}{8} + 6\frac{3}{5}$ → $2\frac{25}{40} + 6\frac{24}{40} = 8\frac{49}{40} = 9\frac{9}{40}$	Add or subtract the whole numbers, then the fractional parts.
Borrowing	$8\frac{3}{5} - 2\frac{11}{15}$ → $8\frac{9}{15} - 2\frac{11}{15}$ → $7\frac{24}{15} - 2\frac{11}{15} = 5\frac{13}{15}$	Borrow from the whole number when the fractional part cannot be subtracted.

CHAPTER 2 ASSIGNMENT

NAME ____________________

CLASS SECTION ____________ DATE ____________

SCORE ________ MAXIMUM __80__ PERCENT ________

---

1. Raise the following fractions to equivalent fractions with the given denominator.

a. $\frac{7}{8} = \frac{\quad}{48}$ **b.** $\frac{11}{25} = \frac{\quad}{75}$ *(2 points)*

2. Reduce each fraction to lowest terms.

a. $\frac{14}{32} =$

b. $\frac{350}{875} =$

(4 points)

3. Convert the mixed numbers to improper fractions.

a. $7\frac{3}{8} =$ **b.** $5\frac{7}{12} =$ *(2 points)*

4. Convert the improper fractions to mixed numbers.

a. $\frac{35}{11} =$ **b.** $\frac{142}{13} =$

(2 points)

5. Add or subtract the following fractions.

a. $\frac{13}{40} + \frac{11}{12} =$ **b.** $\frac{11}{12} - \frac{7}{10} =$

(4 points)

6. Add and/or subtract by the vertical method.

a. $12\frac{1}{4} - 4\frac{3}{10} =$

b. $\frac{4}{9} + 7 - 2\frac{3}{5} =$

(6 points)

7. Three rooms in a home need to be re-carpeted. The recreation room measures $395\frac{1}{2}$ square feet, a hall $200\frac{3}{4}$ square feet, and a bedroom $112\frac{5}{8}$ square feet. What is the total square footage to be carpeted?

(5 points)

8. An advertising flyer will be printed on paper $14\frac{1}{2}$ inches long. If the copy portion takes up $6\frac{7}{8}$ inches, how much space is available for illustrations?

(5 points)

9. In one day of hectic trading, a share of Merck & Co., Inc. stock dropped from $85\frac{3}{8}$ to $79\frac{3}{4}$. How many dollars per share did investors in that stock lose that day?

(5 points)

10. Mrs. Lewis works an irregular part-time schedule on Monday, Wednesday, and Friday mornings. During the past week, she worked $4\frac{1}{4}$ hours on Monday, $3\frac{3}{4}$ hours on Wednesday, and $3\frac{2}{3}$ hours on Friday. How many hours did she work that week?

(5 points)

11. Highway 511 is being resurfaced. Work crews completed $2\frac{2}{5}$ miles on Monday, $4\frac{1}{8}$ miles on Tuesday, $5\frac{3}{4}$ miles on Wednesday, $6\frac{3}{10}$ miles on Thursday, and $3\frac{1}{2}$ miles on Friday. How many miles of highway were resurfaced during the week?

(5 points)

12. Four partners own a company. The first partner owns $\frac{1}{5}$ of the company, the second $\frac{1}{3}$, and the third $\frac{1}{10}$. What portion of the company does the fourth partner own?

(10 points)

13. A real estate investor purchased $25\frac{1}{2}$ acres in January. She sold $12\frac{1}{3}$ acres in April and an additional $7\frac{5}{8}$ acres in August. How many remaining acres does she own?

(10 points)

14. During the last quarter of a football game, the Eagles took possession of the ball. On the first play they gained $6\frac{1}{2}$ yards. On the following plays they gained another $11\frac{3}{4}$ yards, lost $8\frac{2}{3}$ yards, and gained $9\frac{3}{4}$ yards. What was the Eagles' net gain after these plays?

(5 points)

15. On a test, one-fifth of a class achieved an A grade, one-fourth a B grade, and one-third a C. What portion of the class received less than a C grade?

(10 points)

Chapter 3 Fractions: Multiplication and Division

OBJECTIVES

- Multiply and divide fractions
- Find fractional equivalents

SECTIONS

Section 3.1 Multiplication and Division of Fractions

KEY TERMS **Cancellation** **Inversion**

To multiply fractions, multiply numerator by numerator and denominator by denominator.

$$\frac{4}{7} \times \frac{8}{11} = \frac{4 \times 8}{7 \times 11} = \frac{32}{77}$$

$$\frac{3}{5} \times \frac{10}{33} = \frac{3 \times 10}{5 \times 33} = \frac{30}{165} = \frac{2}{11}$$

Use **cancellation** to reduce the terms before multiplying. Reduce pairs of numerators and denominators diagonally as well as vertically to insure that the product will be expressed in lowest terms.

$$\frac{\overset{1}{\cancel{3}}}{\underset{1}{\cancel{5}}} \times \frac{\overset{2}{\cancel{10}}}{\underset{11}{\cancel{33}}} = \frac{2}{11}$$

3 divides 3 and 33
5 divides 5 and 10

Example 1

$$\frac{\overset{1}{\cancel{5}}}{\underset{3}{\cancel{9}}} \times \frac{7}{13} \times \frac{\overset{1}{\cancel{3}}}{\underset{4}{\cancel{20}}} = \frac{7}{3 \times 13 \times 4} = \frac{7}{156}$$

3 divides 9 and 3
5 divides 5 and 20

Division of fractions requires the **inversion** of the divisor: Exchange the numerator with the denominator; then multiply the two fractions.

Convert mixed numbers to improper fractions before multiplying or dividing.

$$\frac{5}{7} \div \frac{4}{9} =$$

$$\frac{5}{7} \times \frac{9}{4} = \frac{45}{28} = 1\frac{17}{28}$$

Example 2

$$\frac{16}{21} \div 3\frac{5}{9} = \frac{16}{21} \div \frac{32}{9} = \frac{\overset{1}{\cancel{16}}}{\underset{7}{\cancel{21}}} \times \frac{\overset{3}{\cancel{9}}}{\underset{2}{\cancel{32}}} = \frac{3}{14}$$

Perform all cancellations **after** inverting the divisor!

PRACTICE ASSIGNMENT 1

1. Multiply the following fractions. Use cancellation to insure that the product will be in lowest terms. Put results into standard form.

   a. $\frac{5}{7} \times \frac{3}{4} =$ **b.** $\frac{5}{6} \times \frac{18}{25} =$

   c. $\frac{8}{15} \times \frac{9}{20} =$ **d.** $7 \times \frac{11}{13} =$

   e. $\frac{7}{16} \times 12 =$

   f. $6 \times \frac{1}{4} \times \frac{15}{16} =$

2. Divide the following fractions. Use cancellation to insure that the quotient will be in lowest terms. Put the final results into standard form.

   a. $\frac{4}{9} \div \frac{5}{12} =$ **b.** $\frac{5}{8} \div \frac{5}{6} =$

   c. $\frac{9}{16} \div 5 =$ **d.** $\frac{7}{11} \div \frac{7}{8} =$

   e. $\frac{5}{12} \div 3\frac{3}{8} =$ **f.** $\frac{9}{10} \div \frac{57}{100} =$

3. Perform the following computations. Put your final answers into standard form.

   a. $7\frac{1}{9} \times 6\frac{3}{4} =$

   b. $1\frac{1}{2} \div 4\frac{5}{6} =$

   c. $6\frac{2}{3} \times \frac{3}{10} \times 5\frac{5}{9} =$

   d. $4\frac{7}{12} \times 2\frac{2}{3} \times 1\frac{5}{6} =$

   e. $8\frac{3}{4} \times 3 \div \frac{3}{4} =$

   f. $5\frac{5}{8} \div 3\frac{3}{4} \times 6\frac{5}{12} =$

Section 3.2 Converting Fractions to Decimals and Decimals to Fractions

All fractions can be converted to decimal form. Fractions having denominators of 10, 100, 1000, etc., are multiples of fractional place values (tenths, hundredths, thousandths, etc.).

$$\frac{3}{10} = 3 \times 0.1 = 0.3$$
$$\frac{57}{100} = 57 \times 0.01 = 0.57$$

Fractions having denominators that are divisors of 10, 100, 1000, etc., can be raised to equivalent fractions with denominators of 10, 100, 1000, etc. Therefore, you can find the decimal forms of these fractions mentally.

$$\frac{1}{5} = \frac{2}{10} = 0.2$$
$$\frac{7}{20} = \frac{35}{100} = 0.35$$

To convert any fraction to decimal form, divide the numerator by the denominator. (Use paper and pencil or a calculator.)

Example 1 Convert $\frac{3}{16}$ to decimal form.

```
     0.1875
16)3.0000
   1 6
   1 40
   1 28
     120
     112
       80
       80
        0
```

key: 3 [÷] 16 [=]
display: 0.1875

Sometimes the decimal form of a fraction does not terminate. When computing with paper and pencil, round it to a desired accuracy. Calculators show the fractional part up to the limit of the display; some round, some do not.

Example 2 Find the decimal form of $\frac{5}{11}$ accurate to the nearest thousandth.

```
     0.4545
11)5.0000
   4 4
     60
     55
      50
      44
       60
       55
        5
```

The paper-and-pencil computation will continue forever if not stopped.

$\frac{5}{11} = 0.454545\ldots = 0.455$ (accurate to the nearest thousandth).

5 [÷] 11 [=] 0.4545454 (on a calculator that displays eight digits and does not round the last digit).

PRACTICE ASSIGNMENT 1

1. Calculating mentally, change the following fractions to decimal form by first converting each to an equivalent fraction with a denominator of 10, 100, 1000, etc. Show each step.

a. $\frac{11}{25} = \quad = $ **b.** $\frac{37}{50} = \quad = $ **c.** $\frac{19}{125} = \quad = $

d. $\frac{63}{250} = \quad = $ **e.** $\frac{33}{500} = \quad = $ **f.** $\frac{17}{1000} = $

2. Use division to convert the following fractions to decimal form. Write all results accurate to six places to the right of the decimal point, rounding where necessary. Compute with paper and pencil or a calculator. On a calculator expand $\frac{a}{b}$ with the sequence a [÷] b [=].

a. $\frac{4}{7} = $ **b.** $\frac{9}{13} = $ **c.** $\frac{17}{30} = $ **d.** $\frac{13}{24} = $

e. $\frac{8}{51} = $ **f.** $\frac{37}{108} = $ **g.** $\frac{58}{71} = $ **h.** $\frac{60}{97} = $

You can convert the fractional part of a terminating decimal number to fraction form by inspection.

The digits to the right of the decimal point become the numerator. Note the place value of the right-most digit; the denominator of this place value becomes the denominator of the fraction. Reduce the fraction to lowest terms.

$$0.35 = \frac{35}{100} = \frac{7}{20}$$

$$0.314 = \frac{314}{1000} = \frac{157}{500}$$

PRACTICE ASSIGNMENT 2

Plan: Write each of the following decimal numbers as a fraction in lowest terms.

1. 0.65 = **2.** 0.171 = **3.** 0.9 = **4.** 0.88 =

5. 0.45 = **6.** 0.503 = **7.** 0.705 = **8.** 0.4007 =

Challenger

Joshua Smith earmarked a sizable portion of his earnings for financial investment, much of it in common stock. On the advice of an investment counselor, Smith purchased the following stock:

Company	Price per Share	Amount Invested
A T & T	$24\frac{1}{4}$	$1,455.00
Digital Equipment Corp.	$88\frac{1}{2}$	4,425.00
Occidental Petroleum	$26\frac{3}{4}$	1,070.00
Procter & Gamble	$76\frac{1}{2}$	1,912.50

How many shares of each company stock did he purchase?

At the end of the year, the stocks in Smith's investment portfolio were listed as follows:

Company	Price per Share
A T & T	$24\frac{3}{8}$
Digital Equipment Corp.	$92\frac{3}{4}$
Occidental Petroleum	$26\frac{1}{2}$
Procter & Gamble	$77\frac{1}{4}$

How much is his investment in each company now worth? How much is his total stock investment now worth?

CHAPTER 3 SUMMARY

Concept	*Example*	*Procedure*
SECTION 3.1 Multiplication and Division of Fractions		
Multiplication of fractions	$\frac{5}{7} \times \frac{3}{8} = \frac{15}{56}$ $\frac{5}{6} \times \frac{3}{8} = \frac{15}{48} = \frac{5}{16}$	Multiply numerators by numerators and denominators by denominators.
Cancelling	$\frac{5}{\cancel{6}_2} \times \frac{\cancel{3}^1}{8} = \frac{5}{2} \times \frac{1}{8} = \frac{5}{16}$	To insure that a product is in lowest terms, "cancel" prior to multiplying.
Division of fractions	$\frac{5}{8} \div \frac{3}{7} =$	
Inverted divisor	$\frac{5}{8} \times \frac{7}{3} = \frac{35}{24} = 1\frac{11}{24}$	Invert the divisor; then multiply.
Multiplication or division of mixed numbers	$6\frac{3}{4} \times 3\frac{5}{8} = \frac{27}{4} \times \frac{29}{8} = \frac{783}{32} = 24\frac{15}{32}$	Convert mixed numbers to improper fractions; then multiply or divide.
SECTION 3.2 Converting Fractions to Decimals and Decimals to Fractions		
Conversion of fractions to decimals	$\frac{7}{10} = 7 \times \frac{1}{10} = 0.7$ $\frac{41}{100} = 41 \times \frac{1}{100} = 0.41$	A fraction with a denominator of 10, 100, 1000, . . . is a multiple of a fractional place value.
	$\frac{3}{5} = \frac{6}{10} = 0.6$ $\frac{111}{125} = \frac{888}{1000} = 0.888$	A fraction whose denominator is a divisor of 10, 100, 1000, . . . can be raised to a fraction with a denominator of 10, 100, 1000, ...
Equivalence of fraction form to a quotient	$\frac{5}{16}$ $5 \div 16 = 0.3125$	To convert any fraction to decimal form, divide the numerator by the denominator.
Nonterminating fractions	$\frac{7}{11}$ $7 \div 11 = 0.636363\ldots$ $= 0.636$ accurate to the nearest thousandth	A fraction whose denominator contains a prime factor other than 2 or 5 will not terminate and must be rounded.
Conversion of terminating decimal fractions to fraction form	$0.42 = \frac{42}{100} = \frac{21}{50}$ $0.125 = \frac{125}{1000} = \frac{1}{8}$	The place value of the right-most digit is the denominator of the fraction. Convert to lowest terms if necessary.

CHAPTER 3 ASSIGNMENT

NAME ______________________

CLASS SECTION ______________ DATE ______________

SCORE __________ MAXIMUM 60 PERCENT __________

1. Multiply and put the results into standard form.

a. $\frac{2}{9} \times \frac{17}{88} =$

b. $\frac{2}{3} \times \frac{3}{8} \times \frac{4}{5} =$

(2 points)

2. Divide and put the results into standard form.

a. $\frac{1}{2} \div \frac{3}{4} =$ **b.** $\frac{5}{12} \div \frac{7}{8} =$

(2 points)

3. Perform the following computations and put the results into standard form.

a. $3\frac{4}{7} \times 2\frac{5}{8} =$ **b.** $9\frac{3}{8} \div 2\frac{1}{12} =$

(2 points)

4. Convert the following fractions to decimal form accurate to six decimal places.

a. $\frac{8}{11} =$ **b.** $\frac{21}{40} =$

(2 points)

5. Convert the following decimal numbers to fractions in lowest terms.

a. 0.72 = **b.** 0.375 =

(2 points)

6. A tax bill passed by Congress was vetoed by the president. To override this veto, both houses of Congress must approve the bill by a two-thirds vote. How many votes of approval will be necessary in

a. The Senate, if all 100 members vote?

b. The House of Representatives, if all 435 members vote? *(6 points)*

7. Monday through Friday, Mary goes to work at a plant located $8\frac{2}{5}$ miles from her home. On Friday, she drives an extra $2\frac{1}{3}$ miles to drop off work orders for the following week. What is the total distance she commutes each week?

(5 points)

8. A wallpaper pattern repeats every $1\frac{1}{6}$ feet. How many times does the pattern appear in a $31\frac{1}{2}$ foot section of wallpaper? *(5 points)*

9. In his will, Uncle Jethro left one-third of his estate to his Labrador retriever. The remainder of the estate was divided equally among his five nieces. What portion of the total estate did each niece receive?

(5 points)

10. Jim estimates that he spends $36\frac{3}{4}$ hours per week on his school program of $16\frac{1}{2}$ credits. On the average, how many hours per credit does he put into his school effort?

(5 points)

11. Henderson Contractors employed 18 workers an average of $42\frac{1}{2}$ hours per week on a job lasting $6\frac{3}{5}$ weeks. How many worker-hours were required for the job? *(8 points)*

1 worker-hour = 1 hour worked by 1 worker
Worker-hours for a job = hours × workers

12. Sandy discovered a new recipe for tuna casserole that serves four people. Sandy wants to convert the recipe to serve six people. Calculate the new amount for each ingredient.

Ingredients	Serves 4	Serves 6
butter	2 tablespoons	
chopped onions	$\frac{1}{3}$ cup	
mushrooms	$\frac{1}{4}$ pound	
ground thyme	$\frac{1}{2}$ teaspoon	
light tuna	$7\frac{3}{4}$ ounces	
rice	$1\frac{1}{3}$ cup	
water	$1\frac{2}{3}$ cup	
chopped parsley	$1\frac{1}{2}$ tablespoons	

(16 points)

Chapter 4 Ratio and Proportion: Finding the Unknown

OBJECTIVES

- Standardize ratios
- Allocate by ratio
- Find the value of the unknown in a simple equation
- Find the value of the unknown in a proportion

SECTIONS

INTRODUCTION

A business may use various criteria to measure its performance and plan for the future. What is the relationship between sales and expenses? How does a company's debt compare to its net worth? Ratios are often used to make such comparisons. Proportions are also useful. How many additional workers should be hired if production is increased? Since a company knows how many workers are involved in current production, it can determine how many more will be needed by setting up a proportion.

Section 4.1 Ratios

KEY TERM **Standardized ratio**

A ratio expresses a relation between two quantities. One summer the ratio of on-time to late departures for a particular airline was 4 to 7. This means that for every four on-time departures, there were seven late departures. A ratio expresses a comparison and may be written as:

number to number
number:number
a fraction

4 to 7
4:7
$\frac{4}{7}$ or 4/7

A **standardized ratio** has whole numbers and, if written as a fraction, is in lowest terms.

Of the 30 desktop computers owned by a small company, 5 are Apples and 25 are IBMs. Compare the number of Apple computers to the number of IBMs. The ratio of Apples to IBMs is 5 to 25 or 5:25. The standardized ratio is 1 to 5 or 1:5. The ratio fraction can be expressed as a decimal number. Since $\frac{1}{5} = 0.2$, the ratio of Apples to IBMs is also 0.2.

5 to 25
5:25
$$\frac{5}{25} = \frac{1}{5} = 0.2$$

Example 1 An entrepreneur wants to market premixed dehydrated coffee. He is experimenting with a sugar-to-coffee particle ratio of .04 to 48.

$$\frac{.04}{48} = \frac{.04 \times 100}{48 \times 100} = \frac{4}{4800} = \frac{1}{1200}$$

The standardized ratio of sugar-to-coffee particles is 1:1200, or 1 particle of sugar to 1,200 particles of coffee.

If the units of measure of the two quantities are not the same, convert one unit to the other and cancel the units in order to obtain the standardized ratio.

Example 2 The ratio of a mechanical drawing is 2 inches of drawing to 15 feet of real distance, or 2 in. to 15 ft. What is the standardized ratio? How many inches of real distance are represented by 1 inch of drawing?

1 ft = 12 in.

15 ft = (15 × 12) in. = 180 in.

$$\frac{2\text{ in.}}{15\text{ ft}} = \frac{2\text{ in.}}{180\text{ in.}} = \frac{2}{180} = \frac{1}{90}$$

The standardized ratio is 1:90.
One inch of drawing represents 90 inches of real distance.

PRACTICE ASSIGNMENT 1

1. Standardize the following ratios.
   a. 8 to 2 **b.** 6 to 9
   c. 0.9 to 0.3 **d.** 0.06 to 0.15
   e. $\frac{7}{8}$ to $\frac{11}{12}$ **f.** $3\frac{1}{4}$ to $2\frac{1}{2}$

2. Standardize the following ratios. Equalize the units of measure.

   12 in. = 1 ft 16 oz = 1 lb
   3 ft = 1 yd 7 days = 1 wk
   4 qt = 1 gal 12 mos = 1 yr.

   a. 8 ft to 4 yd
   b. 15 gal to 32 qt
   c. 30 mo to 5 yr
   d. 12 oz to 1 lb
   e. 20 days to 2 wk
   f. $1\frac{3}{4}$ yr to 8 mo

Put all answers to the following problems into standard form.

3. A bag of fertilizer contains 10.5 lb of nitrate and 6.25 lb of phosphate. What is the ratio of nitrate to phosphate?

4. A small college has 480 female students and 400 male students. What is the ratio of female students to male students?

5. If a room is 11 ft 3 in. in width and $15\frac{1}{2}$ ft in length, what is the ratio of width to length?

Section 4.2 Allocation by Ratio

A ratio is sometimes used to split an amount into two portions. The total number of parts is the sum of the fraction's terms. To find the first portion, divide the amount by the total number of parts and multiply the quotient by the numerator. To find the second portion, divide the amount by the total number of parts and multiply the quotient by the denominator.

Example 1 Two children inherited their parents' estate worth $14,750. The estate is to be apportioned according to the ratio of the children's ages. They are 3 and 7. How much will each child receive?

The total number of parts is $3 + 7 = 10$.

The portion allocated to the younger child will be:

$$\frac{14750}{10} \times 3 = 4425 = \$4{,}425.$$

The portion allocated to the older child will be:

$$\frac{14750}{10} \times 7 = 10325 = \$10{,}325.$$

PRACTICE ASSIGNMENT 1

1. A $25,000 inheritance was divided between two nephews in the ratio of 3:5. How much did each one receive?

a. Nephew A:

b. Nephew B:

2. A shoe company manufactured 1,400 pairs of sandals in a ratio of 3 tan to 2 red. How many pairs are there of each color?

a. Tan:

b. Red:

3. For every four voters in Jackson County, there are five voters in Monroe County. In the two counties, there is a total of 37,926 voters. How many voters are in each county?

a. Jackson County:

b. Monroe County:

4. For every nine right-handed recruits at an army base, there are two left-handed recruits. If there are 1,023 recruits, how many are left-handed? How many are right-handed?

a. Left-handed:

b. Right-handed:

5. A farmer plants 15 acres of corn for every 13 acres of soybeans. If he plants 1,876 acres, how many acres are planted in corn and how many in soybeans?

a. Corn:

b. Soybeans:

Section 4.3 Finding the Unknown in a Simple Equation

KEY TERMS **Equation** **Inverse operations**

What is an equation?

One way to find the value of an unknown is to construct an **equation.** You probably unconsciously use equations every day. Suppose you went to the mall and purchased a compact disc for $12 and a T-shirt for $11. The total amount you spent is an unknown. Simple addition in this case reveals the unknown. Suppose you know that you spent a total of $23, $12 of which you spent on the compact disc. You cannot, however, remember how much you spent on the T-shirt. You use an equation to find that unknown amount. The unknown value is designated by a symbol (often by X). In this chapter, the symbol is U.

$$12 + 11 = U$$

$$12 + U = 23$$

To solve simple equations, use the inverse of an arithmetic operation: $+ \quad - \quad \times \quad \div$.

Think of **inverse operations** as opposite operations.

The four arithmetic operations are:

$$+ \quad - \quad \times \quad \div$$

The four inverse operations are:

$$- \quad + \quad \div \quad \times$$

In all equations the value of the expression to the left of the equal sign must remain equal to the value of the expression to the right of the equal sign. To find the unknown value, apply an inverse operation to both sides.

Example 1

$$U + 8 = 16$$
$$U + 8 - 8 = 16 - 8$$
$$U = 8$$

The inverse operation of $+$ is $-$. Subtract 8 from both sides of the equation.

Example 2

$$U - 4 = 20$$
$$U - 4 + 4 = 20 + 4$$
$$U = 24$$

The inverse operation of $-$ is $+$. Add 4 to both sides of the equation.

Example 3

$$21 \times U = 63$$
$$21 \div 21 \times U = 63 \div 21$$
$$U = 3$$

The inverse operation of $\times$ is $\div$. Divide both sides of the equation by 21.

Example 4

$$U \div 7 = 4$$
$$U \div 7 \times 7 = 4 \times 7$$
$$U = 28$$

The inverse operation of $\div$ is x. Multiply both sides of the equation by 7.

Equations may contain combinations of operations. If an equation involves a multiplication or division and an addition or subtraction, invert the addition or subtraction first.

$$3 \times U - 12 = 15$$
$$U \div 20 + 21 = 26$$

Example 5

$$3 \times U - 12 = 15$$
$$3 \times U - 12 + 12 = 15 + 12$$
$$3 \times U = 27$$

The inverse operation of − is +. Add 12 to both sides of the equation.

$$3 \times U \div 3 = 27 \div 3$$
$$3 \div 3 \times U = 27 \div 3$$
$$U = 9$$

The inverse operation of × is ÷. Divide both sides of the equation by 3.

Example 6

$$U \div 20 + 21 = 26$$
$$U \div 20 + 21 - 21 = 26 - 21$$
$$U \div 20 = 5$$

The inverse operation of + is −. Subtract 21 from both sides of the equation.

$$U \div 20 \times 20 = 5 \times 20$$
$$U = 100$$

The inverse operation of ÷ is ×. Multiply both sides of the equation by 20.

PRACTICE ASSIGNMENT 1

Plan: Find the value of the unknown in each of the following equations. Show the steps in each solution.

1. $U + 7 = 18$

2. $25 + U = 36$

3. $U - 3 = 14$

4. $12 \times U = 6$

5. $U \div 15 = 9$

6. $5 \times U - 3 = 17$

7. $U \div 6 + 32 = 34$

8. $11 \times U - 14 = 52$

9. $4 \times U + 27 = 29$

10. $U \div 12 + 3 = 19$

Sometimes you must simplify one or both sides of an equation before you can solve for the unknown. Numbers or similar products can be combined. If the unknown is on both sides of an equation, it must be placed on one side only. Sometimes it is helpful to reverse the sides of the equation.

Example 7 Simplify and reverse the sides.

$15 - 8 = 3 \times U - 10$ — Simplify: $15 - 8 = 7$

$7 = 3 \times U - 10$ — Reverse the sides.

$3 \times U - 10 = 7$

$3 \times U - 10 + 10 = 7 + 10$ — Add 10 to both sides.

$3 \times U = 17$ — Simplify: $7 + 10 = 17$

$3 \div 3 \times U = 17 \div 3$ — Divide both sides by 3.

$U = \frac{17}{3}$ — Express the answer as a mixed number.

$U = 5\frac{2}{3}$

Example 8 Combine similar products.

$12 \times U + 5 \times U = 75 - 10$ — Combine: $12 \times U + 5 \times U = 17 \times U$

$17 \times U = 75 - 10$ — Simplify: $75 - 10 = 65$

$17 \times U = 65$

$17 \div 17 \times U = 65 \div 17$ — Divide both sides by 17.

$U = \frac{65}{17}$ — Express the answer as a mixed number.

$U = 3\frac{14}{17}$

Example 9 The unknown is on both sides of the equation.

$18 + 7 \times U = 52 - 2 \times U$ — Add $2 \times U$ to both sides.

$18 + 7 \times U + 2 \times U = 52 - 2 \times U + 2 \times U$ — $7 \times U + 2 \times U = 9 \times U$

$18 + 9 \times U = 52$

$18 + 9 \times U - 18 = 52 - 18$ — Subtract 18 from both sides.

$9 \times U = 34$

$9 \div 9 \times U = 34 \div 9$ — Divide both sides by 9.

$U = 34 \div 9$

$U = \frac{34}{9}$ — Show the answer as a mixed number.

$U = 3\frac{7}{9}$

PRACTICE ASSIGNMENT 2

Plan: Find the value of the unknown in each of the following equations. Show the steps in each solution.

1. $14 - 9 = U + 3$

2. $3 \times U + 2 \times U = 47 - 11$

3. $4 \times U + 2 \times U - 7 = 45$

4. $5 \times U + 4 - 3 \times U = 18$

5. $3 \times U - 5 = 7 - U + 6$

6. $4 \times U + 2 + 3 \times U = 18 - U$

7. $7 \times U - 9 - 2 \times U = U + 6$

8. $2 \times U + 11 = 9 \times U + 3 - 5 \times U$

Section 4.4 Finding the Unknown in a Proportion

KEY TERMS **Proportion** **Cross product equation**

A **proportion** is a ratio equation with three given values and one unknown value: Ratio 1 = ratio 2 in which either the left or right ratio includes an unknown value. *U* symbolizes the unknown.

8 is to 3 as 16 is to ?

$8{:}3 = 16{:}U$

$\frac{8}{3} = \frac{16}{U}$ where *U* is the unknown

In a proportion, the product of the outer terms equals the product of the inner terms. This equation is called the **cross product equation.**

Cross Product Equation

Outer Terms **Inner Terms**

$$8 \times U = 3 \times 16$$
$$8 \times U = 48$$
$$8 \times U \div 8 = 48 \div 8$$
$$8 \div 8 \times U = 48 \div 8$$
$$U = 6$$

Example 1 A nut mix contains almonds and filberts in a 2:5 weight ratio. If the mixture contains three pounds of almonds, how many pounds of filberts does it contain?

$$\begin{aligned} \text{parts:pounds} &= \text{parts:pounds} \\ 2:3 &= 5:U \\ 2 \times U &= 3 \times 5 \\ 2 \times U &= 15 \\ U &= 15 \div 2 = 7.5 \end{aligned}$$

PRACTICE ASSIGNMENT 1

1. Find the unknown value in each of the following proportions.

a. 5 is to 4 as 10 is to _____

b. _____ is to 3 as 28 is to 12

c. $\frac{4}{U} = \frac{2}{13}$

d. $\frac{U}{15} = \frac{3}{20}$

e. $6:4 = 3:U$

f. $7.5:3 = U:9.5$

2. At a price of $392 for 25 square yards of carpeting, what will it cost to cover 16 square yards?

3. The profits of a small business are divided between two partners. For every $7 Chambers receives, his partner Lefevre receives $5. If Chambers's share amounts to $50,925, what is Lefevre's share? $36,375

4. The scale of a map is $\frac{3}{8}$ inch to 100 miles. What distance is represented by $2\frac{1}{4}$ inches? 600

5. It took a driver four hours to drive 224 miles. How long will it take the same driver to cover 413 miles under the same driving conditions?

Challenger

At the Limelight Theater, orchestra seats cost $15, and balcony seats cost $8. During one performance, three $8 seats were sold for every two $15 seats sold. Receipts for that performance totaled $5,292. How many seats were sold in the orchestra? How many in the balcony?

CHAPTER 4 SUMMARY

Concept	*Example*	*Procedure*
SECTION 4.1 Ratios		
Ratio	An automobile dealer sells two wagons for every five sedans.	
	2 to 5 2:5 $\frac{2}{5}$	A ratio is a relationship between two quantities expressed as a fraction.
Standardized ratio	0.06:1.4 = 3:70 $\frac{0.06}{1.4} = \frac{6}{140} = \frac{3}{70}$	The fraction has whole number terms and is in lowest terms.
	12 gal:32 qt = 3:2 $\frac{12\text{ gal}}{32\text{ qt}} = \frac{12\text{ gal}}{8\text{ gal}} = \frac{12}{8} = \frac{3}{2}$	Equalize the units of measurement and cancel them.
SECTION 4.2 Allocation by Ratio		
	Distribute $22,000 in the ratio of 3:5.	
	Portion 1 = $\frac{22000 \times 3}{8}$ = 8250	The first portion is the amount divided by the total number of parts, multiplied by the numerator.
	Portion 2 = $\frac{22000 \times 5}{8}$ = 13750	The second portion is the amount divided by the total number of parts, multiplied by the denominator.
SECTION 4.3 Finding the Unknown in a Simple Equation		
	$U - 5 = 13$ $U - 5 + 5 = 13 + 5$ $U = 18$	Apply the inverse operation.
Arithmetic operations		The arithmetic operators are: + − × ÷
Inverse operations		The inverse operators are: − + ÷ ×
	$4 \times U + 5 = 16$ $4 \times U + 5 - 5 = 16 - 5$ $4 \times U = 11$ $4 \div 4 \times U = 11 \div 4$ $U = \frac{11}{4} = 2\frac{3}{4}$	If the equation involves multiplication or division and addition or subtraction, addition or subtraction is inverted first.
SECTION 4.4 Finding the Unknown in a Proportion		
	If 25 sq ft cost $75.80, how much should 38 sq ft cost?	Write the cross product equation and solve for the unknown.
Cross product equation	$25{:}75.8 = 38{:}U$ $25 \times U = 75.8 \times 38$ $U = \frac{75.8 \times 38}{25} = 115.22$	Product of outer terms equals product of inner terms.

NAME ______________________________

CLASS SECTION ______________ DATE ______________

SCORE ____________ MAXIMUM __100__ PERCENT ____________

CHAPTER 4 ASSIGNMENT

1. Standardize the following ratios.

a. 0.07 to 0.5 **b.** 9 qt to 1.5 gal

(2 points)

2. Find the value of the unknown. Show each step.

a. $U - 82 = 112$

b. $6 \times U = 120$

c. $60 \times U + 10 = 550$

d. $U \div 8 + 49 = 54$

e. $2 \times U + 13 = 5 \times U + 8$

f. $12 + 3 \times U - 7 = 6 \times U - 3$

(6 points)

3. Find the unknown in each of the proportions.

a. $\frac{6}{U} = \frac{9}{21}$

b. $7{:}U = 6{:}18$

(2 points)

4. An automobile dealer sold 72 cars with a stick shift and 192 cars with automatic transmission. What is the ratio (in standard form) of cars with a stick shift to cars with an automatic transmission? *(5 points)*

5. A family earning $23,400 of take-home pay spends $4,500 for home insurance, automobile insurance, and health insurance. What is the ratio (in standard form) of insurance dollars to take-home dollars?

(5 points)

6. A laboratory technician must make a solution that contains seven parts alcohol to every three parts water. If the total volume of the solution is 500 cubic centimeters (cc), how many cc of each substance (water and alcohol) will be needed?
Water: ________ Alcohol: ________ *(5 points)*

7. If an estate worth $273,000 is to be divided between a son and a daughter in the ratio of 3:4, how much will each one receive?
Son: ____________ Daughter: ____________

(10 points)

8. A metal alloy weighing 1,570 pounds occupies 40 cubic feet of space. How many pounds of the same alloy will fit in 24 cubic feet of space? *(5 points)*

9. A manufactured article sells for $180. Materials and labor cost $60 and $80, respectively. The rest is profit. What is the ratio in standard form of profit to selling price? *(10 points)*

10. The advertising budget of New Age Appliances is apportioned between television and magazines at 13 to 7. If the total advertising budget is $145,000, how much money is allotted to each medium?
Television: __________ Magazines: __________
(10 points)

11. The vertical scale on a sales graph is $\frac{1}{2}$ inch to 100,000 cars. How many cars are represented by a vertical length of $1\frac{3}{4}$ inches? *(5 points)*

12. For every $7 raised by a charitable organization, $2 must be spent on fund-raising expenses. If the organization has collected $582,750 during its most recent fund-raising drive, how much will be used for expenses? How much will the organization net?
Expenses: __________ Net: __________
(10 points)

13. Company A has assets of $13,000 and liabilities of $8,000. Company B has assets of $20,000 and liabilities of $15,000.

a. What is the assets-to-liabilities ratio of company A?

b. What is the assets-to-liabilities ratio of company B?

c. Which company has more assets compared to its liabilities? *(15 points)*

14. A carnival fun house purchased fourteen mirrors at a total cost of $1,099. How many additional mirrors can be purchased for $471? *(5 points)*

15. A factory produces 150 kitchen tables in four weeks. How many working days (five days per week) are necessary to manufacture 110 tables?

(5 points)

Chapter 5 Electronic Tools

OBJECTIVES

- Compute with electronic calculators
- Compute with electronic spreadsheets

SECTIONS

INTRODUCTION

Not too many years ago, calculators were limited to the electromechanical, desk-model variety found in business offices. Today however, you will find a hand-held, battery- or solar-powered calculator tucked into the pockets, purses, briefcases, and school bags of most people you know. People in a variety of occupations, including students at every level, use hand-held calculators every day. While calculators cannot replace sound mathematical skills, they can reduce the tedium and long hours associated with many computational tasks.

Section 5.1 Introduction to Calculators

KEY TERMS **M-Memory** **Keyboard** **Display**

What Kinds of Calculators Are Available?

Calculators come in a vast array of styles and sizes, equipped with a wide variety of functions, from the most basic to the highly specialized. Prices range from relatively cheap to very expensive. Some calculators print out on paper tape, others display on a small screen, and some have the option of a printed tape, screen display, or both.

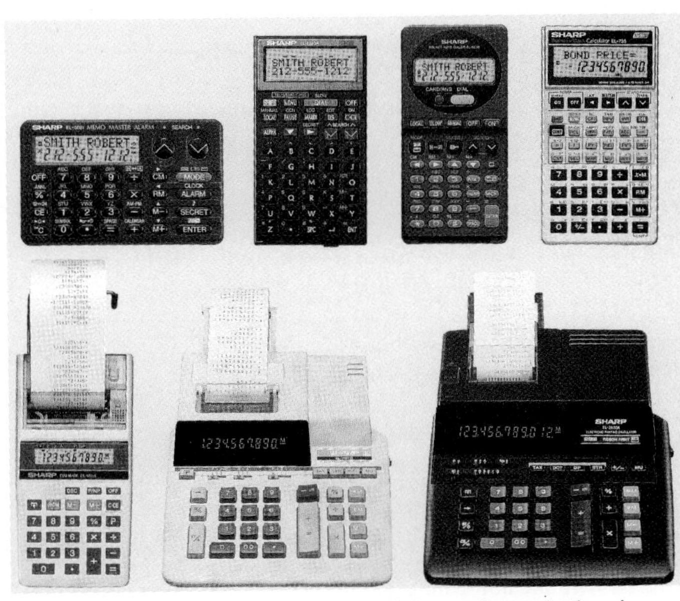

Courtesy of Sharp Electronics. Photo by Dorf & Stanton Communications, Inc.

Desk models continue to be used in business settings. These calculators perform the four arithmetic operations, provide totals and subtotals, and display floating decimal point as well as fixed decimal point numbers. Some models display commas in the results, and many contain one extra memory called M.

The most basic hand-held calculator performs the four arithmetic operations and usually contains **M-memory.** The next category is more expensive and is designed to perform advanced mathematical tasks including algebraic grouping and exponentiation, logarithmic and trigonometric functions. The most expensive hand-held calculators are equipped with special features required in specific fields, such as finance, real estate, and engineering. Many of these calculators may be programmed and are essentially small computers.

What Kind of Calculator Will You Need for This Text?

As long as your calculator contains M-memory, you can perform the computations in this text.

There are some special keys (parenthesis, change-of-sign, and exponentiation keys) that are helpful and will be explained. They are not necessary, however, and alternative techniques for computing will be demonstrated.

What Are the Basic Components of Electronic Calculators?

Calculators have a **keyboard** on which entries are made, a **display** (and/or tape) on which entries and results appear, and memory in which operations are performed. Most calculators have one extra memory called M controlled by three keys.

Add to Memory	[M+]	This key adds the content of the display to the content of M-memory.
Subtract from Memory	[M−]	This key subtracts the content of the display from the content of M-memory.
Memory Recall or Recall Memory	[MR] or [RM]	This key displays the content of M-memory.

Before you begin a keystroke sequence, be sure that all memories are empty to avoid erroneous results. Clear with the clear all [CA] key, by turning the calculator off, or by keying 0 (zero) into M.

Some calculators have an [M=x] key that replaces the content of M-memory with the content of the display. Substituting [M=x] for [M+] at the beginning of a keystroke sequence will properly initialize the memory. You will not need the [M−] key very often. Calculators that have [M+] but not [M−] may feature [+/−], the change-of-sign key. This key reverses the sign of a number in the display from positive to negative (or from negative to positive). In cases where you need [M−], substitute the sequence [+/−] [M+] [+/−] for [M−].

On desk calculators there is often an [M*] key (Memory Total) and an [M◇] key (Memory Subtotal). The [M*] key displays M-memory and clears it; the [M◇] key displays M-memory but does not clear it. For the calculator sequences in this text, you can use either one of these keys, but [M*] is preferable. To keep these sequences uniform, [MR] is always specified.

Section 5.2 Addition and Subtraction

KEY TERMS **Arithmetic logic** **Algebraic logic** **Floating point display** **Editing**

Calculators Compute Correctly if Keyed Correctly

For addition and subtraction, many desk calculators employ arithmetic logic, while most hand-held calculators employ algebraic logic. In this section you will learn keystroke sequences for both types of calculators.

In **arithmetic logic,** enter the [+] or [−] operator after the number; the [=] key is not required to obtain the sum or difference. In **algebraic logic,** enter the [+] or [−] operator between the numbers; press the [=] key to display the sum or difference.

A desk calculator usually displays a maximum of twenty digits, while a hand-held calculator displays a maximum of seven to ten digits.

Computations

37 + 10
37 − 10

Arithmetic (arith) Logic

key: 37 [+] 10 [+]
37 [+] 10 [−]

Algebraic (alg) Logic

key: 37 [+] 10 [=]
37 [−] 10 [=]

How Do Calculator Entries or Displays Differ from Paper-and-Pencil Notation?

If no other digits follow, the right-most zeros after the decimal point may not appear in the display and need not be entered on the keyboard.

7.12100000 can be entered as 7.121 and may display as 7.121.

Numbers having only zeros as fractional digits can be entered as whole numbers. The calculator may supply the decimal point.

37.000 can be entered as 37 and will display as 37 or as 37..

If the whole part of a number is zero, the zero to the left of the decimal point need not be entered; the calculator will supply it.

0.478 can be entered as .478 and will display as 0.478.

Some desk calculators feature an F-marker that allows you to fix the decimal point and view a fixed point display. However, because many calculators do not feature a fixed point display, this text always assumes a **floating point display.** In a floating point display, the decimal point does not always appear in the same position, and the portion of the display to the right of the point is filled with fractional digits.

Many calculators round results; however, some do not. Therefore, the same problem may yield a slightly different answer depending upon the calculator.

Mathematical Answer	8-digit Floating Point Display	12-digit Floating Point Display
0.4545 . . .	0.4545454 0.4545455 rounded	0.45454545454 0.45454545455 rounded
1946.4166 . . .	1946.4166 1946.4167 rounded	1946.41666666 1946.41666667 rounded

Although some desk calculators insert commas into whole number results, in this text, you will perform your own **editing.** How you edit depends on the application. Editing may require:

1. Inserting the dollar sign if the result is money;
2. Inserting commas for better readability;
3. Rounding to the desired number of places (two places, if money);
4. Changing a number to percent form.

	Round to Nearest Tenth	Change to Money Format	Insert Commas	Change to Percent	Change to Percent and Round
Display	534.7894	534.7894	4876392	.0542	.0542
Edited	534.8	$534.79	4,876,392	5.42%	5.4%

The next three examples pertain to money.

Example 1 Compute the sum of 503.74, 73.24, and 14.007.

key alg: 503.74 [+] 73.24 [+] 14.007 [=]
key arith: 503.74 [+] 73.24 [+] 14.007 [+]
display: 590.987
edited: $590.99

Example 2 Compute the difference 503.97 − 73.264.

key alg: 503.97 [−] 73.264 [=]
key arith: 503.97 [+] 73.264 [−]
display: 430.706
edited: $430.71

Example 3 Compute 503.74 − 73.24 + 13.40 − 5.20 + 803.00.

key alg: 503.74 [−] 73.24 [+] 13.4 [−] 5.2 [+] 803 [=]
key arith: 503.74 [+] 73.24 [−] 13.4 [+] 5.2 [−] 803 [+]
display: 1241.7
edited: $1,241.70

PRACTICE ASSIGNMENT

Plan: Use your calculator to find these answers.

1. Compute. Show all decimal places. Do not round or insert dollar signs, but insert commas where appropriate.

   a. 91700 + 551.923 + 8361.44 =

   b. 571.4 + 816 + 400.17 + 26.404 =

   c. 36109.4 + 11968.29 + 40377 + 8236.49 + 22701.55 =

   d. 7050.3 − 9.7294 =

   e. 73.44 − 48.229 − 13.9011 =

   f. 1.868 − 0.1415 − 0.0527 − 0.00054 − 0.00109 =

   g. 219.64 + 308.61 − 153.92 =

   h. 841.3 − 91.62 + 108.366 =

   i. 2854 + 716.3 − 812.4 + 45.25 − 140.9 =

   j. 9 − .0063 + .0135 + .002 − 1.1372 =

2. The children's department of a retail store made the following purchases:

Jan	41,390.57	July	65,045.76
Feb	49,578.45	Aug	66,514.90
Mar	69,873.21	Sept	58,998.64
Apr	60,883.30	Oct	71,209.75
May	62,335.48	Nov	73,541.80
June	60,121.06	Dec	52,208.49

   What were the total purchases for the year?

3. When the New York Stock Exchange opened on Monday morning, the Dow Jones Industrial Average stood at 2041.62. At the close of Monday's trading, the average had gained 16.7 points. On Tuesday the Dow dropped 50.09 points; on Wednesday it dropped another 18.33 points; on Thursday it gained 39.94; and on Friday it lost 14.5 points. What was the Dow Jones average at the end of the week?

Section 5.3 Multiplication and Division

KEY TERM **Order of operations**

Use the same keystroke sequences for multiplication and division on both algebraic and arithmetic calculators. Enter the [×] or [÷] operator between the two numbers and press the [=] key to display the product or quotient.

973 × 84
key: 973 [×] 84 [=]
display: 81732.

1407.3 ÷ 23.8
key: 1407.3 [÷] 23.8 [=]
display: 59.130252

The next three examples pertain to money.

Example 1 Compute the product of 742.3 and 14.3.

742.3 [×] 14.3 [=]
display: 10614.89
edited: $10,614.89

Example 2 Compute the quotient 8005.25 ÷ 3.54.

8005.25 [÷] 3.54 [=]
display: 2261.3701
edited: $2,261.37

Example 3 Compute 742.3 × 0.03 × 14.3 × 8.92 × 17.0.

742.3 [×] .03 [×] 14.3 [×] 8.92 [×] 17 [=]
display: 48289.258
edited: $48,289.26

The **order of operations** is important when you use a calculator. Expressions must be evaluated in the following order:

1. Parenthesized groupings
2. Exponentiation or roots
3. Multiplication or division
4. Addition or subtraction

Unfortunately, not all calculators do this evaluation correctly.

17 + 39 × 3 Multiplication should be performed before addition.

Type of Calculator	Keystroke Sequence	Result	
algebraic (inexpensive model)	17 [+] 39 [×] 3 [=]	168	incorrect
algebraic (better model)	17 [+] 39 [×] 3 [=]	134	correct
arithmetic	17 [+] 39 [+] [×] 3 [=]	168	incorrect
Alternative for calculators that do not perform this computation correctly:			
algebraic	39 [×] 3 [+] 17 [=]	134	correct
arithmetic	39 [×] 3 [=] [+] 17 [+]	134	correct

623 − 189 ÷ 43 Division should be performed before subtraction.

Type of Calculator	Keystroke Sequence	Result	
algebraic (inexpensive model)	623 [−] 189 [÷] 43 [=]	10.093023	incorrect
algebraic (better model)	623 [−] 189 [÷] 43 [=]	618.60465	correct
arithmetic	623 [+] 189 [−] [÷] 43 [=]	10.093023	incorrect
Alternative for calculators that do not perform this computation correctly:			
algebraic	189 [÷] 43 [=] [M+] 623 [−] [MR] [=]	618.60465	correct
arithmetic	189 [÷] 43 [=] [−] 623 [+]	618.60465	correct

Check your calculator to see how it performs these computations.

PRACTICE ASSIGNMENT

Plan: Compute these problems on your calculator.

For problems 1 and 2, show all decimal places; do not round or insert dollar signs, but insert commas where appropriate.

1. Compute.

a. 5729 × 2634 =

b. 17.941 × 8.38 =

c. 0.045 × 0.0055 =

d. 3153 ÷ 16.8 =

e. 836 ÷ 1740 =

f. 531.08 ÷ 0.319 =

g. 0.012 ÷ 0.0073 =

h. 1.75 × 6.5 × 3.75 =

i. 64 × 3.8 × 0.4 × 0.071 =

j. 22.8 × 34 × 1.36 × 4.5 × 2.05 =

2. Remember the order of operations when computing the following problems. Use the keystroke sequences required for your calculator to produce the correct result.

a. 721.4 + 157.88 × 23 =

b. 81.735 + 2.82 ÷ 0.17 =

c. 377.4 + 31.2 × 18.97 × 4.11 =

d. 954.2 − 371 ÷ 5.33 =

e. 583.6 + 80.66 − 14.9 × 38.04 =

3. Complete the following invoice.

INVOICE

The Statler Restaurant Equipment Company

Date: June 1, 199X

Item	Quantity	Description	Unit Price		Extension	
782-B	42	Napkin holders	3	45		
449-K	15	Salad sets	12	39		
913-Z	225	Silverware sets	5	25		
746-B	15	Ice tongs	4	30		
		Total for Merchandise				

4. A baseball player's batting average is calculated by dividing the number of hits by the number of times at bat. In 347 times at bat, "Bubba" Davis had 114 hits. What is his batting average (correct to the nearest thousandth)?

Section 5.4 Spreadsheet Concepts

KEY TERMS **Columns** **Rows** **Cells** **Cell name** **Cell pointer** **Numeric values** **Text** **Formula**

Using spreadsheets may be a new experience for you. To help you, a disk containing partially completed spreadsheets (called templates) is available. The mechanics of creating headers, formatting, and setting column widths are already done for you, so that you can concentrate on the mathematics.

If you do not have access to the necessary equipment, Section 5.5 as well as all spreadsheet assignments may be omitted. You can profit from studying Section 5.4 in either case.

Before the advent of the computer age, business people used paper work sheets containing columns and rows to organize numerical information. Computers entered the scene bringing to business and industry many innovations, among them the electronic spreadsheet. The options for manipulating information to produce a wide variety of reports at the press of a button continues to persuade many businesses to convert to electronic spreadsheets.

The electronic spreadsheet features **columns** identified by letters, and **rows** identified by numbers. **Cells** are formed by the intersection of columns and rows, and are identified by column letter followed by row number. The **cell name** of the cell located at the intersection of column C and row 2 is C2.

Use the arrow keys to move the **cell pointer** from cell to cell. The cell pointer identifies the location by highlighting the cell. If the cell pointer is on cell E4, that cell is highlighted and data or a formula can be entered.

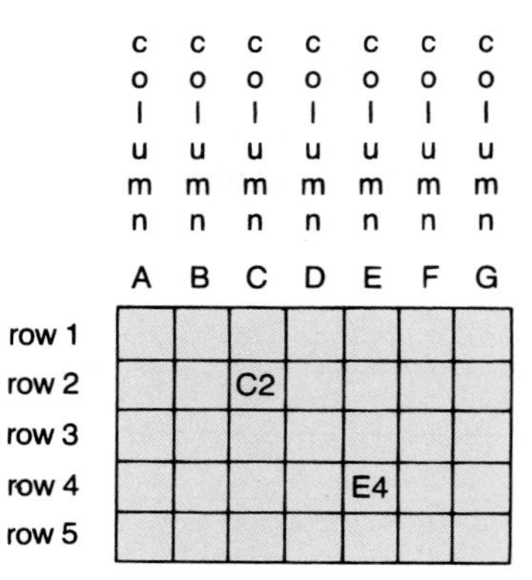

A cell may contain one of the following:

1. **Numeric values.** The left-most character must be a digit, sign, or decimal point.

    5877
    −35.82
    .0225

2. **Text** consisting of headers and other identifying information, sometimes called labels. The left-most character must be a letter. However, numbers can be text if preceded by an apostrophe.

    month
    expenses
    net pay
    '25 credits

3. A **formula** addressing the contents of other cells. A formula computes the content of a cell from values in the addressed cells. The left-most character must be a digit or one of these symbols: + − @ ($ #.

    +C3*C4
    +B12−.5*C16
    @ABS(B6−D10)
    @SUM(A5..A15)

In a formula, symbols denoting multiplication and division operations × and ÷ are * and /, respectively. Exponentiation is denoted by ^. A spreadsheet program (in the absence of parentheses) executes a formula in the following order:

1. Exponentiation
2. Multiplication and division from left to right
3. Addition and subtraction from left to right

+K5*K7 not +K5×K7
+C3/C4 not +C3÷C4

Average

(B1+C2+D3)/3
+B1+C2+D3/3 is incorrect. For an average, the additions must be performed prior to the division.

Parentheses alter the order of operations so that the operations inside parentheses are performed first (in the above order). When there are parentheses within parentheses, the operations inside the innermost parentheses are performed first.

Translating fractional algebraic expressions into spreadsheet language requires extra care. In algebra, the numerator and/or denominator of a fraction constitute groups having implied parentheses. You must insert these parentheses into computer expressions and calculator sequences.

Comparison of Fractional Expressions in Algebra and Spreadsheet Language

Assume that cells A3, B3, C3, and D3 contain numeric values. The initial + sign in the spreadsheet code is often necessary because in most spreadsheet packages an initial letter signifies a text entry; the + sign indicates a numeric entry.

Algebra	Spreadsheet Formula
$\frac{a}{b}$	+A3/B3
$\frac{a+b}{c}$	(A3+B3)/C3
$\frac{a}{b+c}$	+A3/(B3+C3)
$\frac{a+b}{c+d}$	(A3+B3)/(C3+D3)
$\frac{ab}{c}$	+A3*B3/C3
$\frac{a}{bc}$	+A3/(B3*C3)
$\frac{ab}{cd}$	+A3*B3/(C3*D3)

An Example of Composing a Spreadsheet Formula

Algebraic Formula	**Meaning of Formula**
$A = P(1 + RT)$	amount = principal × (1 + rate × time)

Locations of Variables

principal P in B5   rate R in C5   time T in E3

Spreadsheet Formula	
+ B5* (1 + C5*E3)	If the amount A is to be displayed in cell G5, enter the spreadsheet formula into G5.

Section 5.5 Working with Your Templates

KEY TERMS **Cell location** **Cell address** **Relative Address** **Absolute address** **Cell range** **Command menu** **Menu pointer** **Copy command**

In this section you will learn only those skills required to complete the templates on the disk provided with this text. Consult a text on electronic spreadsheets or a software manual for more detailed information. If hardware or software is not available, omit this section and proceed to Challenger (page 60) and Chapter 5 Assignment (page 65).

How to Avoid Formula Pitfalls

Consider the simple formula B2+C5. If the leftmost character is a letter, the formula will enter the cell as text (and as text will not execute). Therefore, precede the formula by a sign (+) indicating that it is not text. The instructions for the spreadsheet assignments will indicate where + signs are needed.

B2+C5 does not execute
+B2+C5 executes
−B2+C5 executes
(B2+C5)/5 executes
@SUM(B2..B10) executes
When a formula begins with a cell address, precede the formula with a + sign.

Relative and Absolute Cell Addresses

A cell name in a formula references a **cell's location,** and is called the **cell address.**

−B7*(K3+.5*K7)

B7, K3 and K7 are cell names, cell locations, and cell addresses.

When a formula is copied from one cell to other cells, the addresses may have to be adjusted. For example, Formula +A1−A3 in column A must become +B1−B3 when it is copied to column B.

B7 is a relative address. It may become B17 if copied ten rows down, or E7 if copied three columns across.

The software performs this adjustment automatically. An adjustable address is called a **relative address.**

An **absolute address** is an address in which either the row or column designation (or both) must not be adjusted. Absolute addresses are designated with dollar signs ($). Column designation, row designation, or both may be preceded by a dollar sign. The instructions for the spreadsheet assignments will indicate where dollar signs are needed.

$B7 contains an absolute column designation. B remains constant.

B$7 contains an absolute row designation. 7 remains constant.

B7 contains absolute column and row designations. Both B and 7 remain constant.

What Is a Cell Range?

A **cell range** defines a rectangular area of one or more cells. Cell range is expressed as

upper left cell..lower right cell.

A cell range that defines a row, column, or part of a row or column is denoted by

left row limit..right row limit

or

upper column limit..lower column limit.

The range C5..L5 defines a row containing ten cells.

The range D4..D13 defines a column containing ten cells.

The range A1..J10 defines a rectangle containing 100 cells.

Command Menu

To bring the main **command menu** to the screen, press /. Use the highlighted **menu pointer** to select a command. Then press [enter]. You can key the first letter of the command instead of using the menu pointer. Each command may bring a sub-menu to the screen. The Copy command will be required for completing most of the spreadsheet assignments.

The Copy Command

The **Copy command** saves time and makes the electronic spreadsheet a powerful business tool. You may copy the contents of one cell to another cell, one cell to a range of cells, or a range of cells to another range of cells.

Copy cells A1..D1 to cells F4..I4.

/
Copy[enter]
inquiry for "from range" or "copy what?"
A1..D1[enter]
inquiry for "to range" or "to where?"
F4..I4[enter]

Key in two ranges, where to copy from and where to copy to ("from range" and "to range" or "copy what?" and "to where?"), following each range with [enter]. When you copy more than one cell with one Copy command, be sure to copy to an equal or greater number of cells.

Another method of defining ranges, applied by experienced users, is highlighting the two ranges. We do not recommend the highlighting method to beginners, however. If you are interested, consult a spreadsheet text or manual.

Key carefully and accurately. If you make a mistake, press the escape key [esc] to cancel the last step. Then proceed with the correct step.

How Do I Bring a Template to the Screen?

Use the file retrieval commands to bring the template to the screen. If, when retrieving, the correct disk drive is not displayed (for example, the program is in drive A: and the templates are in drive B: but the retrieve command displays A:), you can specify B: with the directory method.

The template files are named CHAPn; *n* denotes the number of the chapter in which the exercise appears. The practice template for this chapter is filed under CHAP5.

/
File[enter]
Directory[enter]
inquiry for current directory
B:[enter]

or

/FD
inquiry for current directory
B:[enter]

/
File[enter]
Retrieve[enter]
list of files
CHAP5.wk1[enter]

or

/FR
list of files
CHAP5.wk1[enter]

Saving a Spreadsheet

Use the save command to place your worksheet onto the disk. This should be done not only at the end of a session, but also periodically during each session in order to safeguard against losses of internal memory due to power failures, accidental program exits, or other problems.

/
File[enter]
Save[enter]
inquiry for filename
SPREAD5[enter]

or

/FS
inquiry for filename
SPREAD5[enter]

Save your spreadsheet the first time under the name SPREADn in order to maintain the original CHAPn template. Use a similar procedure for saving the spreadsheet a second, third, or fourth, . . . time. When the program asks if you wish to replace the worksheet already saved under SPREADn, choose the *replace* option. This allows you to save the original template named CHAPn and the latest version of your spreadsheet named SPREADn.

/
File[enter]
Save[enter]
SPREAD5[enter]
replace[enter]

or

/FS
SPREAD5[enter]
R

Printing a Spreadsheet

Be sure the printer is turned on and the paper aligned. Perform any other tasks required to put the printer into ready status.

The Print command has many options not shown here. The range is the range occupied by the spreadsheet. Follow these steps carefully to print your spreadsheet.

/
Print[enter]
Printer[enter]
Range[enter]
A1..H25[enter]
GO[enter]

or

/PPR
A1..H25[enter]
G

Must I Complete a Spreadsheet Exercise in a Single Session?

It is not necessary to complete a spreadsheet exercise in a single session. Save your work before quitting and retrieve it another time. Take plenty of time to become familiar with the practice spreadsheet in this chapter before you move on to the next spreadsheet exercises.

Challenger

Craig Lomax took $2,350 to spend while touring Switzerland. If the currency exchange rate was 1.47 Swiss francs per dollar, how many Swiss francs did Mr. Lomax receive in exchange for his American dollars (correct to the nearest hundredth)? He spent 2,762.35 Swiss francs and had the remainder converted back to American money (at the same rate of exchange). How much money did he have left?

CHAPTER 5 SUMMARY

Concept	*Example*	*Procedure*
SECTION 5.1 Introduction to Calculators		
Extra memory M		
Add to memory	5317.8[M+]	Add display to M-memory.
Subtract from memory	5317.8 [M−]	Subtract display from M-memory.
Recall memory	[RM]	Place memory into display.
SECTION 5.2 Addition and Subtraction		
Arithmetic logic for calculator addition or subtraction	318.984 [+] 507.06 [+] 826.044 397.57 [+] 124.89 [−] 272.68	[+] operator is entered after number. [−] operator is entered after number.
Algebraic logic for calculator addition or subtraction	318.984 [+] 507.06 [=] 826.044 397.57 [−] 124.89 [=] 272.68	[+] operator is entered between numbers. [=] displays sum. [−] operator is entered between numbers. [=] displays difference.
SECTION 5.3 Multiplication and Division		
Calculator multiplication	48.52 [×] 3.4 [=] 164.968	[×] operator is entered between numbers. [=] displays product.
Calculator division	1893.4 [÷] 375 [=] 5.0490666	[÷] operator is entered between numbers. [=] displays quotient.
Order of operations	547 + 391 ÷ 23 = 564	In the absence of parentheses, multiplication or division must be performed before addition or subtraction.
SECTION 5.4 Spreadsheet Concepts		
Electronic spreadsheet		
Column designation	column H	Columns are designated by letters.
Row designation	row 6	Rows are designated by numbers.
Cell	cell H6	The intersections of columns and rows form cells.
Numeric value	5877 −35.82 .0225	The left-most character is a digit, sign, or decimal point.
Text	expenses '.25 × credits	The left-most character is alphabetic or an apostrophe.
Formula	5*E1 +B12−.5*C16 (B1+B2+B3)/3 @AVG(B1..B12) (B1+D3−E4)/3 .5*(C3+F7)	The left-most character is a digit or symbol: + − @ ($ #. The arithmetic operators are: * / + −. Parentheses alter the order of operations.
SECTION 5.5 Working with Your Templates		
Initial sign	+C5−D3	If a formula begins with a cell address, it must be preceded by + sign.
Relative cell address	B6*D6 becomes D7*F7	Cell addresses are adjusted when the formula is moved to another cell.
Absolute cell address	$F8 absolute column F$8 absolute row F8 absolute column and row	Cell addresses are not adjusted.
Cell range	A1..E10	A cell range is a rectangle of cells denoted by upper left cell..lower right cell.
Command menu		Press / to bring a selection of commands to the screen.
Copy command	/ Copy[enter] prompt "from range" B3..C10[enter] prompt "to range" D5..E12[enter]	The copy command copies the contents of one cell to another cell, or the contents of a range of cells to another equal or greater range of cells.
Retrieving a template	/ File[enter] Retrieve[enter] list of files CHAP8.wk1[enter]	The templates are files named CHAPn. Use the file retrieval commands to bring the template to the screen.

CHAPTER 5 SUMMARY—(*Continued*)

Concept	*Example*	*Procedure*
Saving a spreadsheet	/ File[enter] Save[enter] prompt for filename SPREAD8[enter]	The first time, save your spreadsheet under a new name, SPREADn. Thereafter use the replace option.
Printing a spreadsheet	/ Print[enter] Printer[enter] Range[enter] A1..G20[enter] GO[enter]	Turn the printer on and align the paper. Choose the appropriate Print command options and print your spreadsheet.

Chapter 5 Spreadsheet Practice Exercise

Personal Budget

This spreadsheet displays a simple personal budget. Bring your Chapter 5 template to the screen.
It will appear as follows. Remember that you may need to change to the correct disk drive first. For example, if the templates are in drive B:

/
File[enter]
Directory[enter]
inquiry
B:[enter]
or
/FD
inquiry
B:[enter]

/
File[enter]
Retrieve[enter]
list of files
CHAP5.wk1[enter]
or
/FR
list of files
CHAP5.wk1[enter]

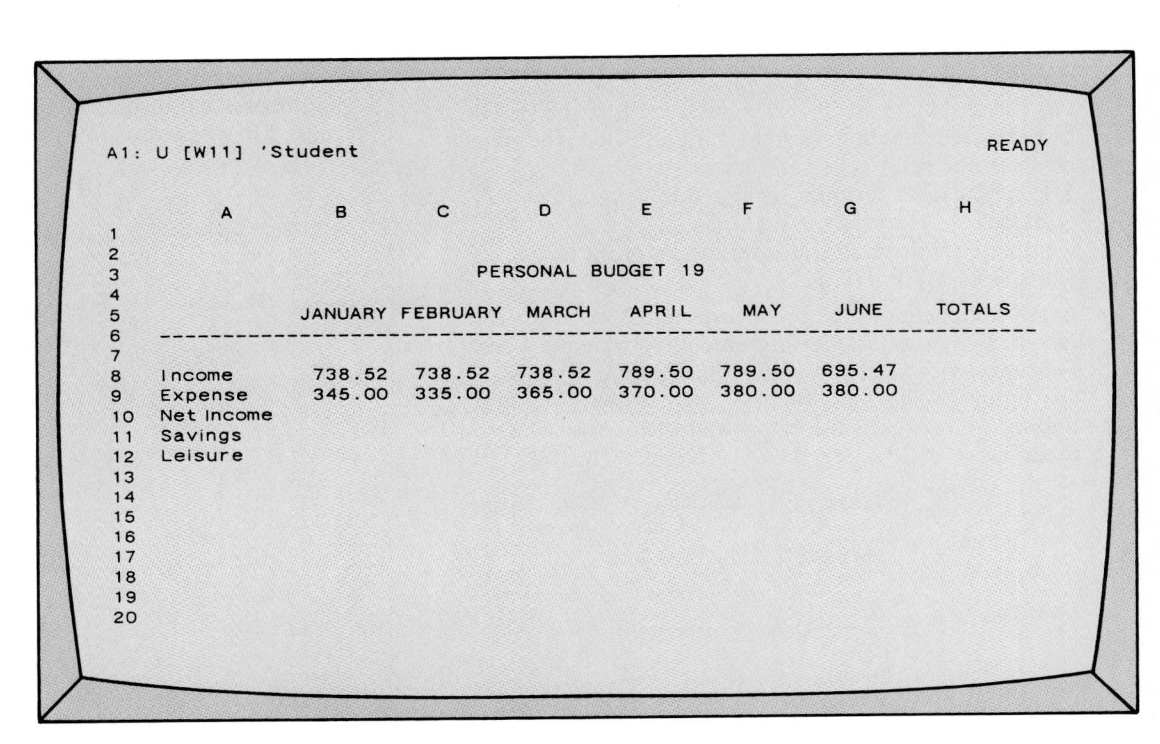

You may enter your name into cell A1 and the last two digits of the year into cell F3. For example, enter 91 as '91. (These entries are not essential.)

Column A contains labels; columns B–G are reserved for data pertaining to the months of January through June, and column H is reserved for totals.

Row 8 contains incomes, row 9 expenses. Row 10 is reserved for net incomes, row 11 for savings, and row 12 for leisure money.

Review the example of composing a spreadsheet formula at the end of Section 5.4. Then complete the exercise.

Be sure to enter all formulas into the computer without spaces between the characters.

Spreadsheet Exercise Steps

1. Row 10: net income
   Net income = income − expense.
   Income is in cell B8; expense in cell B9.
   A spreadsheet formula that begins with a cell address must be preceded by a sign (+).
   The spreadsheet formula is
   + ______ − ______ .
   Complete the formula for net income and store it in cell B10.

2. Row 11: savings
   Assume the savings rate is 50% of net income (50% = .5).
   Net income is in B10.
   The spreadsheet formula is ______ * ______
   Complete the formula for savings and store it in cell B11.

3. Row 12: leisure
   Leisure = net income − savings.
   Net income is in cell B10; savings in cell B11.
   A spreadsheet formula that begins with a cell address must be preceded by a sign (+).
   The spreadsheet formula is
   + ______ − ______ .
   Complete the formula for leisure and store it in B12.
4. Copy the formulas from B10..B12 to C10..G12. (Review the Copy command in Section 5.5.)
5. Column H: totals
   This column will display the totals for January through June.
   For row 8: H8 = B8 + C8 + D8 + E8 + F8 + G8.
   Use the sum function @SUM(range), where the range contains the cells to be summed.
   The spreadsheet formula for row 8 is
   @SUM (______ .. ______) .
   Complete the formula for totals and store it in H8.
   Copy from H8 to H9..H12.
6. After you complete the preceding steps 1–5, verify net incomes, savings, and leisure amounts for January and February with your calculator. Verify total savings. If any figure is incorrect, "debug" your program by first finding what caused the errors and then correcting them.

	January	**February**	**Total Savings**
Net Income	______	______	
Savings	______	______	______
Leisure	______	______	

7. Change the savings rate in cell B11 from 50% to 60%. You can change .5 to .6 by reentering the savings formula. Recopy the altered formula from B11 to C11..G11. All related figures (savings, leisure and totals) will adjust themselves accordingly.

   Verify savings and leisure for January and February. Then verify total savings.

	January	**February**	**Total Savings**
Savings	______	______	______
Leisure	______	______	

Note: In many of the spreadsheet exercises you will find occasional discrepancies of one or two cents caused by the software's rounding procedures.

NAME ______________________________

CLASS SECTION ______________ DATE ______________

SCORE __________ MAXIMUM 100 PERCENT __________

CHAPTER 5 ASSIGNMENT

1. Workers in a laboratory need to know how much magnesium is on hand. One bottle contains 84.3 grams, another bottle 25.69 grams, a third bottle 6.735 grams, a fourth 108.3 grams, and a fifth 2.004 grams. How many grams of magnesium are on hand? *(5 points)*

2. Cathy O'Hare belongs to the Frequent Flyers Club. After flying 50,000 miles, a club member is entitled to a free, round-trip flight to Europe. On five trips Cathy has flown 982, 3086, 1883, 4109, and 5415 miles. How many more miles must she fly to qualify for the free trip? *(5 points)*

3. A student purchased a double hamburger for $1.35, a milk shake for $0.85, an order of french fries for $0.65, and two cookies at $0.23 each. He used a five-dollar bill to pay for his purchase.

a. What was the total cost?

b. What was the correct change? *(10 points)*

4. Profits totaling $794,814 are to be distributed equally among the fifteen owners of a private corporation. However, each owner must pay $2,763 to cover legal expenses. What is the net amount due each owner? *(5 points)*

5. The Elegante Fashion Company pays each of its twenty-eight knitters $7.49 per hour. If each knitter works 37.5 hours per week, what is the total payroll (excluding fringe benefits) for a 39-week season? *(12 points)*

6. Four trucks from Humboldt Trucking Company were loaded beyond maximum allowable limits. Fill in the missing information on the following table.

Truck	Present Load	Maximum	Overload
Truck A	9.2395 tons	8 tons	______
Truck B	12.55	10.765	______
Truck C	10.093	8.07	______
Truck D	15.99	11.1175	______
		Total Overload	______

(5 points)

7. While shopping at a supermarket, Heidi Rostow discovered that the items in her shopping cart cost a total of $58.67, slightly more than she could afford. Before going to the checkout, she put back a bag of Nacho Delights costing $1.79 and a box of Creme de Creme cookies costing $2.45. She also replaced a gourmet frozen dinner costing $4.98 with a generic beans and franks dinner costing $1.69. How much did she spend at the checkout? *(5 points)*

8. An electronics company stocks its radios 28 to a carton. The company received radio orders from four retail stores. One store ordered 67 radios, another 94, a third 82, and the fourth 105. How many cartons must be taken from the warehouse to fill these orders? *(5 points)*

9. A pitcher on the Reds gave up 34 earned runs in 207 innings pitched.

a. What is the average number of earned runs given up per inning? (Round to the nearest thousandth.)

b. What is the pitcher's earned run average for a nine-inning game? (Round to the nearest hundredth.) *(10 points)*

10. Gasoline costs $1.41 per gallon at the full-service pump and $1.33 per gallon at the self-service pump.

a. How much do 18 gallons cost at the full-service pump?

b. How much do 18 gallons cost at the self-service pump?

c. How much less do 18 gallons cost at the self-service pump than at the full-service pump?

d. How many gallons, rounded to the nearest tenth, does $20 buy at the full-service pump?

e. How many gallons, rounded to the nearest tenth, does $20 buy at the self-service pump?

f. How many fewer gallons, rounded to the nearest tenth, does $20 buy at the full-service pump than at the self-service pump? *(18 points)*

11. Three types of spreadsheet entries are: ________________, _______, and __________. *(6 points)*

12. Column identifications on a spreadsheet are _________; row identifications are ___________. *(4 points)*

13. The identification (cell name) of the cell formed by the intersection of column H and row 8 is _______. *(2 points)*

14. The highlighted rectangle indicating the cell that can be accessed is called the cell __________. *(2 points)*

15. If cell C9 contains principal, D10 rate, and E11 days, what is the spreadsheet formula for

$$\frac{\text{principal} \times \text{rate} \times \text{days}}{365}?$$

(6 points)

Chapter 6 Units of Measurement: The Metric System

OBJECTIVE

- Understand standards of measure

SECTIONS

INTRODUCTION

Without standard weights and measures, shopping, trade, industry, and recreation would be in hopeless confusion. While most Americans use the **English inch-pound system** of measure, business and industry in the United States are in the midst of a gradual, if controversial, transition to the standard international **metric system** (SI).

In an economy that is becoming more global every day, a universal system of measure is essential. The English system, with its illogical, historically based units (a foot, for example, was the **length** of the king's foot) must give way to a universal system recognized and accepted by all participants in an international economy. Britain herself adopted metrics in the early 1970s.

During this period of transition many American companies maintain two systems of measure, the English system for goods sold in the U.S. and the metric system for exports. However, keeping two systems is inefficient and expensive, and in time U.S. businesses will drop the English system and use metrics exclusively. In this chapter, you will see how similar quantities are measured in each system.

Section 6.1 Measuring Length

KEY TERMS **English system** **Inch-pound system** **Metric system** **Length** **Meter** **Kilo** **Centi** **Milli**

English

To convert between units, you must memorize a set of seemingly unrelated numerical facts.

1 mile (mi) = 5,280 feet
1 yard (yd) = 3 feet
1 foot (ft) = 12 inches
1 inch (in.)

A person who measures more than six feet is considered tall. How many inches are in six feet?

$$6 \text{ ft} = 6 \times 12 \text{ in.} = 72 \text{ in.}$$

Metric

The metric system, based on the number 10, lends itself more readily to decimal computations.

1 kilometer (km) = 1,000 meters
1 meter (m) = 100 centimeters
1 centimeter (cm) = 10 millimeters
1 millimeter (mm)

The **meter** is the basic unit.

Kilo means one thousand (1000).
Centi means one-hundredth (0.01).
Milli means one-thousandth (0.001).

A person who measures 2 meters is tall. How many centimeters are in two meters?

$$2 \text{ m} = 2 \times 100 \text{ cm} = 200 \text{ cm}$$

A meter is a little longer than a yard, about 1.1 yards.

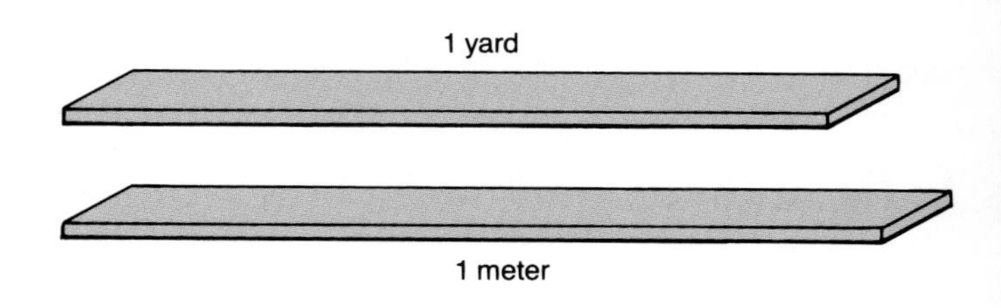

Conversions between Systems

1 mile = 1.6093 kilometers
1 yard = 0.9144 meters
1 inch = 2.5400 centimeters

In the United States, the standard office paper size is $8\frac{1}{2} \times 11$ in. What is this size in centimeters?

$8.5 \times 2.54 = 21.59$
key: 8.5 [×] 2.54 [=]
display: 21.59
edited: 21.6 cm

$11 \times 2.54 = 27.94$
key: 11 [×] 2.54 [=]
display: 27.94
edited: 28 cm

The standard office paper size in centimeters is

$$21.6 \times 28 \text{ cm}.$$

An inch is slightly longer than $2\frac{1}{2}$ centimeters.

1 kilometer = 0.6214 miles
1 meter = 1.0936 yards
1 centimeter = 0.3937 inches

You are driving 100 km/hour. How fast are you driving in mi/hour?

$$100 \times 0.6214 = 62.14$$
$$100 \text{ km/hr} = 62 \text{ mi/hr}.$$

A kilometer is a little more than six-tenths of a mile.

PRACTICE ASSIGNMENT

1. **a.** 18 ft = ______ yd
   b. 5 mi = ______ ft
   c. 21 ft = ______ in.
   d. 8 yd = ______ in.
   e. 132 in. = ______ ft
   f. 42 ft = ______ yd

2. **a.** 14 km = ______ m
   b. 80 mm = ______ cm
   c. 620 m = ______ cm
   d. 4,000 mm = ______ m
   e. 35 mm = ______ cm
   f. 47 cm = ______ m

3. Round the following results to two decimal places.
   a. 125 mi = ______ km
   b. 19 in. = ______ cm
   c. $8\frac{1}{2}$ yd = ______ m
   d. 11 m = ______ ft
   e. 295 km = ______ mi
   f. 37 cm = ______ in.

4. A child grew 5.5 centimeters last year and is now 1.34 meters tall. What was her height in centimeters a year ago?

5. A strip of paper was supposed to measure 6 inches but was cut 4 millimeters too short. How many millimeters long (to the nearest mm) is the strip?

Section 6.2 Measuring Weight

KEY TERMS **Gram** **Weight** **Short ton** **Metric ton**

English

1 short ton = 2,000 pounds
1 pound (lb) = 16 ounces
1 ounce (oz)

Metric

1 metric ton = 1,000 kilograms
1 kilogram (kg) = 1,000 grams
1 gram (g) = 1,000 milligrams
1 milligram (mg)

The **gram** is the basic unit.

Kilo means one thousand (1000).
Milli means one-thousandth (0.001).

What is the **weight** of the contents of a basket containing the following comparable items?

meat	4 lb 9 oz
potatoes	3 lb 4 oz
tomatoes	2 lb 15 oz
cereal	1 lb 7 oz
total	10 lb 35 oz
	= 12 lb 3 oz

meat	2.07 kg	(2 kg 70 g)
potatoes	1.47 kg	(1 kg 470 g)
tomatoes	1.33 kg	(1 kg 330 g)
cereal	0.65 kg	(650 g)
total	5.52 kg	(5 kg 520 g)

A kilogram is a little heavier than a pound, about 2.2 pounds.

1 pound

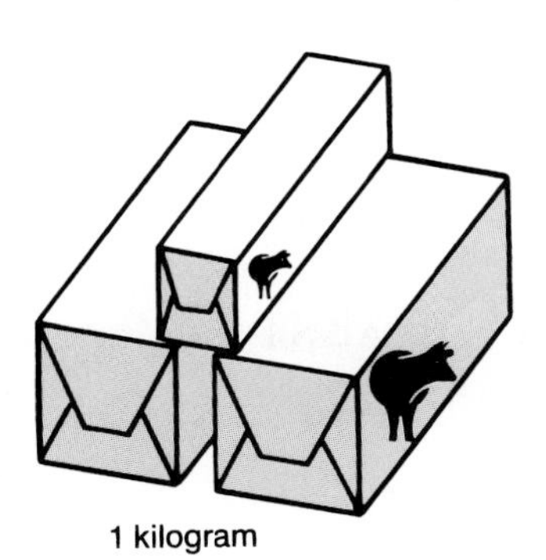
1 kilogram

Conversions between Systems

1 short ton = 0.9072 metric tons
1 pound = 0.4536 kilograms
1 ounce = 28.3500 grams

A truck weighs $2\frac{1}{2}$ tons. How many metric tons does it weigh?

2.5 × 0.9072 = 2.268
key: 2.5 [×] .9072 [=]
display: 2.268
edited: about $2\frac{1}{4}$ metric tons

The **short ton** is a little less than the **metric ton.** ("Ton" usually refers to short ton. There is an English "long ton," however, which is about the same as a metric ton. Thus all tons, short, long, or metric, are comparable.)

1 metric ton = 1.1023 short tons
1 kilogram = 2.2046 pounds
1 gram = 0.0353 ounces

The weight of an ointment is 30.5 grams. How many ounces is that?

30.5 × 0.0353 = 1.07665
key: 30.5 [×] .0253 [=]
display: 1.07665
edited: 1.08 oz

A gram is about the weight of a paper clip.

PRACTICE ASSIGNMENT

1. a. 15 lb = ______ oz
   b. 8 short tons = ______ lb
   c. 48 oz = ______ lb
   d. 68 oz = ______ lb
   e. 850 lb = ______ sh. tons
   f. $2\frac{1}{8}$ lb = ______ oz
2. a. 5.75 kg = ______ g
   b. 9 kg = ______ g
   c. 12 g = ______ mg
   d. 230 kg = ______ g
   e. 622 g = ______ kg
   f. 3700 kg = ______ m. tons
3. Round the following results to two decimal places.
   a. 38 lb = ______ kg
   b. 3 lb = ______ g
   c. 15 oz = ______ g
   d. 325 g = ______ oz
   e. 16 kg = ______ lb
   f. 8 m. tons = ______ sh. tons
4. How many 22-ounce sacks of potting soil can be obtained from a 350-pound supply?
5. How many pounds and ounces (to the nearest ounce) are in a box of detergent weighing 850 grams?

Section 6.3 Measuring Volume

KEY TERMS **Volume** **Liter**

English

1 gallon (gal) = 4 quarts
1 quart (qt) = 2 pints
1 pint (pt) = 2 cups
1 cup (c) = 8 fl ounces
1 fluid ounce (fl oz)

Metric

1 liter (l) = 1,000 milliliters
1 milliliter (ml)

The liter is the basic unit. Milli means one-thousandth (0.001).

What is the **volume** of the following two comparable mixtures?

milk	2 qt 1 pt
water	4 qt
flavoring	$\frac{1}{2}$ pt
total	6 qt $1\frac{1}{2}$ pt

milk	2.37 l	(2 l 370 ml)
water	3.79 l	(3 l 790 ml)
flavoring	0.24 l	(240 ml)
total	6.40 l	(6 l 400 ml)

A **liter** is a little larger than a quart, about 1.06 quarts.

Conversions between Systems

1 gallon = 3.7854 liters
1 quart = 0.9463 liters
1 fluid ounce = 0.0296 liters

You are buying 10 gallons of gasoline. How many liters are you buying?

$$10 \times 3.7854 \text{ l} = 37.854 \text{ l}$$

1 liter = 0.2642 gallons
1 liter = 1.0567 quarts
1 liter = 33.7838 fluid ounces

A large bottle of soda contains 2 liters. How many fluid ounces does it contain?

$2 \text{ l} = 2 \times 33.7838 \text{ fl oz} = 67.5676 \text{ fl oz}$, slightly less than 68 fl oz

A teaspoon holds about five milliliters.

PRACTICE ASSIGNMENT

1. **a.** 12 gal = _______ qt
   b. $5\frac{1}{2}$ qt = _______ pt
   c. 3 pt = _____ c
   d. 5 c = _______ fl oz
   e. 2 qt = _______ fl oz
   f. 34 qt = _______ gal

2. **a.** 32 l = _________ ml
   b. 700 ml = _______ l
   c. 25 l = _________ ml
   d. 3.5 l = _________ ml
   e. 337 ml = _________ l
   f. 18.7 ml = _________ l

3. Round these results to two decimal places.
   a. 6 gal = _________ l
   b. 18 fl oz = _________ ml
   c. 17 qt = _________ l
   d. 17 l = _________ gal
   e. 635 ml = _________ fl oz
   f. 844 ml = ________ qt

4. A deciliter is one-tenth of a liter. A normal cholesterol reading is 180 milligrams per deciliter of serum. How many milligrams of cholesterol would you expect to find in a milliliter of serum?

5. If a tank's capacity is 50 liters and it currently contains 7 gallons of gasoline, how many more gallons (to the nearest tenth) are needed to fill it up?

Section 6.4 Measuring Area and Temperature

KEY TERMS **Area** **Hectare** **Fahrenheit** **Celsius**

In the United States, large land areas are sold and developed by the thousands of acres. In many cases homes are built on lots measured in acres. In much of the rest of the world where not much flat, undeveloped land is left, land is sold in hectares or fractions of hectares. Homes are built on lots measured in square meters. Smaller areas, such as rooms, are measured in square meters; comparable English measurements are square feet or square yards.

English

1 acre = 43,560 square feet (sq ft)

Metric

1 hectare (ha) = 10,000 square meters (sq m)

What is the **area** of a room with the following comparable dimensions?

Length: 15 ft 7 in.
Width: 12 ft 6 in.

15 ft 7 in. = $(15 \times 12 + 7)$ in.
= 187 in.
12 ft 6 in. = $(12 \times 12 + 6)$ in.
= 150 in.
Area = (187×150) sq in.
= 28050 sq in.
= 194 sq ft 114 sq in.

One square foot contains 144 square inches (12×12)

Length: 4.75 m (4 m 75 cm)
Width: 3.81 m (3 m 81 cm)

Area = (4.75×3.81) sq m
= 18.0975 sq m (18 sq m 975 sq cm)

One square meter contains 10,000 square centimeters (100×100)

Conversions between Systems

1 acre = 0.4047 hectares
1 square yard = 0.8361 sq meters
1 square foot = 0.0929 sq meters

1 hectare = 2.4710 acres
1 square meter = 1.1960 sq yards
1 square meter = 10.7639 sq feet

A theme park covering 10,000 acres takes up how many hectares?

10000×0.4047 hectares
= 4047 hectares (close to 4,050 hectares)

A house outside of the U.S. may sit on a 1,000-square meter lot. How is that area expressed in acres?

10,000 sq m = 1.0 hectare
1,000 sq m = 0.1 hectare

0.1×2.471 acres = 0.2471 acres
(not quite a quarter of an acre)

PRACTICE ASSIGNMENT 1

1. **a.** 8 sq ft = ________ sq in.
b. 11 sq yd = ______ sq ft
c. 81 sq ft = _____ sq yd
d. 720 sq in. = _____ sq ft
e. 4.5 sq ft = _______ sq in.
f. 15 sq ft = ______ sq yd

2. **a.** 3.5 ha = _________ sq m
b. 0.87 ha = _________ sq m
c. 450 sq m = _________ ha
d. 1,235 sq m = _________ ha
e. 5 sq m = _________ sq cm
f. 2.5 sq cm = _______ sq mm

3. Round the following results to two decimal places.
a. 31 ha = ________ acres
b. 55 sq m = _________ sq yd
c. 10 sq m = _________ sq ft
d. 175 acres = _________ ha
e. 150 sq ft = _________ sq m
f. 2,500 acres = __________ ha

4. A summer camp is located on 378 acres of land. How many hectares (to the nearest tenth) does the camp occupy?

In the U.S., temperature is usually measured in degrees Fahrenheit (F). The commonly used metric measure for temperature is degrees Celsius (C). To convert from one scale to the other, you must use a formula. It is also helpful to know what is hot and what is cold on both scales.

	Fahrenheit degrees	Celsius degrees
Freezing point of water	32	0
Comfortable outside temperature	70	21 (not quite equal)
Normal body temperature	98.6	37
Boiling point of water	212	100

Temperatures

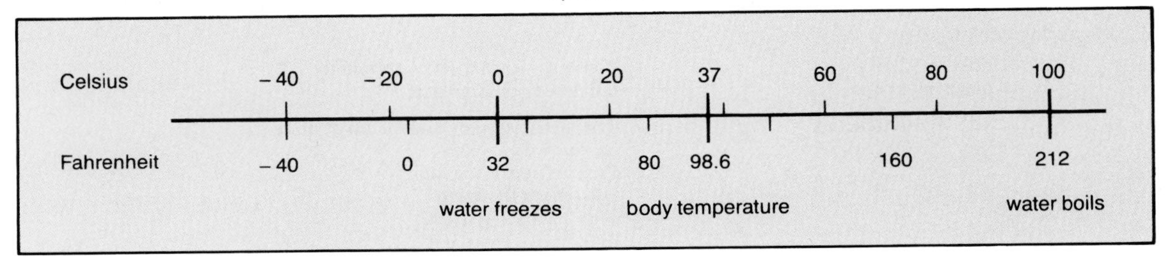

To convert Fahrenheit to Celsius, use the following formula: $C = \frac{5}{9}(F - 32)$.

Example 1 The outside temperature is a hot 88 degrees Fahrenheit. What is the Celsius temperature rounded to the nearest degree?

$$\frac{5}{9} \times (88 - 32) = \frac{5}{9} \times 56$$
$$= 280 \div 9 = 31$$

key alg: 88 [−] 32 [=] [×] 5 [÷] 9 [=]
key arith: 88 [+] 32 [−] [×] 5 [÷] 9 [=]
display: 31.111111
edited: 31 degrees Celsius

The conversion formula can be manipulated to convert Celsius to Fahrenheit.*

$$F = \frac{9}{5}C + 32$$

Example 2 The outside temperature is 10 degrees Celsius. What is the temperature in Fahrenheit? The temperature is between freezing (0 C) and comfortable (21), perhaps a little chilly.

$$\frac{9}{5} \times 10 + 32 = 90 \div 5 + 32$$
$$= 18 + 32 = 50 \text{ degrees Fahrenheit}$$

*

$C = \frac{5}{9}(F - 32)$	Reverse the sides
$\frac{5}{9}(F - 32) = C$	Multiply by $\frac{9}{5}$
$\frac{9}{5} \times \frac{5}{9}(F - 32) = \frac{9}{5}C$	Simplify
$F - 32 = \frac{9}{5}C$	Add 32
$F - 32 + 32 = \frac{9}{5}C + 32$	Simplify
$F = \frac{9}{5}C + 32$	

PRACTICE ASSIGNMENT 2

Plan: Round to the nearest tenth.

1. 82 degrees F = ________ degrees C
2. 101.6 degrees F = ________ degrees C
3. 15 degrees F = ________ degrees C
4. 17 degrees C = ________ degrees F
5. 7 degrees C = ________ degrees F
6. 22.4 degrees C = ________ degrees F

Challenger

How heavy is water? A milliliter of water weighs one gram. How many pounds and ounces (to the nearest ounce) do 50 gallons of water weigh?

1 gal = 3.7854 l = 3785.4 ml
3785.4 ml weigh 3785.4 grams
3785.4 g = 3785.4 × 0.0353 oz = 133.62462 oz
50 gal weigh 50 × 133.62462 oz = 6681.231 oz
6681.231 oz ÷ 16 = 417.57694 = 417 lb 9 oz

CHAPTER 6 SUMMARY

Concept	*Example*	*Procedure*	*Formula*
SECTION 6.1 Measuring Length			
English system	1 mile = 5,280 feet 1 quart = 4 cups	You must learn a set of seemingly unrelated numeric facts.	
Metric system	1 km = 1000 meters 1 liter = 1000 ml	The metric system is made up of units based on decimals. Prefixes adjust the units: kilo means thousand centi means hundredth milli means thousandth	
Length	A person is 6 feet tall.	The English system units are miles, yards, feet, inches.	
	A person is 1.82 meters tall.	The basic metric unit is the meter.	
SECTION 6.2 Measuring Weight			
Weight	A bag of potatoes weighs 5 pounds.	The English system units are short tons, pounds, ounces.	
	A bag of potatoes weighs 2.268 kg.	The basic metric unit is the gram.	
SECTION 6.3 Measuring Volume			
Volume	A can of soda contains 12 fl oz.	The English system units are gallons, quarts, pints, cups, fluid ounces.	
	A can of soda contains 355 ml.	The basic metric unit is the liter.	
SECTION 6.4 Measuring Area and Temperature			
Area	A room measures 120 square feet.	The English system units are square miles, square yards, square feet, square inches.	
	A room measures 11.148 square meters.	The metric system units are square kilometers, square meters, square centimeters.	
	A housing lot measures 1 acre. A housing lot measures 0.4 hectares.	The English system unit is the acre. The metric system unit is the hectare.	
Temperature	The temperature in a house is 68 degrees Fahrenheit.	The English system unit is degrees Fahrenheit.	$C = \frac{5}{9}(F - 32)$ alg: F[−]32[=][×]5[÷]9[=] arith: F[+]32[−][×]5[÷]9[=] 5/9*(F − 32)
	The temperature in a house is 20 degrees Celsius.	The metric system unit is degrees Celsius.	$F = \frac{9}{5}C + 32$ alg: 9[÷]5[×]C[+]32[=] arith: 9[÷]5[×]C[=][+]32[+] 9/5*C + 32

NAME ______________________________

CLASS SECTION ______________ DATE ______________

SCORE __________ MAXIMUM 85 PERCENT __________

CHAPTER 6 ASSIGNMENT

1. If a turtle travels 14 feet per hour, how many yards will it travel in 5 hours? *(5 points)*

2. If a car maintains a speed of 75 kilometers per hour, how many meters will it travel in a half hour? *(5 points)*

3. A baby weighed 7 pounds 13 ounces at birth. He gained 1 pound 12 ounces his first month. What was his weight at 1 month? *(5 points)*

4. If a bag of dog food is labeled 3.5 kg, how many grams does it weigh? *(5 points)*

5. How many fluid ounces are in a container holding 1 gallon plus 2 quarts plus 1 cup? *(10 points)*

6. How many liters of vegetable oil can be obtained by adding 350 milliliters of soybean oil to 225 milliliters of peanut oil? *(5 points)*

7. How many acres (to the nearest hundredth) is a lot that measures 75 × 150 feet? *(5 points)*

8. Two small farms are combined. What is their total area in hectares if one is 19,500 square meters and the other 14,650 square meters? *(5 points)*

9. How many square yards (whole number) of carpeting will cover a 12 × 17-foot room? *(5 points)*

10. How many cubic feet of sand (to the nearest tenth) will fill a depression 27 inches wide by $3\frac{1}{2}$ feet long by 11 inches deep? *(10 points)*

11. If a farm produces 1,600 kilograms of grain per hectare, how many metric tons will 7.2 hectares produce? *(5 points)*

12. How many milliliters (to the nearest ml) of cough syrup are in a bottle labeled 4.7 fl oz? *(5 points)*

13. If the thermometer outside reads 63.5 degrees F., what is the corresponding reading (to the nearest tenth of a degree) on the Celsius scale? *(5 points)*

14. A child's temperature is 38.2 degrees C. What is the temperature (to the nearest tenth of a degree) on the Fahrenheit scale? *(5 points)*

15. If a driver averages 73 kilometers per hour, how many miles (to the nearest tenth) can he drive in 15 minutes? *(5 points)*

PART

2 PERCENTAGE AND APPLICATIONS

Chapter 7 Percentage

OBJECTIVES

- Compute bases, percentages, and rates
- Find increases or decreases in terms of percent
- Learn the meaning of percentage points

SECTIONS

INTRODUCTION

The major networks announce that two large banks have dropped interest rates for their best customers to eight percent, that three states have increased sales tax to six percent, that a large corporation has asked its executives to take a five percent cut in pay, that unemployment has risen by two-tenths of a percent, and that the president's approval rating has dropped by two percentage points. What does it all mean?

Section 7.1 Equivalence of Percent and Fractions

A percent is a ratio whose fraction contains a denominator of 100. The term *percent* literally means *per hundred.*

> 7/100
> 7% seven percent

To convert a decimal number to percent, move the decimal point two places to the right, thereby multiplying by 100.

> $0.384 = 38.4\%$

To convert a fraction to percent, divide the numerator by the denominator; then move the decimal point two places to the right.

> $$\frac{7}{15} = 7 \div 15 = 0.4666666 = 46.7\%$$

Example 1 Conversions to percent form:

$$8 = 800\% \quad 1.35 = 135\%$$

$$\frac{3}{4} = 0.75 = 75\%$$

$$1\frac{2}{5} = 1.4 = 140\%$$

To convert percent to a pure decimal number, move the decimal point two places to the left, thereby dividing by 100.

$$72.5\% = 0.725$$

To convert percent to a fraction, place the number over a denominator of 100 and simplify the fraction.

$$35\% = \frac{35}{100} = \frac{7}{20}$$

Example 2 Conversions from percent form:

$$18.34\% = 0.1834 \quad 100\% = 1 \quad 0.5\% = 0.005 \quad 37\% = \frac{37}{100}$$

$$8\frac{1}{2}\% = \frac{8.5}{100} = \frac{8.5 \times 2}{100 \times 2} = \frac{17}{200}$$

$$\frac{3}{8}\% = \frac{\frac{3}{8}}{100} = \frac{3}{8} \div \frac{100}{1} = \frac{3}{8} \times \frac{1}{100} = \frac{3}{800}$$

PRACTICE ASSIGNMENT

1. Convert each of the following to percent form.

a. 0.65 = **b.** 0.07 =

c. 0.85338 = **d.** 2.4 =

e. $\frac{23}{100}$ = **f.** $3\frac{4}{5}$ =

2. Convert each of the following to a fraction in lowest terms or to a mixed number with the fractional part in lowest terms.

a. 59% = **b.** 20% =

c. 2% = **d.** $33\frac{1}{3}\%$ =

e. $14\frac{2}{5}\%$ = **f.** 680% =

3. Complete the following table.

Percent Form	Decimal Number	Fraction in Lowest Terms
20%	______	______
______	______	$\frac{7}{20}$
______	0.06	______
$1\frac{1}{5}\%$	______	______
______	14.45	______
______	______	$8\frac{3}{10}$

Section 7.2 Finding the Percentage

KEY TERMS **Base** **Rate** **Percentage**

Base, rate, and percentage are the building blocks for all percentage problems. The **base** is a whole unit, often an initial or total value. It is 100 percent. The **rate** is the number of percent that defines a portion (or multiple) of the base. The **percentage** is the amount represented by the rate. For every rate there is a corresponding percentage. The fundamental equation is: Percentage = **B**ase × **R**ate.

What is $\underbrace{35\%}_{\text{rate}}$ of $\underbrace{200}_{\text{base}}$?

Percentage = Base × Rate

$P = B \times R$

35% of 200 =

200 × 0.35 = 70

$B \times R = P$

The three forms of the percentage equation follow. You may want to memorize them.

Find P by covering P.	$\frac{P}{BR}$	$P = BR$	This is the fundamental equation.
Find R by covering R.	$\frac{P}{BR}$	$R = \frac{P}{B}$	In the two related equations, percentage is the numerator of a fraction.
Find B by covering B.	$\frac{P}{BR}$	$B = \frac{P}{R}$	

You may also derive the related equations from the fundamental equation. When you solve for the rate, R is the unknown. To isolate R, divide both sides of the fundamental equation by B. (Use the inverse operation technique explained in chapter 4, section 4.3, Finding the Unknown in a Simple Equation.)

The fundamental equation.	$P = B \times R$
Reverse the two sides of the equation.	$B \times R = P$
Divide each side by B.	$B \times R \div B = P \div B$
Exchange the multiplication and division.	$B \div B \times R = P \div B$
Simplify.	$1 \times R = P \div B$
Simplify; write the division as a fraction.	$R = P/B$

When you solve for the base, B is the unknown. To isolate B, divide both sides of the fundamental equation by R.

The fundamental equation.	$P = B \times R$
Reverse the two sides of the equation.	$B \times R = P$
Divide each side by R.	$B \times R \div R = P \div R$
Simplify.	$B \times 1 = P \div R$
Simplify; write the division as a fraction.	$B = P/R$

Example 1 Payroll expenses for the Palmer Company include 18.65% for fringe benefits. If monthly payroll expenses amount to $750,436, what is the cost of fringe benefits?

18.65% = 0.1865
Percentage = base × rate = 750436 × 0.1865
= 139956.31

key: 750436 [×] .1865 [=]
display: 139956.31
edited: $139,956.31

PRACTICE ASSIGNMENT

Plan: Each problem contains a base and a rate. Compute the percentage.

1. Compute the following percentages, rounded to four places where necessary.

   a. 75% of 256 =

   b. 9% of 574 =

   c. 115.65% of 3000 =

   d. $8\frac{3}{4}$% of 342.5 =

   e. 0.35% of 210.15 =

   f. $\frac{1}{2}$% of 240 =

2. A shopper purchased a refrigerator priced at $845 and made a down payment of 18.5%. What was the amount of the down payment?

3. Paul Kessler owes $673.82 in back taxes. In addition, he must pay $9\frac{1}{2}$% in interest. What is the interest on the taxes owed?

4. A meteorite weighing 1520 kilograms (kg) consists of 91.5% iron and 7.2% nickel based on weight.

a. What is the weight of the iron computed to the nearest tenth of a kg?

b. What is the weight of the nickel computed to the nearest tenth of a kg?

c. What is the weight of other miscellaneous ingredients computed to the nearest tenth of a kg?

Section 7.3 Finding the Rate

A cafeteria lunch, priced at $2.45, rose by $0.65. What is the percent of increase? In this case, you must compute the rate, given the base and percentage.

$$\text{Rate} = \frac{\text{percentage}}{\text{base}}$$

Convert the rate from pure decimals to percent.

What percent of $\underbrace{2.45}_{\text{base}}$ is $\underbrace{0.65?}_{\text{percentage}}$

$$0.65 \div 2.45 = 0.2653061$$
$$P \div B = R$$

$$0.2653061 = 26.531\%$$

Example 1 In an election the total precinct vote was 3,012, of which 1,860 votes were cast for candidate X. What percent of the precinct vote did candidate X receive?

$$\text{Rate} = \text{percentage} \div \text{base} = 1860 \div 3012 = 0.6175298$$

In many situations accuracy to the nearest tenth of a percent is sufficient.

key: 1860 [÷] 3012 [=]
display: 0.6175298
edited: 61.8%

PRACTICE ASSIGNMENT

Plan: Each problem contains a base and a percentage. Compute the rate. Round rates to the nearest tenth of a percent.

1. **a.** What percent of 1,500 is 225?

b. What percent of 357.2 is 61.0812?

c. 24 is what percent of 132?

d. 350 is what percent of 25?

e. 0.005 is what percent of 0.1?

f. $\frac{4}{5}$ is what percent of $\frac{3}{10}$?

2. After deductions and income adjustments, a family paid a federal income tax of $1,221 on a gross income of $27,400. What was their effective federal income tax rate that year?

3. In a small town with a population of 6,435, there are 388 people who are 75 years old or older. What percent of the population is at least 75 years old?

4. Of 5,826 full-time students enrolled in a college, 3,719 are enrolled in either the School of Mathematics and Science or the School of Business Administration. What percent of the full-time student body is studying mathematics, science, or business administration?

5. A man pawned a gold watch for $10.50. The pawn shop sold the watch for $38.99. What percent profit (based on cost) did the pawn shop make on this sale?

Section 7.4 Finding the Base

Toy department sales were eight percent of store sales. What were the store's total sales if the toy department's sales amounted to $10,342? In this case, compute the base, given the rate and percentage.

$$\text{Base} = \frac{\text{percentage}}{\text{rate}}$$

8% of what is 10342?
rate — percentage

10342 ÷ 0.08 = 129275
$P \div R = B$

Example 1 A sales agent earned $277.55 in commissions at a rate of 6.25%. How much did the agent sell?

$$\begin{aligned} 6.25\% &= 0.0625 \\ \text{Base} &= \text{percentage} \div \text{rate} \\ &= 277.55 \div 0.0625 \\ &= 4440.80 \end{aligned}$$

key: 277.55 [÷] .0625 [=]
display: 4440.8
edited: $4,440.80

PRACTICE ASSIGNMENT

Plan: Each problem contains a rate and a percentage. Compute the base.

1. Round answers to nearest tenth.

a. 312 is 65% of what number?

b. 76 is 62.5% of what?

c. 15 is 2.5% of what?

d. 95% of _______ is 646.

e. 150% of _______ is 117.

f. $66\frac{2}{3}$% of _________ is 74.53.

2. An investor received a check for $26.19, which represented a 4.2% dividend based on the investment's value. How much was the investment worth?

3. The quality control department of Blitzen Electronics rejected 5 defective radios, which represented 0.8% of the total output for one day. How many radios were produced that day?

4. The gross profit on a pair of men's shoes was 15% of the cost. If the gross profit was $3.30, what was the cost of the shoes?

5. A homeowner pays 5.45% of the assessed value of the home for property taxes. If taxes are $2,740, what is the assessed value of the home?

Section 7.5 Quick Mental Computations

By now you know that you can compute some problems quickly in your head. Here are some shortcuts to help you.

To find 1% of a base, divide by 100:

$$1\% \text{ of } 36459 = 364.59.$$

To find 2%, 3%, etc., find 1% and multiply by 2, 3, etc.:

$$2\% \text{ of } 36459 = 2 \times 364.59 = 729.18.$$

To find 10% of a base, divide by 10:

$$10\% \text{ of } 36459 = 3645.9.$$

To find 20%, 30%, etc., find 10% and multiply by 2, 3, etc.:

$$20\% \text{ of } 36459 = 2 \times 3645.9 = 7291.8.$$

To find 50%, find one-half:

$$50\% \text{ of } 10800 = \frac{1}{2} \times 10800 = 5400.$$

To find 25%, find one-fourth:

$$25\% \text{ of } 10800 = \frac{1}{4} \times 10800 = 2700.$$

To find $33\frac{1}{3}\%$, find one-third:

$$33\frac{1}{3}\% \text{ of } 6600 = \frac{1}{3} \times 6600 = 2200.$$

To find $66\frac{2}{3}\%$, find two-thirds:

$$66\frac{2}{3}\% \text{ of } 6600 = \frac{2}{3} \times 6600 = 4400.$$

Use the reverse technique to find the base, given the percentage for 1%, 2%, 10%, 20%, 25%, 50%, $33\frac{1}{3}\%$, $66\frac{2}{3}\%$, etc.

Example 1 47 is 1% of what?
Since $1\% = \frac{1}{100}$, the answer is $100 \times 47 = 4700$.

Example 2 362 is 20% of what?
Since $20\% = \frac{1}{5}$, the answer is $5 \times 362 = 1810$.

PRACTICE ASSIGNMENT

Plan: Use the suggested shortcuts to solve these problems.

1. Find the following:

- **a.** 1% of 8,920 =
- **b.** 10% of $3,895.50 =
- **c.** 20% of $214.30 =
- **d.** 2% of 7,213 =
- **e.** 50% of $8,116.44 =
- **f.** 25% of $6,372 =
- **g.** $33\frac{1}{3}\%$ of 217.5 =
- **h.** 5% of 682.18 =

2. Complete the following:

- **a.** 918 is 10% of:
- **b.** $1.75 is 1% of:
- **c.** $891.50 is 20% of:
- **d.** 2.036 is 40% of:
- **e.** $593.72 is 25% of:
- **f.** 1,304 is 50% of:
- **g.** $.75 is 5% of:
- **h.** $48.60 is 200% of:

Section 7.6 Increase or Decrease Problems

In business it is often necessary to examine an increase or decrease in value, regarding the old value as the base. This may involve finding the old value, the new value, or the rate of increase or decrease.

How Can the New Value Be Found?

If you know the old value and the rate of increase, you can find the new value. The old value is the base. Add the rate of increase to 100 percent. The increased rate is the rate corresponding to the new value. Find the percentage.

Base	=	100%
Rate of increase	=	6%
Rate	=	106%

If you know the old value and the rate of decrease, you can find the new value. The old value is the base. Subtract the rate of decrease from 100 percent. The decreased rate is the rate corresponding to the new value. Find the percentage.

Base	=	100%
Rate of decrease	=	6%
Rate	=	94%

Example 1 An employee whose annual salary is $21,044 receives a raise of 3.5%. What will the new salary be?

$$100\% + 3.5\% = 103.5\% = 1.035$$
$$\begin{aligned} \text{Percentage} &= \text{base} \times \text{rate} \\ &= 21044 \times 1.035 \\ &= 21780.54 \end{aligned}$$

key: 21044 [×] 1.035 [=]
display: 21780.54
edited: $21,780.54

Example 2 A school ordered 214 calculators for a first-semester mathematics course. A decrease in enrollment of 8% is expected for the following semester. How many calculators will the school order for the second semester?

$$100\% - 8\% = 92\% = 0.92$$
$$\begin{aligned} \text{Percentage} &= \text{base} \times \text{rate} \\ &= 214 \times 0.92 \\ &= 196.88 \end{aligned}$$

key: 214 [×] .92 [=]
display: 196.88
edited: 197

PRACTICE ASSIGNMENT 1

Plan: In each problem, a value has increased or decreased by a certain percent. Find the new value by multiplying the old value (the base) by the increased or decreased rate.

1. A supermarket that sold milk for $1.79 per gallon increased its price by 5%. What is the new price?

2. During a one-week period, the price of gold declined by 9.5%. If the price was $490 per ounce at the beginning of the week, what was the price at the end of the week?

3. A company paid $75,000 for office equipment last year. The value of this equipment has depreciated by $8\frac{1}{3}$%. What is the equipment worth now?

4. Orders for recreational vehicles increased 11.8% over the previous year. If 46,500 vehicles were ordered in the previous year, how many were ordered in the current year?

5. A Japanese car manufacturer agreed to reduce the number of cars exported to the United States by $9\frac{1}{2}$%. If 63,800 cars were exported last year, how many should be exported to the United States this year?

6. A pair of boots that sold for $68.50 now sells for 9% more. What is the new price?

How Can the Old Value Be Found?

If you know the new value and the rate of increase or decrease, you can find the old value. The old value is the base. The new value is the percentage. Obtain the increased or decreased rate and find the base.

$$\text{Base} = \frac{\text{new value}}{\text{increased or decreased rate}}$$

Example 3 A plant's production was 4% greater this year than last year. What was last year's production if this year's output was 27,325 units?

$$\begin{aligned} 100\% + 4\% &= 104\% = 1.04 \\ \text{Base} &= \text{percentage} \div \text{rate} \\ &= 27325 \div 1.04 \\ &= 26274.038 \end{aligned}$$

key: 27325 [÷] 1.04 [=]
display: 26274.038
edited: 26,274

Example 4 A credit card company found that the number of credit cardholders with a balance of over $1,000 declined by 8.75% since the previous year. If there were 33,684 such customers this year, how many were there last year?

$$\begin{aligned} 100\% - 8.75\% &= 91.25\% = 0.9125 \\ \text{Base} &= \text{percentage} \div \text{rate} \\ &= 33684 \div 0.9125 \\ &= 36913.972 \end{aligned}$$

key: 33684 [÷] .9125 [=]
display: 36913.972
edited: 36,914

PRACTICE ASSIGNMENT 2

Plan: In each problem, a value has increased or decreased by a certain percent. Find the old value by dividing the new value (the percentage) by the increased or decreased rate.

1. The Lundgrens' telephone bill was $27.95 this month, 8% higher than last month's bill. How much was last month's bill?

2. A suburban school district is building a new school and will increase its teaching staff by 23.2%. After the increase, there will be 616 teachers. How many teachers were there before the increase?

3. A coal mine produced 98,450 tons of coal this year. This is 6.8% less than the amount mined last year. How much coal was mined last year to the nearest ton?

4. A set of antique furniture is valued at $540,000. This represents an increase of 1400% over what is was worth 25 years ago. What was the value of the furniture 25 years ago?

5. After six weeks of dieting, Valerie lost 8.6% of her original weight and was down to 112.5 lb. What was her original weight to the nearest tenth of a pound?

6. Sales in the electronics department of Michaels and Company increased 6.2% during the past year. If this year's sales were $643,050, what were last year's sales to the nearest dollar?

Section 7.7 Finding the Rate of Increase or Decrease

In cases where you know both old and new values, you can calculate the percent that the old value increased or decreased. Obtain this rate by dividing the increase or decrease by the old value and converting the result to percent form.

$$\text{Rate of increase or decrease} = \frac{\text{increase or decrease}}{\text{old value}}$$

Example 1 During the summer, the price of gasoline increased from $1.34 per gallon to $1.44 per gallon. What was the percent of increase?

$$\begin{aligned}\text{Increase} &= \text{new value} - \text{old value}\\ &= 1.44 - 1.34 = 0.10\end{aligned}$$

$$\begin{aligned}\text{Rate of increase} &= \text{increase} \div \text{old value}\\ &= 0.10 \div 1.34 = 0.0746268\end{aligned}$$

key alg: 1.44 [−] 1.34 [=] [÷] 1.34 [=]
key arith: 1.44 [+] 1.34 [−] [÷] 1.34 [=]
display: 0.0746268
edited: 7.5%

Example 2 A men's fashion store sells a particular sport jacket for $79.95. During a promotional sale the price was reduced to $64.95. By what percent did the price decrease?

$$\begin{aligned}\text{Decrease} &= \text{old value} - \text{new value}\\ &= 79.95 - 64.95 = 15\end{aligned}$$

$$\begin{aligned}\text{Rate of decrease} &= \text{decrease} \div \text{old value}\\ &= 15 \div 79.95 = 0.1876172\end{aligned}$$

key alg: 79.95 [−] 64.95 [=] [÷] 79.95 [=]
key arith: 79.95 [+] 64.95 [−] [÷] 79.95 [=]
display: 0.1876172
edited: 18.8%

PRACTICE ASSIGNMENT

Plan: In each problem, a value has increased or decreased, and both old and new values are given. Find the rate of increase or decrease by dividing the increase or decrease by the old value. Where necessary, round the rate of increase or decrease to the nearest tenth of a percent.

1. From one year to the next, a family's taxable income increased from $21,000 to $23,500. What was the rate of increase?

2. During the stock market crash of 1987, the Dow Jones Industrial Average dropped 506 points from approximately 2,550. What was the rate of decline?

3. In the produce section of a supermarket, twelve cases of tomatoes were in stock on Monday morning. By Sunday evening, two cases were left. What percent of the tomatoes were sold during the week?

4. Last year the construction industry reported 279,352 new housing starts. This year there were only 207,828. What is the rate of decrease in housing starts?

5. The price of a share of Superconductors Unlimited stock rose from $27\frac{1}{4}$ to $39\frac{3}{4}$. What was the rate of increase?

6. November sales in a toy store totaled $587,345. Christmas shopping boosted December sales to $882,179. What was the percent of increase?

Section 7.8 Percentage Points

If a state raises its sales tax rate from 6 percent to 7 percent, the tax on $100 worth of taxable goods will rise from $6 to $7, an increase of $1, one-sixth, or 16.7 percent. However, the news media might report that the sales tax rose by one "percentage point," from 6 percent to 7 percent.

A report that the number of Republicans in the House of Representatives dropped by three percentage points may mean that Republicans in the House decreased from 44 percent to 41 percent. Since the number of representatives remains the same (435), the base is constant, and the decrease in percent from the old value can be computed as:

$$\frac{44 - 41}{44} = \frac{3}{44} = 0.0681818 = 6.8\%.$$

The number of Republicans in the House dropped three percentage points since last term.

Percentage Point Approach
Drop of 3 percentage points

The number of Republicans in the House is 93.2% of the number of Republicans during the previous term.

Percent Approach
Decrease of 6.8%

An increase or decrease of x percentage points implies a difference of x percent. Percentage points are expressed as numbers without the percent sign. If a change is stated in percentage points and the base remains the same, compute the rate of change in percent with the following formula:

$$\text{Rate of increase or decrease} = \frac{\text{change in percentage points}}{100 \times \text{old rate}}$$

Example 1 In a computer room, 15% of the computers are defective. After extensive maintenance, the number of defectives decreased by five percentage points. What is the percent of decrease?

$$\frac{\text{change in percentage points}}{100 \times \text{old rate}} = \frac{5}{15} = \frac{1}{3} = 33.3\%$$

If the rate of change is stated as percent, compute the change in percentage points with the following formula (assuming no change in base):

Percentage point increase or decrease
= 100 × old rate × rate of increase or decrease

Example 2 In an accounting class, 20% of students scored an A on the first exam. On the next exam, the number of As increased by 5%. Assume that the same number of students took the second exam as took the first. What was the percentage point increase?

$$\begin{aligned} 100 \times \text{old rate} \times \text{rate of increase} &= 20 \times 0.05 \\ &= 1 \text{ percentage point} \end{aligned}$$

PRACTICE ASSIGNMENT

Plan: In each problem, a value has increased or decreased. In some problems, the change in percent must be converted to a change in percentage points; in others the change in percentage points must be converted to a change in percent (rounded to the nearest tenth of a percent).

1. Sixty-nine percent of a graduating class obtained employment one month after graduation. The number of graduates employed after two months rose by three percentage points. By what percent did the number of employed graduates increase?

2. Five years ago a 500-bed hospital had an average of 30 empty beds. Last year the rate of empty beds increased by 120%. How is this increase expressed as percentage points?

3. Ken Rafferty earned $29,700 annually, of which 31% was deducted from each paycheck. This year his salary remained the same, while his deductions rose by one percentage point. By what percent did his net income decrease?

4. The tax on a property assessed at $45,000 was $945. The following year the tax increased to $2,035. What was the percentage point increase?

Challenger

Rent, utilities, maintenance, office supplies, insurance, etc., are expenses over and above the cost of merchandise. While these overhead expenses (*overhead* for short) affect a business as a whole, portions are usually allocated to departments. Some companies base their allocation of overhead expenses on department sales.

1. Compute the ratio of department sales to total sales for each department; then convert this ratio to percent form.
2. For each department, find the percentage of total overhead that corresponds to the rate found in step 1. Round ratios to the nearest hundredth of a percent and money amounts to the nearest dollar.

Martin's Department Store allocates overhead by sales. If the August overhead was $4,350, how much was allocated to each department?

Department	Sales	Distribution Ratio	Overhead Allocation
Ladies' wear	9702	________%	________
Menswear	4260	________%	________
Toys	1745	________%	________
Appliances	2193	________%	________
Sporting goods	2759	________%	________
Furniture	3491	________%	________
Total sales	24150		

CHAPTER 7 SUMMARY

Concept	*Example*	*Procedure*	*Formula*
SECTION 7.1 Equivalence of Percent and Fractions			
Percent	$7\% = \frac{7}{100} = 0.07$	Percent is a ratio whose fraction contains a denominator of 100.	
Conversion from pure numeric to percent form	$0.384 = 38.4\%$ $1.35 = 135\%$ $\frac{7}{15} = 0.4666\ldots$ $= 46.67\%$	Move decimal point two places to the right, thereby multiplying by 100.	
Conversion from percent form to pure numeric	$68.4\% = 0.684$ $3\frac{3}{4}\% = 3.75\%$ $= 0.0375$	Move decimal point two places to the left, thereby dividing by 100.	
SECTION 7.2 Finding the Percentage			
Fundamental percentage formula	What is $8\frac{1}{2}\%$ of 2016? $2016 \times 0.085 = 171.36$	Percentage = base × rate	$P = BR$ B[×]R[=] B*R
SECTION 7.3 Finding the Rate			
Finding the rate	22 is what percent of 500? $22 \div 500 = 0.044 = 4.4\%$	Rate = percentage ÷ base	$R = \frac{P}{B}$ P[÷]B[=] P/B
SECTION 7.4 Finding the Base			
Finding the base	$17\frac{1}{4}$ is 32% of what? $17.25 \div 0.32 = 53.90625$	Base = percentage ÷ rate	$B = \frac{P}{R}$ P[÷]R[=] P/R
SECTION 7.5 Quick Mental Computations			
Mental Shortcuts	50% of 348 = 348 ÷ 2 = 174 25% of 972 = 972 ÷ 4 = 243		
SECTION 7.6 Increase or Decrease Problems			
Finding the new value after an increase or decrease	A storeroom contained 1803 lb of flour. The content is increased by 2%. $1803 \times 1.02 = 1839.06$	Add the rate of increase or subtract the rate of decrease from 100%. Find the percentage using the new rate.	Percentage formula
Finding the old value prior to an increase or decrease	There are 1,850 lb left after the content is decreased by 3%. What was the prior content? $1850 \div 0.97 = 1907.2164$	Add the rate of increase or subtract the rate of decrease from 100%. Find the base using the new rate.	Base formula
SECTION 7.7 Finding the Rate of Increase or Decrease			
Finding the rate of increase or decrease	The price of a gallon of gas increased from $1.02 to $1.05. What was the rate of increase? $\frac{1.05 - 1.02}{1.02} = \frac{0.03}{1.02}$ $= 0.0294117 = 2.94\%$	Increase or decrease is the percentage. Find the rate using the old value as base.	$R = \frac{\text{new} - \text{old}}{\text{old}}$ alg: new[−]old[=][÷]old[=] arith: new[+]old[−][÷]old[=] (new-old)/old

CHAPTER 7 SUMMARY—(*Continued*)

Concept	*Example*	*Procedure*	*Formula*
SECTION 7.8 Percentage Points			
Percentage point	The unemployment rate decreased by one percentage point (from 6.5% to 5.5%).	The percentage point increase or decrease is the change in the rate expressed without the percent sign.	
Conversion of percentage points to percent (same base)	A 30% tax rate is increased by four percentage points. What is the increase in percent? $\frac{4}{30} = \frac{2}{15} = 0.1333\ldots$ $= 13.33\%$		Rate of increase or decrease $= \frac{\text{change in percentage points}}{100 \times \text{old rate}}$
Conversion of percent increase or decrease to percentage points (same base)	Voter turn-out increased by 5% over last election's 46%. By how many percentage points did voter turn-out increase (assuming the same number of eligible voters)? $46 \times 0.05 = 2.3$ percentage points		Percentage points $= 100 \times \text{old rate} \times \text{rate of increase or decrease}$

NAME ______________________________

CLASS SECTION ______________ DATE ______________

SCORE ____________ MAXIMUM __170__ PERCENT ____________

CHAPTER 7 ASSIGNMENT

1. Convert the following decimal numbers to percent form.

a. 0.40 =

b. 0.02 =

c. 1.592 =

d. $\frac{3}{200}$ =

e. $2\frac{3}{4}$ =

(5 points)

2. Convert to a fraction in lowest terms.

a. 25% =

b. $37\frac{1}{2}$% =

c. $56\frac{1}{4}$% =

d. 115% =

e. 45.8% =

(5 points)

3. A hardware store sold a power drill at a special sale for 82% of cost. What was the selling price if the cost was $83.50? *(5 points)*

4. Of a store's total inventory valued at $498,750, $133,262 worth was destroyed in a fire. What percent of the inventory was destroyed? *(5 points)*

5. Stanley Harmon donates 7% of his weekly paycheck to his church. His contribution last week was $34.09. What was the amount of his paycheck? *(5 points)*

6. An investor owns 200 shares of a stock valued at $62\frac{3}{8}$ per share. If the stock pays a 6.3% dividend, what is the dividend per share? *(5 points)*

7. An agent received a commission of $1,938.13 for selling merchandise worth $21,500. What was the rate of the agent's commission? *(5 points)*

8. Of a company's travel expense budget of $544,750, 51.6% was spent on airplane fares and 27.4% on hotel accommodations.

a. How much was spent for air transportation?

b. How much was spent for hotels?

(10 points)

9. A literary agent received $105.30, a $6\frac{1}{2}$% commission, for selling an article to a national magazine. How much did the article sell for? *(5 points)*

10. Paolo's Department Store had sales totaling $187,607.50. The sporting goods department had sales of $57,376.12. What percent of total sales was contributed by this department? *(5 points)*

11. Income taxes of 18% of a family's taxable income amounted to $3,170. What was the taxable income? *(5 points)*

12. A 2.15-ounce tube of ointment contains 4.7% benzocaine. The remainder consists of inert ingredients. How many ounces (to the nearest hundredth) of inert ingredients are in the ointment? *(10 points)*

13. The selling price of a dress was 135% of its purchase price. If the dress sold for $38.99, what was its purchase price? *(5 points)*

14. Kevin returned a book to the public library fifteen days after it was due. He paid a fine of $.05 per day for each day the book was overdue. If the book cost $12.50, the library fine is what percent of the cost? *(5 points)*

15. Tiffany Clarke earns a gross annual salary of $24,600. If 28.3% of her salary is withheld for taxes and other deductions, what is the amount of each monthly paycheck? *(10 points)*

16. Eighteen percent of the selling price of a personal computer was gross profit. If the gross profit was $205.20, what was the selling price? *(5 points)*

17. The Farm and Garden Equipment Company issued sales quotas for its five regions. At the end of the quarter, sales were as follows:

Region	Quota	Sales	Percent
Northeast	$254,750	$189,438.70	________
Southeast	342,175	356,019.25	________
Midwest	505,500	421,394.85	________
Northwest	386,525	373,578.15	________
Southwest	655,500	692,067.50	________

To the nearest tenth of a percent, calculate the percent of each quota filled by actual sales. *(10 points)*

18. When April Santana started her jogging program, she was able to cover 2.5 miles. Now she jogs 4.7 miles. What is her rate of increase? *(5 points)*

19. The new sales manager of a tool company is dissatisfied with last year's sales of $239,000 and wants to increase sales by 7% this year. What will this year's sales figures have to be to reach this goal? *(5 points)*

20. A major league baseball team won 96 games last year. This year it won $16\frac{2}{3}$% fewer games. How many games did the team win this year? *(5 points)*

21. Subscribers to Popular Sports magazine may buy a one-year subscription for $13.92, which is 42% less than the newsstand price for twelve monthly issues.

a. What is the price of twelve issues purchased at a newsstand?

b. What is the newsstand price of one issue?

c. What is the subscription price of one issue? *(15 points)*

22. Last year Mancuso and Dworsky, attorneys-at-law, received a total of $795,000 in fees; they paid $284,500 in expenses. The previous year the law firm collected a total of $719,700 in fees and paid $266,250 in expenses. What was the percent increase in profit? *(10 points)*

23. A monthly salary of $2,104.17 included a 4% increase.

a. What was the monthly salary before the increase?

b. What was the annual salary before the increase?

c. What was the annual salary after the increase? *(15 points)*

24. Last year the rate of return on a one-year investment of $1,000 was 7.4%. This year the rate of return on the same investment declined by one percentage point. By what percent has the rate declined? *(5 points)*

25. A marketing company surveyed 150 people for their soft drink preferences. Thirty-two percent preferred Polar-Cola to other drinks. After this group took a comparative taste test, the rate of preference for Polar-Cola increased by twenty-five percent. What was the percentage point increase in people preferring Polar-Cola? *(5 points)*

Chapter 8 Trade Discounts

OBJECTIVES

- Compute trade discounts and net prices
- Apply a series of discounts and find the equivalent single rate

SECTIONS

INTRODUCTION

When businesses purchase products from each other, they offer a price reduction—discount—to the buyer. This discount, called a trade discount, generates large-scale sales to regular customers at prices below those paid by the general public (list price).

Section 8.1 Single Trade Discounts

KEY TERMS **List price** **Net price** **Complement rate**

A trade discount is expressed as a percent of the **list price;** the list price less the discount amount is the **net price.**

List price	=	815.50
Discount	=	10%
815.5 × 0.1	=	81.55
Net price	=	733.95

Example 1 Linda's Dress Shop ordered merchandise that lists for $3,470.50. The supplier gives a 15% trade discount. What is the amount of discount and the net price?

List price = 3470.50
Discount = 15%
Discount amount = 3470.50 × 0.15 = 520.58

Net price = list price − discount amount
= 3470.50 − 520.58
= 2949.92

key:	3470.5 [×] .15 [=]
display:	520.575
rounded:	520.58 **discount amount**
key alg:	3470.5 [−] 520.58 [=]
key arith:	3470.5 [+] 520.58 [−]
display:	2949.92
edited:	$2,949.92 **net price**

Another method for computing the net price is to multiply the list price by the complement of the discount rate. The complement of a rate, called the **complement rate,** is the difference between 100 percent and the rate. To compute the discount amount, subtract the net price from the list price.

100% − 10% = 90%
100% − rate = complement rate
List price × complement = net
815.50 × 0.9 = 733.95

List price	815.50
− Net price	733.95
= Discount amount	81.55

Example 2 The Beckworth Company is entitled to a trade discount of 18%. How much will the company pay on an order that lists for $893.85, and how much will be deducted?

List price = 893.85
Complement of 18% = 82%
Net price = 893.85 × 0.82 = 732.96

Discount amount = list price − net price
= 893.85 − 732.96
= 160.89

key: 893.85 [×] .82 [=]
display: 732.957
rounded: 732.96 **net price**
key alg: 893.85 [−] 732.96 [=]
key arith: 893.85 [+] 732.96 [−]
display: 160.89
edited: $160.89 **discount amount**

PRACTICE ASSIGNMENT

Plan: Compute the discount amount and subtract it from the list price to obtain the net price. When using the complement method, first multiply the list price by the complement rate to compute the net price; then subtract the net price from the list price to obtain the discount amount.

1. Find the discount amount and net price.

	List Price	Discount Rate	Discount Amount	Net Price
a.	$579.00	10%	______	______
b.	805.00	15%	______	______
c.	1,720.25	11%	______	______
d.	602.00	$12\frac{1}{2}$%	______	______
e.	942.50	8.9%	______	______
f.	3,175.32	$10\frac{3}{4}$%	______	______

2. Use the complement method to find the net price; then find the discount amount.

	List Price	Discount Rate	Complement Rate	Net Price	Discount Amount
a.	$3,475.50	15%	%	______	______
b.	887.50	8%	%	______	______
c.	1,800.00	$12\frac{1}{2}$%	%	______	______
d.	89.70	9.5%	%	______	______
e.	2,160.00	14.8%	%	______	______
f.	3,062.80	$16\frac{2}{3}$%	%	______	______

3. A retailer receives a trade discount of 15% on china. How much was deducted from the list price on an order listing for $7,375?

4. Stanley's Sound Shop ordered records from Miracle Music at a total list price of $421.50. If a trade discount of $9\frac{1}{2}$% is offered, what is the net price?

5. An electronics importer sells digitalized stereo television sets for $579.99 with a 22% trade discount to wholesalers.

a. What is the trade discount for wholesalers?

b. What is the wholesale price?

Section 8.2 Multiple Trade Discounts

KEY TERMS **Series of discounts** **Net cost factor**

Trade discounts are not uniform. Customers in better marketing positions may receive greater discounts. Seasons and locations sometimes influence discount policy. In some cases more than one discount may be offered to some customers. While the list price remains the same, net prices vary by the addition or removal of discounts.

Multiple discounts are usually shown in descending order. For example, the **series of discounts** 20 percent, 10 percent, and 5 percent may be shown as 20/10/5. The most favored customers qualify for all three discounts; the next favored customers enjoy a 20 percent and a 10 percent discount; other customers receive a 20 percent discount.

List price	=	1850.00
20%	=	370.00
First net	=	1480.00
10%	=	148.00
Second net	=	1332.00
5%	=	66.60
Final net	=	1265.40

Example 1 J & M Tires receives trade discounts of 15/8/5 on steel-belted radials. On an order with a list price of $2,800, what is the final net? What is the total discount amount?

List price	= 2800.00
First discount = 15%	
First discount amount = 2800 × 0.15	= 420.00
First net	= 2380.00
Second discount = 8%	
Second discount amount = 2380 × 0.08	= 190.40
Second net	= 2189.60
Third discount = 5%	
Third discount amount = 2189.60 × 0.05	= 109.48
Final net	= 2080.12

Total discount amount = 2800 − 2080.12 = 719.88

You may use the complement method for computing the final net price. The **net cost factor** is the product of the complement rates. Multiply the list price by the net cost factor to find the net price.

Complement of 20%	= 80%
Complement of 10%	= 90%
Complement of 5%	= 95%
Net cost factor	=
0.8 × 0.9 × 0.95	= 0.684

Net = 1850 × 0.684 = 1265.40

Example 2 A tractor manufacturer lists lawn mowers for $769.40, subject to trade discounts of 15/10/5. What are the net price and total deduction for a customer who receives all discounts?

Complement rate of 15% = 100% − 15% = 85% = 0.85
Complement rate of 10% = 100% − 10% = 90% = 0.90
Complement rate of 5% = 100% − 5% = 95% = 0.95
Net cost factor = 0.85 × 0.9 × 0.95 = 0.72675

Net price = list price × net cost factor
= 769.40 × 0.72675
= 559.16

Amount of discount = list price − net price
= 769.40 − 559.16
= 210.24

key:	.85 [×] .9 [×] .95 [=]
display:	0.72675 **net cost factor**
key:	[×] 769.4 [=]
display:	559.16145
rounded:	559.16 **net price**
key alg:	769.4 [−] 559.16 [=]
key arith:	769.4 [+] 559.16 [−]
display:	210.24
edited:	$210.24 **discount amount**

PRACTICE ASSIGNMENT

Plan: You can start with the list price and compute the discount amounts and nets successively, changing the base to the previous net at each step. However, you may find the complement method more convenient. First find the complement rates and the net cost factor. Then multiply the list price by the net cost factor to find the final net. Either way, remember that the price reduction is the difference between the list price and the final net.

1. Complete the following table.

	List Price	First Discount	Net Price after First Discount	Second Discount	Net Price after Second Discount
a.	$2,650.00	20%	______	$12\frac{1}{2}$%	______
b.	208.00	10%	______	7.5%	______
c.	1,900.00	22.4%	______	10%	______
d.	4,800.00	18%	______	7%	______
e.	1,740.50	20%	______	5.5%	______
f.	928.00	$13\frac{1}{2}$%	______	6%	______

2. Show the complement rates in percent form and in decimal form.

	Rate	Complement Rate Percent Form	Complement Rate Decimal Form
a.	17%	______%	0.83
b.	6%	______%	______
c.	$20\frac{1}{2}$%	______%	______
d.	$33\frac{1}{3}$%	______%	______
e.	19.9%	______%	______
f.	$16\frac{3}{8}$%	______%	______

3. Complete the following table.

	List Price	Trade Discounts	Net Cost Factor	Net Price	Discount Amount
a.	$ 790.00	15%	______	$______	$______
b.	908.00	17.5%	______	______	______
c.	650.00	10/15	______	______	______
d.	3,550.00	20/12	______	______	______
e.	7,290.00	30/10/10	______	______	______
f.	5,000.00	20/15/8	______	______	______

4. Beback Garden Supply Company lists edgers for $69.85, subject to trade discounts of 20/15/8. What is the lowest net price?

5. Jimmy's Bargain Store receives trade discounts of 20/15/7 on alarm clocks listed at $6.45. What is the amount of discount per clock?

6. A department store ordered 600 sweaters listed at $18.50 each. After discounts of 20/5/5, how much will the store pay for the shipment?

Section 8.3 Single Discount Equivalent

What Single Rate Results in the Same Savings as a Series of Rates?

In some cases you may want to find the one rate that results in the same savings as a series of discount rates. (It is *not* found by computing the sum of the rates.) This rate is called the single discount equivalent, which is the complement of the net cost factor.

Discount series: 20/10/5

Net cost factor =
$0.8 \times 0.9 \times 0.95 = 0.684$

$1.000 - 0.684 = 0.316$

Single discount equivalent $= 31.6\%$

Example 1 Pack Corporation lists its low-end personal computer and printer for $3,800 with possible trade discounts of 20/15/10. What is the net price and the single discount equivalent for a customer who receives all three discounts?

Complement rate of 20% = 100% − 20% = 80% = 0.8
Complement rate of 15% = 100% − 15% = 85% = 0.85
Complement rate of 10% = 100% − 10% = 90% = 0.9
Net cost factor = $0.8 \times 0.85 \times 0.9 = 0.612$
Net price = $3800 \times 0.612 = 2325.60$
Single discount equivalent = $1 - 0.612$
= 0.388 = 38.8%

key:	.8 [×] .85 [×] .9 [=]
display:	0.612 **net cost factor**
key:	[×] 3800 [=]
display:	2325.6
edited:	$2,325.60 **net price**
key alg:	1 [−] .612 [=]
key arith:	1 [+] .612 [−]
display:	0.388
edited:	38.8% **single discount equivalent**

PRACTICE ASSIGNMENT

Plan: Given a list price and a series of discounts, find the complement rates and the net cost factor. As before, multiply the list price by the net cost factor to find the net price. The single discount equivalent is the complement of the net cost factor.

1. Compute the single discount equivalents for the following series of discounts.

Discounts	Net Cost Factor in Decimals	Net Cost Factor in Percent	Single Discount Equivalent in Decimals	Single Discount Equivalent in Percent
a. 15/15	______	______%	______	______%
b. 25/17	______	______%	______	______%
c. 12/9	______	______%	______	______%
d. 25/20/10	______	______%	______	______%
e. 20/13/10	______	______%	______	______%
f. 12/8/4	______	______%	______	______%

2. Find the single discount equivalents and decide which one would result in a lower net price.

Discount 1	Single Discount Equivalent	Discount 2	Single Discount Equivalent	Lower Net Price 1 or 2
a. 20/10	______%	15/15	______%	1
b. 15/7.5	______%	18/4.5	______%	2
c. 20/10/10	______%	25/10/5	______%	2
d. 25/15/8	______%	23/15/10	______%	1
e. 15/10	______%	16/8	______%	1
f. 10/10/10	______%	15/15	______%	2

3. Daniel Jobbers offers discounts of 12/10/10 on men's suits. Central Square Wholesale grants discounts of 15/9/8 on the same merchandise. Whose rate is better for the customer?

4. Lopez Bait and Tackle Shop placed an order for lures with a total list price of $487.50. The shop is entitled to trade discounts of 15/5/5.

a. What is the single discount equivalent?

b. What is the final net price?

5. A department store receives discounts of 15/10/8 on a tricycle. The tricycle lists at $57.75.

a. What is the single discount equivalent?

b. If the store orders forty tricycles, what is the final net price?

Challenger

Wilson High School has budgeted $50,000 for laboratory equipment. Foster Science Company grants a 20% educational discount, which will reduce its bid price to $55,450. In order to win the contract, Foster Science is considering a second discount.

a. What would the single discount equivalent have to be for Foster Science to receive the order (accurate to the nearest hundredth of a percent)?

b. What would the net cost factor have to be?

c. What would the second discount have to be?

CHAPTER 8 SUMMARY

Concept	*Example*	*Procedure*	*Formula*
SECTION 8.1 Single Trade Discounts			
Trade discount	List price = $712.50 Discount = 8%	Trade discounts are offered as percent of list price. Use the percentage formula to compute the discount amount.	
Computation of discount amount and net price	List price = 712.50 712.5 × 0.08 = 57.00 Net price = 655.50	Net price = list price − discount amount	da = list × d list[×]d[=] list*d net = list − da alg: list[−]da[=] arith: list[+]da[−] list-da
Computation of net price by the complement method	Complement rate = 100% − 8% = 92% = 0.92 or 1.00 − 0.08 = 0.92	The complement rate is the difference between 100% and the rate.	comp(R) = 1 − R alg: 1[−]R[=] arith: 1[+]R[−] 1-R
	Net price = 712.50 × 0.92 = 655.50	Net price = list price × complement rate	net = list × comp(R) list[×]compR[=] list*(1-R)
	Discount amount = 712.50 − 655.50 = 57.00	Discount amount = list price − net price	da = list − net alg: list[−]net[=] arith: list[+]net[−] list-net
SECTION 8.2 Multiple Trade Discounts			
Multiple discounts	List price = $930.00 Discounts are 10/5	Compute the first net using the list price as base.	da1 = list × d1 net1 = list − da1
Computation of multiple nets	930.00 1st disc. = 930 × 0.1 = 93.00 1st net = 837.00 2nd disc. = 837 × 0.05 = 41.85 2nd net = 795.15	Compute the next net using the previous net as base. The net cost factor is the product of the complement rates.	da2 = net1 × d2 net2 = net1 − da2 Net cost factor = comp(d1) × comp(d2) × . . . compd1[×]compd2[×] . . . [=] (1-d1)*(1-d2)* . . .
Computation of final net by complement method	Final net = 930 × 0.9 × 0.95 = 795.15	To obtain the final net, multiply the list price by the net cost factor.	Final net = list price × ncf list[×]ncf[=] list*ncf
SECTION 8.3 Single Discount Equivalent			
Single discount equivalent	What is the single discount equivalent of 20/10/5? 0.8 × 0.9 × 0.95 = 0.684 1.000 − 0.684 0.316 = 31.6%	Compute the net cost factor. The single discount equivalent is the complement of the net cost factor converted to percent form.	Single discount equivalent = 100 × comp(ncf) alg: 1[−]ncf[=][×]100[=] arith: 1[+]ncf[−][×]100[=] 100*(1-ncf)

Chapter 8 Spreadsheet Exercise

Trade Discount Analysis

When completed, this spreadsheet will display the single discount equivalent, net cost factor, net price, and discount amount that correspond to various discount series.

Bring your chapter 8 template onto the screen. Remember that you may need to change to the correct disk drive first. For example, if the templates are in Drive B:

/
File[enter]
Directory[enter]
Inquiry
B:[enter]

or

/FD
Inquiry
B:[enter]

/
File[enter]
Retrieve[enter]
List of files
CHAP8.wk1[enter]

or

/FR
List of files
CHAP8.wk1[enter]

It will appear as follows:

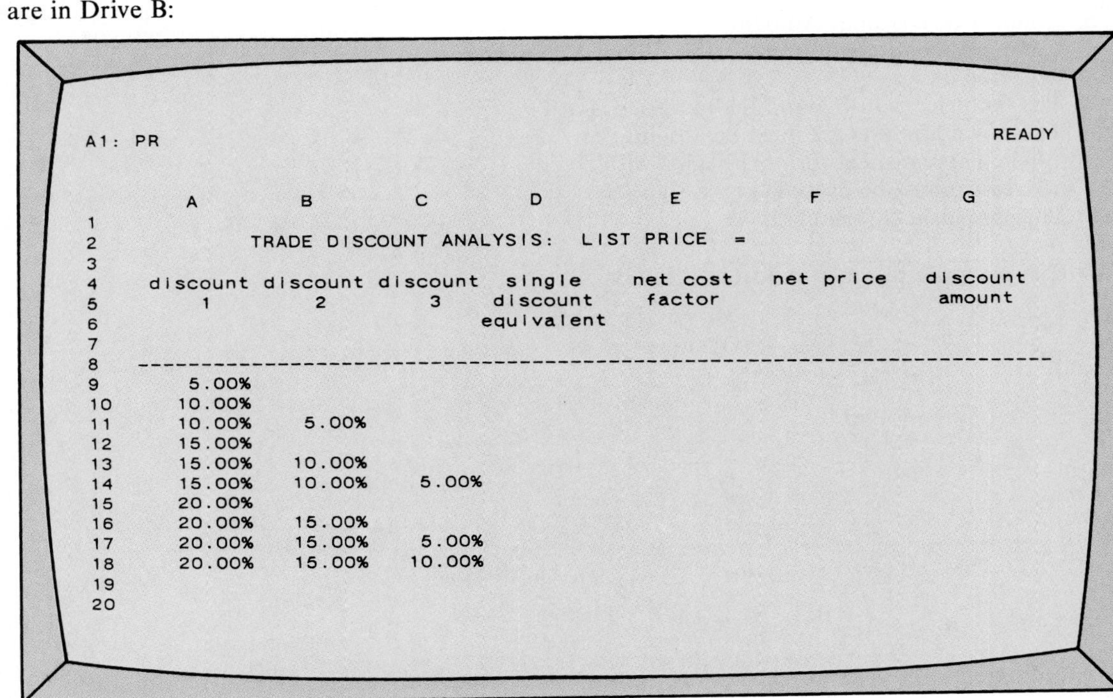

Columns A, B, and C contain ten different discount combinations, beginning with a single discount of 5% in row 9 and ending with the series 20/15/10 in row 18.
Be sure to enter all formulas into the computer without spaces between the characters.

Spreadsheet Exercise Steps

1. Enter 500 as the list price into F2.
2. Column D: Single Discount Equivalent
   The single discount equivalent equals
   1 − net cost factor.
   The net cost factor will be in E9.
   The spreadsheet formula is _____ − _______
   Complete the formula and store it in D9.
   (This cell will not display correctly until step 3 is completed.)
3. Column E: Net Cost Factor
   The net cost factor equals
   (1 − discount 1) × (1 − discount 2) ×
   (1 − discount 3).
   The three discounts are in A9, B9, and C9.
   The spreadsheet formula is
   (_____ − _______)*(_____ − _______)
   *(_____ − _______)
   Complete the formula and store it in E9.

4. Column F: Net Price
   Compute the net price from the list price and net cost factor.
   The list price is in F2, and the net cost factor is in E9.
   The row address in F2 must be absolute; therefore, the 2 must be preceded by a dollar sign.
   When a spreadsheet formula begins with a cell address it must be preceded by a sign (+).
   The spreadsheet formula is
   + _____ $ _____ * ______
   Complete the formula and store it in F9.
5. Column G: Discount Amount
   Compute the discount amount from the list price and net price.
   The list price is in F2, and the net price is in F9.
   The row address in F2 must be absolute.
   When a spreadsheet formula begins with a cell address it must be preceded by a sign (+).
   The spreadsheet formula is
   + _____ $ _____ − ______
   Complete the formula and store it in G9.
6. Copy the formulas from D9..G9 to D10..G18. (Review the Copy command in Section 5.5.)

In steps 7 through 10, verify your results with a calculator.

7. What single discount is equivalent to discounts of 15% and 10%?
8. What is the discount amount for the series 15/10/5?
9. What is the discount amount for the series 20/15/5 if the list price is $1,272?
10. What is the net price for the series 20/15/10 if the list price is $1,280.50?

CHAPTER 8 ASSIGNMENT

NAME ______________________________

CLASS SECTION ______________ DATE ______________

SCORE __________ MAXIMUM 150 PERCENT __________

1. Complete the following table.

Quantity	Stock Number/Description	Unit	Unit Price	Amount
5	1401/Heater	pc	$109.50	________
3	705/Spark plugs	dz	15.00	________
4	2365/Carburator	pc	65.45	________
16	2326/Shock absorber	pc	43.75	________
8	8215/Seat covers, plastic	pair	32.15	________
	Total			________
	Trade discount 18%			________
	Net price			________

(20 points)

2. Grandma's Chicken is entitled to a trade discount of $11\frac{1}{4}$% on all plastic ware purchased from a franchise supply company. An order lists for $639.75.
What amount can be deducted?
What is the net price? *(10 points)*

3. Silicorp Semiconductors offers trade discounts of 20/12/9 on pocket calculators to its most favored customers. What single discount is equivalent to this series? *(5 points)*

4. The list price of a bookcase is $487, subject to trade discounts of 35/15. What is the lowest wholesale price? *(5 points)*

5. A sporting goods store ordered 35 basketballs listing for $8.95 each. If it receives a trade discount of 16%, what is the net price? *(5 points)*

6. The list price of a patio set is $375, subject to trade discounts of 20/12.5.

a. What is the single discount equivalent?

b. Compute the savings and the net price from the single discount equivalent.
Savings:
Net: *(20 points)*

7. Inez Interiors plans to purchase 250 cans of varnish at $4.75 per can. The supplier promises trade discounts of 25/5. What will the discount amount be on this order? *(10 points)*

8. Trapper Auto Parts is entitled to trade discounts of 15/10/4 on windshield wipers listed at $12.80 per pair. On an order of 65 pairs, what is the total discount amount? *(10 points)*

9. University Medical Center plans to order 500 boxes of microscope slides. Lindstrom Medical Supplies offers slides at $5.50 per box, less discounts of 15/10. Science-Tek offers the same quality slides at $5.25 per box, less 20% discount.

a. What are the two single discount equivalents?
Lindstrom:
Science-Tek:

b. What are the two nets?
Lindstrom:
Science-Tek: *(20 points)*

10. A wholesaler grants trade discounts of $10\frac{1}{2}\%$ to Gordon's Pharmacy, a local drug store, and $14\frac{1}{2}\%$ to Presto, a chain of pharmacies. If Gordon's orders 17 cases of shaving cream listing for $18.35 per case and Presto orders 150 cases of the same product, how much will each pharmacy be billed? How much less per case will Presto pay than Gordon's?

a. Gordon's Pharmacy:

b. Presto:

c. Difference per case: *(15 points)*

11. Campus Sugar Shop enjoys standard trade discounts of 12/8 on imported taffy listing at $16.25 per carton. On orders of 100 cartons or more, an additional discount of 5% is granted. How much per carton can the store save by placing the larger order? If 125 cartons are ordered, what are the total savings due to the additional discount?

a. Savings per carton:

b. Savings on 125 cartons: *(10 points)*

12. First-Rate Office Supplies offers discounts of 15/15 on large volume stationery, while its competitor, Bradford Office Supplies, offers 20/10 on similar purchases.

a. Compute the two single discount equivalents.
First-Rate:
Bradford:

b. What is the difference in net price on annual purchases totaling $25,000? *(20 points)*

Chapter 9 Cash Discounts and the Invoice

OBJECTIVES

- Interpret sales terms
- Understand and compute cash discounts
- Apply partial payment rendered within a discounting period

SECTIONS

INTRODUCTION

To encourage quick settlement of commercial accounts, some merchants offer small price reductions called cash discounts. Cash discounts are not uniform and do not necessarily require payment on the day of purchase. A cash discount is applied to the net price after any trade discounts have been applied. All details of a sales transaction including terms and conditions are recorded on an invoice.

Section 9.1 Ordinary Sales Terms

KEY TERMS **Invoice** **Discounting period**

The sales terms listed on an **invoice** indicate whether a cash discount is offered and, if so, under what conditions.

Sales terms of 2/10,n/30 mean that 2 percent can be deducted if the invoice is paid within 10 days of the invoice date, the **discounting period.**

2	/	**10**	,	**n**	/	**30**
Discount 2%		Period 10 days		Full payment		30 days

Prior net	=	733.95
Cash discount	=	2%
733.95 × 0.02	=	14.68
New net price	=	719.27

Complement method: 733.95 × 0.98 = 719.27

In another example, terms of 3/15,n/45 mean that 3 percent can be deducted if the invoice is paid within 15 days of the invoice date. The number after the n/ indicates in how many days full payment is expected. If the bill is not paid within 15 days, no cash discount is allowed and the prior net amount is due within 45 days of the invoice date.

Example 1 A paint store received an invoice in the amount of $1,280.45 dated August 5 with terms of 2/15,n/45. Determine the discount amount and the new net if payment is rendered on August 10.

A cash discount of 2 percent may be taken if payment is rendered within 15 days of August 5; that is, by August 20. If payment is rendered on August 10, 2 percent may be deducted.

Discount amount = 2% of previous net
= 1280.45 × 0.02
= 25.61

New net = previous net − discount amount
= 1280.45 − 25.61
= 1254.84

key: 1280.45 [×] .02 [=]
display: 25.609
rounded: 25.61 **discount amount**
key alg: 1280.45 [−] 25.61 [=]
key arith: 1280.45 [+] 25.61 [−]
display: 1254.84
edited: $1,254.84 **net price**

The complement method may be more convenient.

New net = 98% of previous net
= 1280.45 × 0.98
= 1254.84

Discount amount = previous net − new net
= 1280.45 − 1254.84
= 25.61

key: 1280.45 [×] .98 [=]
display: 1254.841
rounded: 1254.84 **net price**
key alg: 1280.45 [−] 1254.84 [=]
key arith: 1280.45 [+] 1254.84 [−]
display: 25.61
edited: $25.61 **discount amount**

Staggered terms may be offered, such as 2/10,1/30,n/60. A two percent discount is offered for payment within 10 days; one percent for payment within 30 days. If payment is not made within the discounting period, no cash discount is allowed and the full amount is due within 60 days of the invoice date.

Invoice date: July 17
Payment date: August 1

Payment is not in time for the 2% discount, but is in time for the 1% discount.

PRACTICE ASSIGNMENT

Plan: Compute the end of the discounting period: Invoice date + number of days in discounting period. If payment is not rendered within the discounting period, full payment is required. If payment is rendered within the discounting period, reduce it by the discount amount. In the case of staggered terms, test each term in succession and apply the appropriate discount rate.

1. Determine the net in accordance with the following terms.

	Invoice Date	Net Prior to Cash Discount	Terms	Payment Date	Net
a.	Aug 16	$ 880.45	2/10,n/30	Sept 24	________
b.	Mar 6	1,203.15	3/15,n/30	Mar 21	________
c.	Apr 21	950.70	3/20,n/60	May 7	________
d.	July 30	1,880.00	2/15,n/45	Aug 13	________
e.	Aug 24	395.00	2/10,1/30,n/60	Sept 3	________
f.	May 22	2,060.50	3/10,2/20,n/45	June 16	________

2. Jablonski's Pharmacy received an invoice totaling $547.50 for first aid supplies. The invoice was dated August 17 and specified terms of 5/10,n/60. If payment is made on September 1, how much must be paid?

3. A supermarket received a shipment of cereal on July 14. The invoice, dated July 9, totaled $675 with terms of 2/10,n/30. How much did the supermarket pay if payment was rendered on July 19?

4. Good Riddance Pest Control received an invoice dated January 30 in the amount of $1,537 for spraying equipment. Terms of the invoice were 3/15,n/60. Good Riddance paid the invoice on February 10. What amount was paid?

5. A contractor bought a load of bricks. He received an invoice totaling $749.75, dated June 1, with terms of 3/15,2/25,n/45. If the invoice was paid on June 23, what was the net payment?

Section 9.2 Other Sales Terms

KEY TERMS **Receipt-of-goods** **End-of-month**

Receipt-of-goods (ROG) terms are advantageous to a buyer when a time lapse occurs between the invoice date and the delivery date.

A term of 2/10 ROG means that a cash discount of two percent may be taken if payment is rendered within ten days after the merchandise has been received.

2 /	**10**	**ROG**
Discount 2%	Period 10 days	After goods received

Example 1 A buyer of stereo equipment received an invoice dated September 7 with terms of 3/10 ROG for merchandise delivered on September 21. The invoice amount was $1,032. If payment is rendered by September 30, what is the new net?

A cash discount of 3% applies to payment rendered within 10 days after September 21. The end of the discounting period is September 21 + 10, or October 1. Payment rendered on September 30 is within the discounting period.

Net price = 97% of previous net
= $1032 \times 0.97 = 1001.04$

key: 1032 [×] .97 [=]
display: 1001.04
edited: $1,001.04

End-of-month (EOM) terms insure that most payments are received in the beginning of a month.

A sales term of 2/10 EOM means that a cash discount of two percent applies to payment rendered within 10 days after the end of the month in which an invoice is issued.

2 /	**10**	**EOM**
Discount / 2%	Period 10 days	After end of month

Example 2 A shoe outlet received an invoice dated June 21 with terms of 4/10 EOM. What is the new net if payment is rendered by July 9 on a prior net of $590?

A cash discount of 4% applies to payment rendered within 10 days after June 30, or by July 10. Since payment is rendered within the discounting period, the net is 96% of $590 = 590 × 0.96 = 566.40.

key: 590 [×] .96 [=]
display: 566.4
edited: $566.40

An additional month is usually granted on invoices dated from the 26th to the end of the month. For example, if terms are 2/10 EOM on an invoice dated July 26, the discounting period ends on September 10.

PRACTICE ASSIGNMENT

Plan: In the case of ROG terms, the end of the discounting period is:

delivery date + number of days in discounting period.

In the case of EOM terms, the end of the discounting period is:

end-of-month + number of days in discounting period.
Grant one additional month on invoices dated from the 26th to the end of the month.

If payment is not rendered within the discounting period, full payment is required.
If payment is rendered within the discounting period, reduce it by the discount amount.

1. Determine the net in accordance with the following terms.

	Delivery Date	Net Prior to Cash Discount	Terms	Payment Date	Net
a.	Aug 10	$ 690.00	2/10 ROG	Aug 17	$ ________
b.	Apr 27	1,140.50	3/15 ROG	May 18	________
c.	Dec 21	2,075.00	2/15 ROG	Jan 10	________
d.	July 30	988.13	4/10 ROG	Aug 16	________
e.	Nov 25	427.90	3/20 ROG	Dec 10	________
f.	May 22	989.00	1/25 ROG	June 12	________

2. Determine the net in accordance with the following terms.

	Invoice Date	Net Prior to Cash Discount	Terms	Payment Date	Net
a.	Sept 15	$ 290.00	2/10 EOM	Oct 20	$ ________
b.	May 29	725.25	3/15 EOM	July 11	________
c.	Oct 18	804.00	3/5 EOM	Nov 4	________
d.	Dec 12	2,115.50	2/15 EOM	Jan 16	________
e.	June 30	695.90	3/20,n/60 EOM	Aug 5	________
f.	Nov 9	1,785.70	2/10,n/45 EOM	Dec 12	________

3. On May 17 Harmony Music ordered several guitars. One week later the music store received an invoice dated May 22 in the amount of $945 with terms of 4/5 ROG. The guitars were delivered on June 12. What amount must be remitted if payment is rendered on June 20?

4. Thoreau Book Outlet received an invoice dated March 15 attached to a shipment of paperbacks. The invoice amount was $375 with terms of 3/15,n/45 EOM. How much should be remitted if the book store renders payment on April 2?

5. First State Bank ordered a copier costing $2,645 for a branch office. The invoice, dated April 29, showed terms of 3/10,n/60 EOM. How much must be remitted if payment is rendered on June 1?

6. A department store ordered lighting fixtures costing $965. The invoice, dated October 25 with terms of 2/10 ROG, was received on October 27. The lighting fixtures arrived on November 5. What remittance should be rendered on November 12?

Section 9.3 Partial Payment within the Discounting Period

Occasionally, a buyer may lack the cash necessary to pay the entire amount. If a partial payment is rendered within the discounting period, by how much will the obligation be reduced?

Example 1 Yvonne's Fashion Boutique received an invoice for $3,640, dated June 5, with terms of 3/15,n/45. A payment of $2,500 was remitted by June 20. How much credit should the buyer receive?

Within the discounting period, 97% of the obligation is sufficient payment. (97% is the complement of the 3% discount rate.) Therefore, a payment of $2,500 represents 97% of the credit received. Credit received = base. Find the base.

$$B = P \div R = 2500 \div 0.97 = 2577.32$$

Outstanding balance = invoice amount − credit received
= 3640.00 − 2577.32
= 1062.68

key:	2500 [÷] .97 [=]	
display:	2577.3196	
rounded:	2577.32	**credit received**
key alg:	3640 [−] 2577.32 [=]	
key arith:	3640 [+] 2577.32 [−]	
display:	1062.68	
edited:	$1,062.68	**balance**

PRACTICE ASSIGNMENT

Plan: In each of these problems a cash discount is allowed, but only partial payment is rendered within the discounting period. If the payment is taken as percentage and the complement of the discount rate as rate, the base becomes the credit received. The outstanding balance is the difference between the full invoice amount and the credit received.

1. A beauty salon received an invoice in the amount of $612.88 for supplies. Terms are 3/15,n/30, and the invoice is dated November 5. If a partial payment of $300 is rendered on November 18, how much will be credited toward the salon's account?

2. Torrelli's Tackle Shop received an invoice in the amount of $1,478.30 dated April 29 with terms of 2/10,n/30. If a partial payment of $700 is rendered on May 8, what is the remaining balance? $764.01 On what date is this balance due?

3. The Hercules Health Club received an invoice dated March 30 in the amount of $1,287.90 for some weight-lifting equipment. Terms of the invoice were 2/10,n/45 EOM, and a partial payment of $750 was rendered on May 8. By how much has the club's obligation been reduced?

4. Chez Francois, a new restaurant, ordered tablecloths and napkins costing $782.75. The invoice date was May 17 and the terms were 1/10 ROG. If the order was received on May 21 and payment of $275 was rendered on May 30, how much does the restaurant still owe?

5. Arkin's Veterinary Clinic received an invoice in the amount of $817.05 for drugs and supplies. The terms of the invoice dated March 30 were 3/10,1/20,n/30. If a payment of $500 is rendered on April 11, what is the remaining obligation?

Section 9.4 Other Items on the Invoice

KEY TERMS **Credits** **Freight charges**

After trade and cash discounts (if any) are deducted from the list price, other adjustments may be necessary, such as credits for damaged or returned goods and transportation charges.

Transportation charges may vary and are subject to negotiation. FOB (free on board) means that the seller will place the merchandise on the transportation vessel. FOB destination (e.g., FOB Buffalo) means that the seller will pay the transportation cost to the city of destination. Billing forms may differ, and invoices may vary.

	Total list price
−	Trade discounts
=	Net after trade discounts
−	Cash discount
=	Net after cash discount
−	Credits
=	Net after credits
+	Freight charges
=	Final total

PRACTICE ASSIGNMENT

Complete both columns of the following invoice.

INVOICE

NTT COMPUTER MANUFACTURING COMPANY

Route 121
Green Valley Industrial Road
Deforst, GA 30300

Ship to:	Springfield Data Processing 125 Medical Center Drive Springfield, GA 31329
Bill to:	Same

Order No.: 3422	Date: Oct. 20, 199X
Terms:	2/30, n/60 ROG
Freight:	paid by buyer
Invoice No.:	751

Description	Stock No.	Units	Unit Price	Extension
Monitors, monochrome	112-M	5	512.60	2563.00
CPU 640 K	83-A	5	1300.00	6500.00
Printer cables	111-O	20	67.80	1356.00
Total List Price			10419.00	10419.00
– Trade Discounts 15/5				
Net after Trade Discounts				
– Cash Discount			0.00	
Net after Cash Discounts				
– Credit, Invoice No. 712			513.18	513.18
Net after Credits				
+ Freight, Dixie Express			1128.00	1128.00
Net Price				

Challenger

Buddy's Sight and Sound received an invoice dated July 31 attached to a shipment of 50 radios. The amount of the invoice was $867.50, the terms were 3/10,2/20,n/45. However, since only 45 radios had been ordered, Buddy's returned the extra merchandise immediately. On August 17, the company paid a partial payment of $400. Disregarding freight charges and assuming full credit for the returned merchandise, what was Buddy's remaining obligation?

CHAPTER 9 SUMMARY

Concept	*Example*	*Procedure*	*Formula*
SECTION 9.1 Ordinary Sales Terms			
Ordinary sales terms	2/10,n/30	The buyer is entitled to the discount shown if payment is made within the number of days specified (discounting period).	
Discount amount	Invoice date: July 5 Previous net = $618.10 Cash discount = 618.1 × 0.02 = 12.36 If invoice is paid by July 15, net = 618.10 − 12.36 = 605.74 Otherwise, net = $618.10	A cash discount is a percent of the prior net. Use the percentage formula to compute the discount amount. (The complement method may also be used.)	da = prev. net × d prev.net[×]d[=] prev.net*d New net = prev. net − da alg: prev.net[−]da[=] arith: prev.net[+]da[−] prev.net-da
SECTION 9.2 Other Sales Terms			
Receipt-of-goods terms	3/10 ROG Invoice date: July 5 Goods received on July 25 Cash discount is 3% if invoice is paid by August 4 (computation as in 9.1).	The buyer is entitled to the discount shown if payment is made within the discounting period. The discounting period begins when goods are received (rather than on the invoice date).	
End-of-month terms	4/8 EOM Invoice date: July 5 Cash discount is 4% if invoice is paid by August 8 (computation as in 9.1).	The buyer is entitled to the discount shown if payment is made within the discounting period. The discounting period begins at the end of the month. An additional month may be granted when the invoice date is close to the end of the month (26th or later).	
SECTION 9.3 Partial Payment within the Discounting Period			
Partial payment within discounting period	The discounting period for an invoice amounting to $3,195.25 ends on June 10. Partial payment of $2,000.00 is made on June 9. How much is credited to the buyer, and what is the balance? Credit = 2000 ÷ 0.98 = 2040.82	Credit for partial payment = $\frac{\text{Partial payment}}{\text{Complement of discount rate}}$	Credit = $\frac{\text{payment}}{\text{comp}(d)}$ payment[÷]compd[=] payment/(1-d)
	Balance = 3195.25 − 2040.82 = 1154.43	Balance = previous net − credit	Balance = previous net − credit alg: prev.net[−]credit[=] arith: prev.net[+]credit[−] prev.net-credit
SECTION 9.4 Other Items on the Invoice			
Invoice		Total list price − Trade discounts − Cash discount − Credits + Freight charges = Final total net	

Chapter 9 **Spreadsheet Exercise**

Determining Net Prices

If a cash discount is offered, net price depends on whether or not payment is rendered within the discount period. This spreadsheet calculates the end of the discount period, determines whether or not payment was on time, and computes net price accordingly.

The DATEVALUE function converts dates to date serial numbers, thereby permitting computations with dates. The IF function enables the program to recognize whether payment was received within the discount period.

Bring your chapter 9 template onto the screen. Remember that you may need to change to the correct disk drive first by accessing File, Directory.

/
File[enter]
Retrieve[enter]
list of files
CHAP9.wk1[enter]

or

/FR
list of files
CHAP9.wk1[enter]

It will appear as follows:

```
A1: PR [W10]                                                              READY

        A        B        C        D         E        F   G        H           I
1                               DETERMINING NET PRICES
2
3    invoice          invoice  delivery  payment     r   d     end-of-       net
4    amount           date     date      date        a   a     discount      price
5                              if                    t   y     period
6                              rog                   e   s
7
8  ------------------------------------------------------------------------------
9   1,480.35         05/01/89            05/10/89  2.0% 10
10    850.25         05/15/89            05/25/89  3.0%  8
11  1,040.25         06/05/91            07/05/91  3.5%  5
12    345.18 rog              08/14/89   08/24/89  1.0% 15
13    505.10         08/21/90            09/04/90  4.0%  5
14  1,290.00 rog              04/20/90   04/30/90  1.5% 10
15  1,585.00         09/12/91            10/06/91  3.5%  8
16  2,575.60         10/17/91            10/25/91  2.5% 10
17    805.14 rog              01/18/91   02/28/91  3.0% 10
18  1,777.41         02/21/91            02/28/91  2.0% 15
19
20
```

Column A contains ten invoice amounts in rows 9 to 18. The cells in column B are either blank or contain the term *rog.* If a cell in column B is blank, the corresponding cell in column C contains the invoice date. If a cell in column B contains *rog,* the corresponding cell in column D contains the delivery date.
Column E contains the payment dates.
Column F contains the discount rates.
Column G contains the number of days in the discount period.

Spreadsheet Exercise Steps

1. Column H: End of Discount Period
The end of the discount period is

invoice date + number of days in the term

or

delivery date + number of days in the term.

Invoice dates, where given, begin in C9.
Delivery dates, where given, begin in D9.
The number of days in the term is in G9.
The IF function, @IF(condition,*x*,*y*), produces the value *x* if the condition is true or the value *y* if the condition is false.

The condition is whether the term in B9 is blank or not.
If it is blank, use the invoice date; otherwise, use the delivery date.
The spreadsheet formula is:

@IF(B9=" ",
@DATEVALUE(______) + ______ ,
@DATEVALUE(______) + ______)

Make sure to enter a space between the two quotation marks by pressing the spacebar. Otherwise, there should be no spaces between the characters.
All parentheses and @ signs must be correctly entered.
Complete the formula and store it in H9.

2. Column I: Net Price
If the payment date is greater than the date at the end of the discount period, net price = invoice amount.
Otherwise, net price =
invoice amount * (1-discount rate).
The invoice amount is in A9.
The payment date is in E9.
The discount rate is in F9.
The end of the discount period is in H9.
The IF function @ IF(condition,*x*,*y*) becomes

@ IF(@ DATEVALUE(E9)>@ DATEVALUE(H9),
______ , ______ * (_____ − ______))

Complete the formula and store it in I9.

3. Copy the formulas from H9..I9 to H10..I18.

4. In each case, verify whether the discount (or lack of it) was justified.

Row	End of Discount Period	Payment Date	Write *on time* or *late*	Net Price Verified by Calculator
9	______	______	______	______
10	______	______	______	______
11	______	______	______	______
12	______	______	______	______
13	______	______	______	______
14	______	______	______	______
15	______	______	______	______
16	______	______	______	______
17	______	______	______	______
18	______	______	______	______

NAME ______________________________

CLASS SECTION ______________ DATE ______________

SCORE ____________ MAXIMUM 125 PERCENT ____________

CHAPTER 9 ASSIGNMENT

1. A wholesale furniture company allows a cash discount of 4/15,n/60 to some of its customers. If an invoice is dated November 2, what is the last date on which payment can be rendered in order to qualify for the discount? *(5 points)*

2. On April 8, Banion Contractors received an invoice dated April 5. The terms were 3/10 ROG, and the amount was $2,780 for a shipment of building materials received on April 15. If payment was rendered on April 21, what was the amount remitted? *(5 points)*

3. The Kennedy Square Dental Clinic was invoiced $637.50 on August 15 for supplies received on August 31. Terms were 3/15 EOM. If payment was rendered on September 30, how much was remitted? *(5 points)*

4. Miller Restaurant Supply lists terms of 2/10,1/25,n/90 on its invoices. If an invoice totaling $2,750 is dated September 18, how much should the supplier receive if payment is rendered:

a. September 25?

b. October 5? *(10 points)*

5. An invoice of $2,960.30 is dated April 18 with terms of 2/15 EOM.

a. What is the last date of the discounting period?

b. If payment is rendered by May 10, what is the net? *(10 points)*

6. Riding Wheels received an invoice for bicycles in the amount of $2,847. The invoice date was May 25 and terms were 3/15,n/45 for bicycles delivered on May 31. A partial payment of $1,500 was rendered on June 12. What was the outstanding balance after the partial payment? *(5 points)*

7. Person-to-Person Employment Agency ordered stationery and received an invoice dated January 31 in the amount of $315.60. Terms were 2/15,n/30.

a. What is the last date on which payment can be made to qualify for the discount?

b. What is the net amount if payment is rendered by that date? *(10 points)*

8. On August 15, Schuman's clothing store ordered men's overcoats. An invoice dated August 22 in the amount of $1,750 was received on August 25. Terms were 2/15 ROG. The shipment was received on September 10.

a. What is the last day payment can be made to take advantage of the discount?

b. How much money will the discount save? *(10 points)*

9. Farber's Fashions received an invoice dated March 1 in the amount of $2,535 for sewing machines. Terms of the sale are 3/10,2/20,1/25,n/60. What is the final date on which payment can be rendered in order to qualify for each of the three discounts? What is the amount of each discount?

Discount	Final Date	Amount
a. First	________	________
b. Second	________	________
c. Third	________	________

(15 points)

10. Donaldson's Uniform Shop was billed $682.50 for an order of nurse's uniforms. Terms of the invoice, dated November 7, were 3/10,2/20,n/45. How much credit can Donaldson's receive if a payment of $350 is rendered on:

a. November 13?

b. November 26? *(10 points)*

11. Complete the following invoice, including the blanks at the bottom. Assume the year is not a leap year.

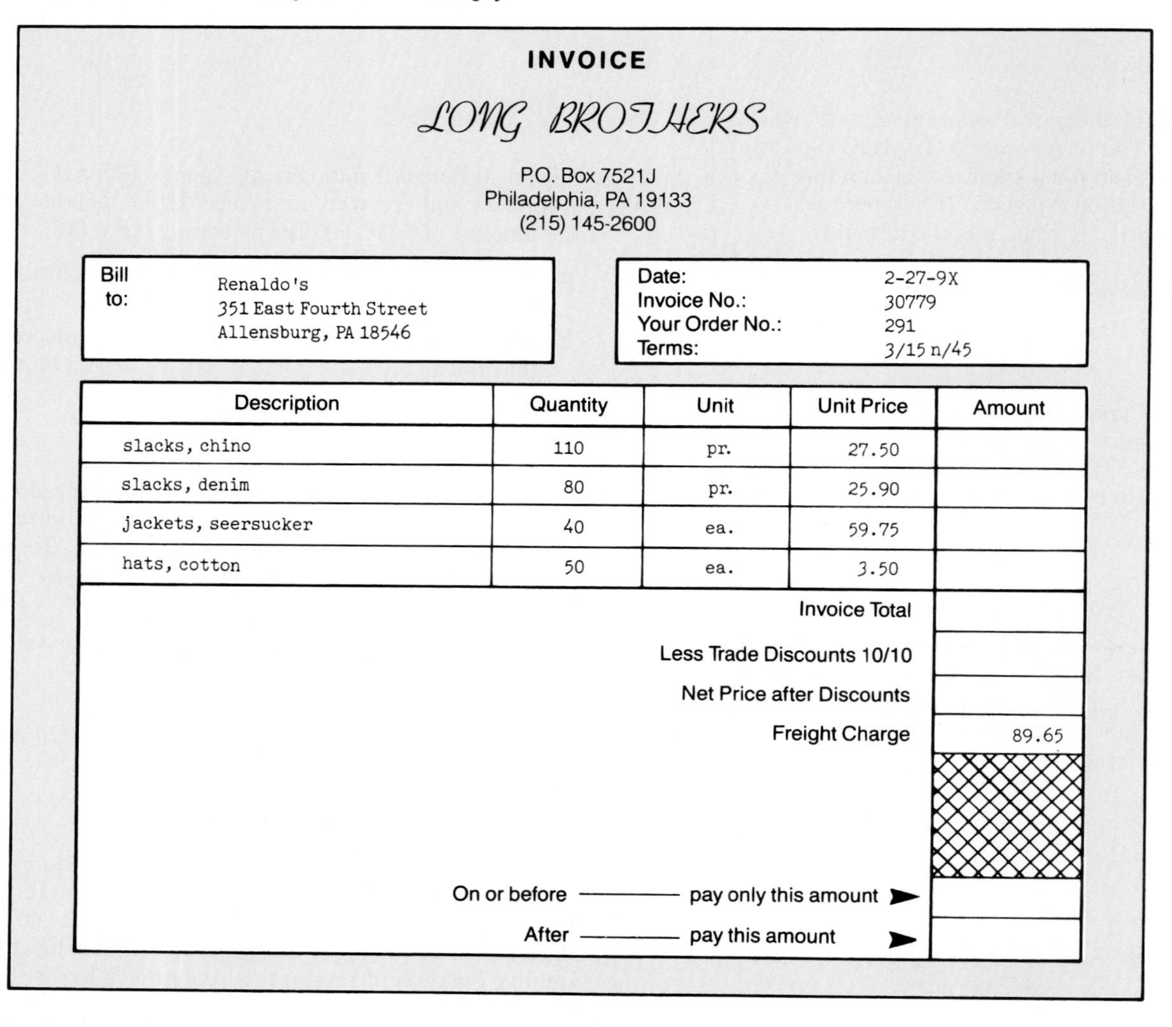

INVOICE

LONG BROTHERS

P.O. Box 7521J
Philadelphia, PA 19133
(215) 145-2600

Bill to: Renaldo's
351 East Fourth Street
Allensburg, PA 18546

Date: 2-27-9X
Invoice No.: 30779
Your Order No.: 291
Terms: 3/15 n/45

Description	Quantity	Unit	Unit Price	Amount
slacks, chino	110	pr.	27.50	
slacks, denim	80	pr.	25.90	
jackets, seersucker	40	ea.	59.75	
hats, cotton	50	ea.	3.50	
			Invoice Total	
			Less Trade Discounts 10/10	
			Net Price after Discounts	
			Freight Charge	89.65
			On or before ——— pay only this amount ➤	
			After ——— pay this amount ➤	

On what date is reduced payment due?

What is the amount of reduced payment?
On what date is full payment due?
What is the amount of full payment?

(40 points)

Chapter 10 Markup

OBJECTIVES

- Compute markup amounts and rates based on cost
- Compute markup amounts and rates based on selling price
- Find the cost or selling price when markup is based on cost or selling price
- Convert a markup rate based on cost to a markup rate based on selling price and the reverse

SECTIONS

INTRODUCTION

Retailers are in business to make money. When they sell products, they must set selling prices that not only cover their costs, but also generate a profit. The difference between cost and selling price is markup.

Section 10.1 Markup Based on Cost

KEY TERMS **Cost** **Markup** **Selling Price**

The prices retail firms charge their customers reflect three components: **cost,** expenses, and a return on investment called net profit. Gross profit, also called **markup,** is the sum of expenses and net profit. **Selling price** is the sum of cost and markup.

selling price
cost + expenses + net profit
gross profit
markup
cost + markup

Markup is often expressed as a percent of cost.

Cost is the base or 100 percent.

Markup is the percentage that corresponds to the markup rate.

Selling price is the percentage that corresponds to the rate (100% + markup rate).

Markup Based on Cost		
	R	
Cost	100%	B \$550.00
+ Markup	35%	P \$192.50
= Selling price	135%	P \$742.50

Example 1 The cost of a small appliance to the Southport Company is \$57.82; the company applies a markup rate of 21 percent. What is the markup? What is the selling price?

$$\text{Markup} = P = B \times R = \text{cost} \times \text{markup rate}$$
$$= 57.82 \times 0.21 = 12.14$$

$$\text{Selling price} = \text{cost} + \text{markup}$$
$$= 57.82 + 12.14$$
$$= 69.96$$

key: 57.82 [×] .21 [=]
display: 12.1422
rounded: 12.14 **markup**
key alg: 57.82 [+] 12.14 [=]
key arith: 57.82 [+] 12.14 [+]
display: 69.96
edited: \$69.96 **selling price**

Example 2 What is the markup and the rate of markup on a desk lamp that sells for $18.25 and costs Krantz Lighting $11.57?

Markup = selling price − cost
= 18.25 − 11.57 = 6.68

Markup rate = $P \div B$ = markup ÷ cost
= 6.68 ÷ 11.57
= 57.7%

key alg: 18.25 [−] 11.57 [=]
key arith: 18.25 [+] 11.57 [−]
display: 6.68 **markup**
key: [÷] 11.57 [=]
display: 0.5773553
edited: 57.7% **markup rate**

Example 3 A department store sells a refrigerator for $683.25, which includes a markup of 15.75 percent. What is the cost of the refrigerator to the department store?

Markup rate = 15.75%

Rate of selling price = 115.75%

Cost = $B = P \div R$
= selling price ÷ rate of selling price

683.25 ÷ 1.1575 = 590.28077 = 590.28

key: 683.25 [÷] 1.1575 [=]
display: 590.28077
edited: $590.28

PRACTICE ASSIGNMENT

Plan: The variables are cost, markup, markup rate, and selling price. Find the missing variable, either a rate, a percentage (markup or selling price), or the base (cost). In this assignment, markup is always based on cost.

1. Find the markup and selling price.

	Cost	Rate of Markup	Markup	Selling Price
a.	$ 29.00	20%	$ ________	$ ________
b.	48.50	42%	$ ________	$ ________
c.	2,059.00	100%	$ ________	$ ________
d.	91.44	$66\frac{2}{3}$%	$ ________	$ ________
e.	370.80	19.9%	$ ________	$ ________
f.	261.35	120%	$ ________	$ ________

2. Find the markup rate. Round to the nearest tenth of a percent where necessary.

	Cost	Selling Price	Rate of Markup
a.	$ 18.00	$ 24.00	________
b.	90.25	121.84	________
c.	135.75	237.56	________
d.	68.93	99.95	________
e.	189.75	379.50	________
f.	207.80	446.77	________

3. Find the cost and selling price.

	Rate of Markup	Markup	Cost	Selling Price
a.	30%	$ 4.71	$ ________	________
b.	$66\frac{2}{3}$%	28.14	$ ________	________
c.	39%	198.12	$ ________	________
d.	26.5%	24.91	$ ________	________
e.	105%	364.35	$ ________	________
f.	46.75%	114.07	$ ________	________

4. A discount chain store sells sets of furniture for $428.65 after a markup of 12.5%. What is the cost to the discount chain?

5. What is the rate of markup on a loaf of bread that sells for $1.24 and costs the grocery store $1.20?

6. The markup on a pair of jeans is $6.57; the markup rate is 45 percent. What is the selling price?

Section 10.2 Markup Based on Selling Price

Markup can be expressed as a percent of selling price.

Cost is the percentage that corresponds to the rate (100% − markup rate).

Markup is the percentage that corresponds to the markup rate.

Selling price is the base or 100 percent.

Markup Based on Selling Price		
	R	
Cost	65%	P $149.50
+ Markup	35%	P $80.50
= Selling price	100%	B $230.00

Example 1 Viking, Inc. sells a set of exercise equipment for $258.75. What is the markup if the markup rate is 17.5 percent? What is the cost to Viking?

Markup = $P = B \times R$ = selling price × markup rate
= 258.75 × 0.175 = 45.28

Cost = selling price − markup
= 258.75 − 45.28
= 213.47

key: 258.75 [×] .175 [=]
display: 45.28125
rounded: 45.28 **markup**
key alg: 258.75 [−] 45.28 [=]
key arith: 258.75 [+] 45.28 [−]
display: 213.47
edited: $213.47 **cost**

Example 2 A VCR retails for $315.80. What is the markup rate on this item if the cost to the retailer is $270?

Markup = selling price − cost
= 315.80 − 270.00 = 45.80

Markup rate = $R = P \div B$
= markup ÷ selling price
= 45.80 ÷ 315.80
= 14.5%

key alg: 315.8 [−] 270 [=]
key arith: 315.8 [+] 270 [−]
display: 45.8
edited: 45.80 **markup**
key: [÷] 315.8 [=]
display: 0.1450284
edited: 14.5% **markup rate**

Example 3 The Boulder Company sells a line of telescopes marked up at 23.5 percent. If the markup on a telescope is $550, what is the cost to the company?

Selling price = $B = P \div B$ = markup ÷ markup rate
= 550 ÷ 0.235 = 2340.43

Cost = selling price − markup
= 2340.43 − 550
= 1790.43

key: 550 [÷] .235 [=]
display: 2340.4255
rounded: 2340.43 **selling price**
key alg: 2340.43 [−] 550 [=]
key arith: 2340.43 [+] 550 [−]
display: 1790.43
edited: $1,790.43 **cost**

PRACTICE ASSIGNMENT

Plan: The variables are cost, markup, markup rate, and selling price. Find the missing variable, either a rate, a percentage (cost or markup), or the base (selling price). In this assignment, markup is always based on selling price.

1. Find the markup and cost.

Rate of Markup	Selling Price	Markup	Cost
a. 15%	$ 88.00	$______	$______
b. 24%	106.50	$______	$______
c. $12\frac{1}{2}$%	59.50	$______	$______
d. 17.9%	47.59	$______	$______
e. $33\frac{1}{3}$%	1,453.95	$______	$______
f. $22\frac{3}{4}$%	1,260.00	$______	$______

2. Find the markup rate. Round to the nearest one-tenth of a percent where necessary.

	Cost	Selling Price	Rate of Markup
a.	$ 18.00	$ 27.00	____%
b.	1,050.00	1,689.95	____%
c.	848.00	1,290.50	____%
d.	2,245.00	3,490.00	____%
e.	97.75	139.99	____%
f.	3,800.00	6,999.95	____%

3. Find the selling price and cost.

	Markup	Rate of Markup	Selling Price	Cost
a.	$ 17.25	15%	$______	$______
b.	15.57	18%	$______	$______
c.	147.81	$19\frac{1}{2}\%$	$______	$______
d.	71.91	28.2%	$______	$______
e.	1,040.17	41%	$______	$______
f.	68.04	$16\frac{2}{3}\%$	$______	$______

4. A toy store sells a model airplane for $24.80. If the markup rate is 27.5%, what is the markup on this item?

5. Winter Wonderland marks up ice skates $64.80 per pair, a rate of 27%. What is the selling price?

6. A food processor is marked up 24%. If the selling price is $62.79, what is the cost?

Section 10.3 Conversion of Markup Rates

A markup rate of 20 percent based on cost is not equivalent to a markup rate of 20 percent based on selling price.

Markup Based on Cost		
Cost	100%	$150
+ Markup	20%	$ 30
= Selling price	120%	$180

Markup Based on Selling Price		
Cost	83.3%	$150
+ Markup	16.7%	$ 30
= Selling price	100%	$180

Markup may be based on cost or sales. Consumers like to know the markup rate based on cost. However, since sales data is more accessible than cost data, trade statistics are often calculated from a sales base.

To compare markup rates, each rate must refer to the same base. Convert a markup rate from one base to another with the following formulas:

$$\text{Rate based on sales} = \frac{\text{rate based on cost}}{1 + \text{rate based on cost}}$$

$$\text{Rate based on cost} = \frac{\text{rate based on sales}}{1 - \text{rate based on sales}}$$

Example 1 On the average, the Food Tiger grocery chain marks up food by 6.25 percent based on cost. For comparisons, the industry usually lists markups based on sales. What is the corresponding markup rate based on selling price?

$$\begin{aligned}\text{Rate based on selling price} &= \frac{\text{rate based on cost}}{1 + \text{rate based on cost}}\\ &= \frac{0.0625}{1.0625}\\ &= 0.0588235\\ &= 5.9\%\end{aligned}$$

Example 2 An industry magazine lists markups based on selling price. A consumer advocacy group, however, prefers to see the markup rates based on cost. If the average markup rate on camping equipment is listed at 12.5 percent, how will the consumer group list it?

$$100\% - 12.5\% = 87.5\% = 0.875$$

$$\text{Rate based on cost} = \frac{\text{rate based on selling price}}{1 - \text{rate based on selling price}}$$

$$= \frac{0.125}{0.875}$$

$$= 0.1428571$$

$$= 14.3\%$$

PRACTICE ASSIGNMENT

Plan: If the markup rate is based on cost, find the corresponding markup rate based on selling price. If the markup rate is based on selling price, find the corresponding markup rate based on cost. Round resulting markup rates to the nearest tenth of a percent.

1. *Consumer Report* magazine surveys the retail trade and lists the following markup rates based on cost. Show the rates based on selling price.

	Based on Cost	Based on Selling Price
Food	8%	____%
Automobiles	18%	____%
Clothing	23%	____%
Cosmetics	58%	____%
Home improvements	31%	____%
Jewelry	110%	____%

2. *Computer World* lists markup rates based on selling price for different brands of personal computers. Show the markup rates based on cost.

	Based on Selling Price	Based on Cost
UBM	11.0%	____%
Orange	18.5%	____%
Handy	8.5%	____%
Dexta	9.9%	____%
Lang	16.0%	____%
Yamata	19.4%	____%

3. A gift shop marks up a set of candle holders 40% based on selling price. What is the rate of markup based on cost?

4. Rizzo's Pharmacy marks up bottles of sun block 37% based on cost. What is the markup rate based on selling price?

Challenger

Dollar Saver is a discount store that marks up portable televisions 9% based on selling price. An order of forty-five TVs costs the store $5,467.50 less trade discounts of 20/15. What is the markup on each set?

CHAPTER 10 SUMMARY

Concept	*Example*	*Procedure*	*Formula*
SECTION 10.1 Markup Based on Cost			
Markup		Selling price = cost + markup	
Markup based on cost	Cost 100% 550.00 + Markup 35% 192.50 = Selling price 135% 742.50	Cost is base. Markup and selling price are percentages related to cost.	
Finding markup given cost and markup rate	Cost = \$600.00 Markup rate = 15% What is the markup? 600 × 0.15 = 90	Find percentage given base and rate.	Markup = BR = cost × rate cost[×]rate[=] cost*rate
Finding markup rate given cost and markup	Cost = \$62.50 Markup = \$10.00 What is the markup rate? 10 ÷ 62.5 = 0.16 = 16%	Find rate given base and percentage.	Markup rate = $\frac{P}{B} = \frac{\text{markup}}{\text{cost}}$ markup[÷]cost[=] markup/cost
Finding cost given markup and markup rate	Markup = \$17.85 Markup rate = 18% What is the cost? 17.85 ÷ 0.18 = 99.17	Find base given rate and percentage.	Cost = $\frac{P}{R} = \frac{\text{markup}}{\text{rate}}$ markup[÷]rate[=] markup/rate
SECTION 10.2 Markup Based on Selling Price			
Markup based on selling price	Cost 65% 149.50 + Markup 35% 80.50 = Selling price 100% 230.00	Selling price is base. Markup and cost are percentages related to selling price.	
Finding markup given selling price and markup rate	Selling price = \$490.35 Markup rate = 15% What is the markup? 490.35 × 0.15 = 73.55	Find percentage given base and rate.	Markup = BR = selling price × rate selling price[×]rate[=] selling price*rate
Finding markup rate given selling price and markup	Selling price = \$65.40 Markup = \$17.50 What is the markup rate? 17.5 ÷ 65.4 = 0.267584 = 26.76%	Find rate given base and percentage.	Markup rate = $\frac{P}{B} = \frac{\text{markup}}{\text{selling price}}$ markup[÷]selling price[=] markup/selling price
Finding selling price given markup and markup rate	Markup = \$73.20 Markup rate = 12.5% What is the selling price? 73.2 ÷ 0.125 = 585.60	Find base given rate and percentage.	Selling price = $\frac{P}{R} = \frac{\text{markup}}{\text{rate}}$ markup[÷]rate[=] markup/rate
SECTION 10.3 Conversion of Markup Rates			
Conversion of markup rate based on cost to markup rate based on selling price	Rate based on cost = 18% Rate based on sales $= \frac{0.18}{1.18}$ = 0.1525423 = 15.25%		Rate based on sales = $\frac{\text{rate based on cost}}{1 + \text{rate based on cost}}$ rate based on cost[÷] 1+ rate based on cost[=] rate based on cost/ (1+rate based on cost)
Conversion of markup rate based on selling price to markup rate based on cost	Rate based on sales = 20% Rate based on cost $= \frac{0.2}{0.8}$ = 0.25 = 25%		Rate based on cost = $\frac{\text{rate based on sales}}{1 - \text{rate based on sales}}$ rate based on sales[÷] 1− rate based on sales[=] rate based on sales/ (1−rate based on sales)

Chapter 10 Spreadsheet Exercise

Comparison of Markup Rates

For a particular cost and markup, the markup rate based on selling price is lower than the markup rate based on cost. However, the method of computing the markup rate does not affect the selling price. This spreadsheet displays both rates for a comparison.

Bring your chapter 10 template onto the screen. It will appear as follows:

```
A1: PR [W8]                                                        READY

          A          B          C       D        E         F         G
1                            COMPARISON OF MARKUP RATES
2
3                                 COST    =
4   ----------------------------------------------------------------------
5       markup               markup              cost     markup    selling
6        rate                 rate                          in       price
7      based on             based on                      dollars
8        cost               selling
9                            price
10  ----------------------------------------------------------------------
11       5.0%
12      10.0%
13      15.0%
14      20.0%
15      25.0%
16      30.0%
17      35.0%
18      40.0%
19      45.0%
20      50.0%
```

Column A contains twenty different markup rates based on cost, beginning with 5% in row 11 and ending with 100% in row 30.

Spreadsheet Exercise Steps

1. Enter 500 as cost into E3.
2. Column C: Markup Rate Based on Selling Price
   Compute the markup rate based on selling price from the markup rate based on cost.
   The rate based on cost is in A11.
   When a spreadsheet formula begins with a cell address, it must be preceded by a sign (+).
   The spreadsheet formula is + ______ /
   (_____ + ______).
   Complete the formula, store in C11, and copy from C11 to C12..C30.
3. Column E: Cost
   The cost is replicated from cell E3 into the cells of this column; the row address must be absolute.
   (When you vary cost, change it in E3 only.) Place +E$3 into E11.
4. Column F: Markup in Dollars
   Compute the markup from the cost and from the markup rate based on cost. The cost is in E11, and the markup rate based on cost is in A11. When a spreadsheet formula begins with a cell address, it must be preceded by a sign (+).
   The spreadsheet formula is + ______ * ______ .
   Complete the formula and store it in F11.
5. Column G: Selling Price
   Compute the selling price from cost and markup.
   Cost is in E11 and markup in F11.
   When a spreadsheet formula begins with a cell address, it must be preceded by a sign (+).
   The spreadsheet formula is
   + ______ + ______ .
   Complete the formula and store it in G11.

6. Copy the formulas from E11..G11 to E12..G30.
7. What is the markup when the markup rate based on cost is 15%?
8. What is the selling price when the markup rate based on cost is 20% and the cost is $850?
9. What is the selling price when the markup rate based on selling price is 31% and the cost is $681.50?
10. What is the markup rate based on cost if the markup rate based on sales is 31%?

NAME ____________________

CLASS SECTION ____________ DATE ____________

SCORE ________ MAXIMUM __70__ PERCENT ________

CHAPTER 10 ASSIGNMENT

1. A merchant marks up merchandise 31.5% based on cost. What is the selling price of an item costing $43.50? *(5 points)*

2. A camera that sells for $171.25 carries a markup of 21% based on selling price. What is the cost to the store? *(5 points)*

3. A health food store marks up garlic tablets $7.59 per bottle, a rate of 85% based on cost. What is the selling price? *(5 points)*

4. A tennis racket is advertised in a tennis magazine for $151.90 by a mail-order company. If the mail-order company buys the racket for $120, what is the markup rate based on the selling price? *(5 points)*

5. Show the markup and the two markup rates.

Item	Cost	Selling Price	Markup in Dollars	Markup % Based on Cost	Markup % Based on Selling Price
a. Pen	$ 0.10	$ 0.58	$______	______%	______%
b. Desk	202.80	299.00	$______	______%	______%
c. Radio	174.20	259.90	$______	______%	______%
d. Washer	268.00	387.35	$______	______%	______%
e. Boat	2,653.29	3,154.00	$______	______%	______%

(15 points)

6. Chic 'n Comfy Shoes sells boots for $95.75 per pair. If the markup rate is 37% based on cost, what is the markup? *(5 points)*

7. A furniture store marks up all merchandise 40% based on selling price. If a bedroom suite is priced at $8,379.50, what is the markup? *(5 points)*

8. Video Town marks up blank video cassettes $1.62 each, a rate of 30% based on selling price. How much does Video Town pay for each cassette? *(5 points)*

9. A pair of gold earrings is sold with a markup of $650, 105% of cost. What is the cost of the earrings? *(5 points)*

10. Which rate results in a higher markup?

a. 45% markup based on cost

b. 35% markup based on selling price *(5 points)*

11. Mother Goose Village purchases teddy bears for $11.39 and marks them up $5.36. What is the rate of markup based on selling price? *(5 points)*

12. The Left Bank, an art supply store, sells camel's hair paint brushes for $10.60 each, $2.65 higher than cost. What is the markup rate based on cost? *(5 points)*

PART

3 INTEREST

Chapter 11 Simple Interest

OBJECTIVE

- Compute a charge for the use of borrowed money

SECTIONS

INTRODUCTION

The survival of most businesses depends upon the ability to borrow money at a reasonable interest rate. Interest is the charge paid for the use of borrowed money. The expression *simple interest* means that interest is calculated once for the entire term of the loan. At the end of the loan period, the borrower repays the loan plus the interest.

Section 11.1 The Simple Interest Equation

KEY TERMS **Principal** **Interest** **Term**

The money loaned is the **principal.** The rate quoted for computing the **interest** is the annual rate, the percent of the principal charged for one year. (During the past two decades, interest rates fluctuated from a high of nearly 20 percent to a low of about 6 percent.)

Principal = loan = \$1,000
Interest rate = 6% (per year)
Term of loan = $1\frac{1}{2}$ years
$I = 1000 \times 0.06 \times 1.5 = 90.00$

The time period of a loan is the **term.** Since the term is not always one year, it must be factored into the interest formula. The simple interest formula is:

$$\text{Interest} = \text{Principal} \times \text{Rate} \times \text{Time}$$
$$I = P \times R \times T$$
$$I = PRT$$

The time T must be a fraction of a year (or a whole number of years). Therefore, if the term is a number of months, it must be converted to one of the fractions shown to the right. On a calculator, the fractions $\frac{1}{4}$, $\frac{1}{2}$, and $\frac{3}{4}$ can be handled most conveniently as decimals 0.25, 0.5, and 0.75. Manipulate the other fractions with multiplication by the numerator and division by the denominator.

Number of Months	T
1	$\frac{1}{12}$
2	$\frac{2}{12} = \frac{1}{6}$
3	$\frac{3}{12} = \frac{1}{4} = 0.75$
4	$\frac{4}{12} = \frac{1}{3}$
5	$\frac{5}{12}$
6	$\frac{6}{12} = \frac{1}{2} = 0.5$
7	$\frac{7}{12}$
8	$\frac{8}{12} = \frac{2}{3}$
9	$\frac{9}{12} = \frac{3}{4} = 0.75$
10	$\frac{10}{12} = \frac{5}{6}$
11	$\frac{11}{12}$

Example 1 A loan of $2,500 is to be repaid in $1\frac{3}{4}$ years at 9.4 percent simple interest. What is the interest charge?

$$9.4\% = 0.094$$

$$1\frac{3}{4} = 1.75$$

$$I = P \times R \times T$$
$$= 2500 \times 0.094 \times 1.75$$
$$= 411.25$$

key: 2500 [×] .094 [×] 1.75 [=]
display: 411.25
edited: $411.25

Example 2 Blake Salazar borrowed $650 at $7\frac{1}{2}$ percent interest for five months. How much interest will he pay?

$$7\frac{1}{2}\% = 7.5\% = 0.075$$

$$I = P \times R \times T$$
$$= 650 \times 0.075 \times \frac{5}{12}$$
$$= 20.3125 = 20.31$$

key: 650 [×] .075 [×] 5 [÷] 12 [=]
display: 20.3125
edited: $20.31

PRACTICE ASSIGNMENT

Plan: In all of these problems, you must find simple interest. If the term is not a whole number of years, express the time T as a fraction. Apply the simple interest formula $I = PRT$.

Round the interest to the nearest cent where necessary. Add the interest to the principal to calculate the total amount to be repaid.*

1.

	Principal	Rate	Term	Interest
a.	$ 400.00	8%	2 years	$______
b.	925.00	6%	$1\frac{1}{2}$ years	$______
c.	2,630.00	$8\frac{1}{4}$%	$1\frac{3}{4}$ years	$______
d.	804.50	6.85%	15 months	$______
e.	1,595.00	9.35%	$1\frac{1}{3}$ years	$______
f.	12,000.00	$8\frac{2}{3}$%	33 months	$______

2.

	Principal	Rate	Term	Interest
a.	$ 790.00	7%	6 months	$______
b.	163.25	$6\frac{1}{2}$%	3 months	$______
c.	1,075.00	8.6%	2 months	$______
d.	968.00	$9\frac{1}{4}$%	11 months	$______
e.	4,585.00	10.15%	9 months	$______
f.	3,900.00	8.075%	5 months	$______

3. First Standard Bank charges its preferred borrowers 8.5% interest. A preferred borrower received a loan of $6,000 and repaid it in ten months. How much interest was charged for this loan?

4. Ian Randolph borrowed $1,800 from a friend to furnish his apartment. He repaid the loan plus 3% interest in seven months. How much did he repay?

*To be precise, the principal is repaid and the interest in paid. However, in common usage, repayment refers to principal plus interest.

5. When Ralph and Cathy Bruno got married, they pooled their money ($1,700 and $2,350, respectively) and purchased a six-month certificate of deposit (CD). If the CD yielded 7.35% interest, how much was it worth on the due date?

Section 11.2 Ordinary Interest

KEY TERM **360-day year**

When computing ordinary interest, assume a year of 360 days. Many loans are of short-term duration, months or days. Ordinary interest was used to simplify calculation when paper-and-pencil computing prevailed. If days per month was assumed to be 30 (called approximate time) and days per year 360, the fractions used for computing the interest could usually be reduced. For example, ordinary interest on $915 for two months at 4 percent could be calculated as follows:

$$\frac{915}{1} \times \frac{4}{100} \times \frac{60}{360} = \frac{\overset{61}{\cancel{\overset{305}{\cancel{915}}}}}{1} \times \frac{1}{\underset{5}{\cancel{25}}} \times \frac{1}{\underset{2}{\cancel{6}}} = \frac{61}{1} \times \frac{1}{5} \times \frac{1}{2} = \frac{61}{10} = 6.10$$

Computing interest for any number of days, or using a **360-day year** with the exact number of days in each month (called exact time) generated fractions that did not always reduce so neatly. Nevertheless, computing ordinary interest with exact time was a common practice called *Bankers' Rule.*

When using electronic devices, however, there is no need to expand the interest formula to fractions. Any number of days divided by 360 or 365 will produce a fairly precise decimal result. But in certain business fields, ordinary interest is still charged, perhaps because it yields more interest.

Example What is the interest on a loan of $915 negotiated for 75 days at $6\frac{1}{2}$ percent ordinary interest?

$6\frac{1}{2}\% = 6.5\% = 0.065$

The term is 75 days. $T = \frac{75}{360}$

$$I = P \times R \times T = 915 \times 0.065 \times \frac{75}{360}$$

$$= 12.390625 = 12.39$$

key: 915 [×] .065 [×] 75 [÷] 360 [=]
display: 12.390625
edited: $12.39

PRACTICE ASSIGNMENT

Plan: In all of these problems you must find ordinary interest. Express the time T as $\frac{d}{360}$ where d is the number of days. The simple interest expression PRT is now $PR \times \frac{d}{360}$. Round the interest to the nearest cent where necessary.

1.

	Principal	Rate	Term	Interest
a.	$ 815.00	7%	50 days	$________
b.	260.00	$6\frac{1}{2}\%$	100 days	$________
c.	933.33	7.8%	90 days	$________
d.	2,010.00	12.45%	300 days	$________
e.	15,210.00	$8\frac{5}{8}\%$	60 days	$________
f.	10,150.00	11.015%	30 days	$________

2. **a.** What is the interest on a loan of $1,800 for 236 days at $8\frac{1}{2}$%?
   b. What is the interest on a loan of $1,800 for 236 days at $9\frac{1}{2}$%?

3. A corporation borrowed $725,000 for 45 days at 8.05% interest. How much money will the corporation have to repay?

4. Toy-Mart borrowed $25,000 for 30 days at 7.85% interest.
   a. How much interest will be paid?
   b. What will the total repayment be?

5. Easy Credit Corporation loaned $875 to Tina Hansen on November 18. She repaid the loan with interest 89 days later. If the loan company charged 18% interest, how much interest did she pay?

Section 11.3 Exact Interest

KEY TERM **365-day year**

When computing exact interest, assume a year of 365 days. A computer or calculator divides by 365 (or 366) as easily as by 360. Therefore, many lending institutions compute exact interest. While some institutions may compute with 366 days in a leap year, traditionally "exact interest" implies 365 days.

Example A loan of $915 is negotiated for 75 days at $6\frac{1}{2}$ percent exact interest.

$$6\frac{1}{2}\% = 6.5\% = 0.065$$

The term is 75 days. $T = \frac{75}{365}$

$$I = P \times R \times T = 915 \times 0.065 \times \frac{75}{365}$$

$$= 12.22089 = 12.22$$

Compare with ordinary interest of $12.39. When the other variables remain constant, ordinary interest is always greater than exact interest.

key: 915 [×] .065 [×] 75 [÷] 365 [=]
display: 12.22089
edited: $12.22

PRACTICE ASSIGNMENT

Plan: In all of these problems you must find exact interest. Express the time T as $\frac{d}{365}$ where d is the number of days. The simple interest expression PRT is now $PR \times \frac{d}{365}$. Round the interest to the nearest cent where necessary.

1. This problem contains the same exercises as problem 1 in the section 11.2 practice assignment. Copy the ordinary interest and compute the exact interest. Observe the differences.

	Principal	Rate	Term	Exact Interest	Ordinary Interest
a.	$ 815.00	7%	50 days	$______	$______
b.	260.00	$6\frac{1}{2}$%	100 days	$______	$______
c.	933.33	7.8%	90 days	$______	$______
d.	2,010.00	12.45%	300 days	$______	$______
e.	15,210.00	$8\frac{5}{8}$%	60 days	$______	$______
f.	10,150.00	11.015%	30 days	$______	$______

2. A loan of $1,278 was negotiated for a term of 162 days. How much interest will be charged if the rate is 9.3%?

3. A recording studio borrowed $9,150 for 75 days at $6\frac{1}{2}$% interest. What is the amount of interest on this loan?

4. Stephanie Olivier needed $3,000 to open a beauty salon. She negotiated a loan in this amount at $12\frac{3}{4}$% interest, to be repaid in 277 days. How much interest must she pay?

5. A builder purchased lumber costing $7,450 on credit at $7\frac{1}{2}$% interest. If he purchased the lumber on July 7 and paid off his account 78 days later, what was the total repayment?

Section 11.4 One Percent Combinations

Before the advent of electronic devices, shortcuts were devised to simplify paper-and-pencil calculation, for example, reducing fractions. Compute 5 percent ordinary interest for sixty days on $1,000 without a calculator:

$$PRT = \frac{1000}{1} \times \frac{5}{100} \times \frac{60}{360}$$

$$= \frac{\overset{5}{\cancel{10}}}{1} \times \frac{5}{1} \times \frac{1}{\underset{3}{\cancel{6}}}$$

$$= \frac{25}{3} = 8\frac{1}{3} = 8.33$$

Interest is computed with paper and pencil only in very simple cases. However, occasional calculation by hand promotes a broader understanding of arithmetic concepts. An interesting paper-and-pencil shortcut takes advantage of rates and terms that surprisingly yield ordinary interest of one-hundredth of the principal.

When the product of the whole number in the rate multiplied by the number of days equals 360, ordinary interest equals one-hundredth of the principal. Division by 100 can be done mentally.

Rate = 6%
Term = 60 days

$$I = PRT = P \times \frac{6}{100} \times \frac{60}{360}$$
$$= P \times \frac{360}{100 \times 360}$$
$$= P \times \frac{1}{100}$$

This shortcut is called the 6%-60-day method, but other combinations of rates and terms are just as applicable.

For any of the following combinations, ordinary interest equals principal/100.

Rate	Days
18%	20 days
12%	30 days
9%	40 days
6%	60 days
4%	90 days
3%	120 days

Furthermore, when rate and/or time are convenient multiples, you can find ordinary interest with relatively simple multiplication or division operations.

Ordinary interest on $654.13 borrowed for 60 days at 6% equals 654.13/100 = 6.54.

Therefore, ordinary interest on $654.13 borrowed for 120 days at 6% equals 6.54 × 2 = 13.08.

Example A loan of $1,500 with 12 percent interest was repaid in 40 days. How can you compute the ordinary interest without using a calculator?

When $R = 12\%$ and $T = 30$ days,

$$I = \frac{P}{100} = \frac{1500}{100} = 15$$

Interest at 12% for 30 days = 15
Interest at 12% for 10 days = 15 ÷ 3 = 5
Interest at 12% for 40 days = 15 + 5 = 20

PRACTICE ASSIGNMENT

Plan: Find the ordinary interest for each of the following problems. To practice the methods described in this section, perform these exercises without using a calculator. In each of these problems, the rate and term form a one percent combination or some variation. In the case of variations, solutions are outlined.

1. $735.00 at 6% for 60 days = $________
2. $1,580.00 at 6% for 30 days = $________ ÷ 2 = $________
3. $822.50 at 6% for 120 days = $________ × 2 = $________
4. $4,675.00 at 12% for 60 days = $________ × 2 = $________

5. $312.00 at 6% for 90 days	=		$________	4% for 90 days
		+	________	2% for 90 days
		=	$________	6% for 90 days
6. $986.60 at 5% for 60 days	=		$________	6% for 60 days
		−	________	1% for 60 days
		=	$________	5% for 60 days
7. $5,800.00 at 9% for 40 days	=		$________	
8. $297.00 at 4% for 90 days	=		$________	
9. $1,250.00 at 4% for 120 days	=		$________	3% for 120 days
		+	________	1% for 120 days
		=	$________	4% for 120 days
10. $2,105.00 at 18% for 80 days	=		$________	18% for 20 days
		×	4	
	=		$________	18% for 80 days

Challenger

Lyn Blanchard was determined to earn a quick profit by trading stock of Thirdwave Industries. First, Blanchard withdrew $3,500 from a savings account that paid 5.75 percent simple interest. Then she borrowed $2,200 at 15 percent simple interest from a finance company. She purchased as many shares as she could with $5,700 at $7\frac{1}{8}$ per share. She paid the broker's commission of 5 percent on the purchase out of her petty cash fund.

After three months Blanchard sold the stock at $10\frac{3}{4}$ per share, paying another 5 percent commission. Considering brokerage commissions, interest charges, and loss of earnings from the savings account, what was her net gain?

CHAPTER 11 SUMMARY

Concept	*Example*	*Procedure*	*Formula*
SECTION 11.1 The Simple Interest Equation			
Simple interest equation	What is simple interest on $1,250 at $7\frac{1}{2}$% for 3 months? 1250 × 0.075 × 0.25 = 23.4375 = 23.44	Convert the months to a portion of a year in decimal form. Multiply principal × rate × time.	$I = PRT$ P[×]R[×]T[=] P*R*T
	What is simple interest on $1,250 at $7\frac{1}{2}$% for 5 months? 1250 × 0.075 × 5 ÷ 12 = 39.0625 = 39.06	Convert the months to a fraction of a year. $T = \frac{m}{12}$	$I = PRT = PR\frac{m}{12}$ P[×]R[×]m[÷]12[=] P*R*m/12
SECTION 11.2 Ordinary Interest			
Ordinary Interest	What is ordinary interest on $1,500 at $6\frac{3}{4}$% for 100 days? 1500 × 0.0675 × 100 ÷ 360 = 28.125 = 28.13	Use a 360-day year. $T = \frac{d}{360}$	$I = PRT = PR\frac{d}{360}$ P[×]R[×]d[÷]360[=] P*R*d/360
SECTION 11.3 Exact Interest			
Exact interest	What is exact interest on $1,500 at $6\frac{3}{4}$% for 100 days? 1500 × 0.0675 × 100 ÷ 365 = 27.739726 = 27.74	Use a 365-day year. $T = \frac{d}{365}$	$I = PRT = PR\frac{d}{365}$ P[×]R[×]d[÷]365[=] P*R*d/365
SECTION 11.4 One Percent Combinations			
One percent combinations	P = $657.50 R = 9% T = 40 days I(ord) = 657.5 ÷ 100 = 6.58	Whenever $100\,R \times d = 360$, $I = \frac{P}{100}$ (Use this shortcut for paper-and-pencil computation of ordinary interest.)	

Chapter 11 Spreadsheet Exercise

Interest Table

In the past, business mathematics texts devoted hundreds of pages to mathematical tables with detailed explanations of how to use them. While computer programs have virtually eliminated the need for tables, some, such as interest tables, remain useful for observing gains at varying rates and terms.

This spreadsheet computes and displays a portion of a simple interest table. It shows the ordinary and exact interest on a principal of $1,000 for 1 day to 11 months at rates varying from five percent to nine percent. (Each month is assumed to contain 30 days.) Producing this table prior to the advent of computers would have required considerable time and effort.

Bring your chapter 11 template onto the screen. It will appear as follows:

```
A1: PR [W4]                                                              READY

     A     B     C     D     E     F     G     H     I     J     K     L
1          SIMPLE INTEREST ON 1000 DOLLARS ROUNDED TO THE NEAREST CENT
2    ----------------------------------------------------------------------
3
4          ordinary I = 1000 x R x d/360          exact I = 1000 x R x d/365
5
6    R     5.0%  6.0%  7.0%  8.0%  9.0%        5.0%  6.0%  7.0%  8.0%  9.0%
7
8    days
9
10     1
11     2
12     3
13     4
14     5
15     6
16     7
17     8
18     9
19    10
20    11
```

Columns B–F will display ordinary interest at rates from 5% to 9%. Columns H–L will display the corresponding exact interest. Rows 10–39 will display interest for 1–30 days; rows 45–55 will display interest for 1–11 months.

Spreadsheet Exercise Steps

1. Begin with the ordinary simple interest formula.
   The rate is in B6; the row address must be absolute (the 6 must be preceded by $).
   The days are in A10; the column address must be absolute (the A must be preceded by $).
   The spreadsheet formula is
   ________ * _____ $ _____ * $ _______ /
   _______.
   Complete the formula and store it in B10; copy from B10 to B10..F39.

2. Monthly interest = 30 × daily interest (assuming every month contains 30 days).
   The daily interest is in B10.
   The spreadsheet formula for the monthly interest is
   ______ * _______.
   Complete the formula and store it in B45; copy from B45 to B45..F55.
   You have now completed the ordinary interest portion.

3. Exact interest = $\frac{72}{73}$ × ordinary interest (because $\frac{360}{365} = \frac{72}{73}$).
Recall that exact interest is somewhat smaller than ordinary interest.
The ordinary interest is in B10.
The spreadsheet formula for the exact interest is
______ / ______ * _______ .
Complete the formula and store it in H10.
Copy the formula from H10 to H10..L39 and from H10 to H45..L55.

4. Verify the ordinary interest for 11 days at 7% with your calculator. __________

5. Verify the exact interest for 11 days at 7% with your calculator. __________

6. What is the ordinary interest on a loan of $1,000 at 6% for 23 days? __________

7. How much must be repaid after 7 months for a loan of $1,000 at 8% exact interest? ____________

8. How much must be repaid after 10 months and 18 days for a loan of $1,000 at 9% exact interest?
____

NAME ____________________

CLASS SECTION ____________ DATE ____________

SCORE ________ MAXIMUM __110__ PERCENT ________

CHAPTER 11 ASSIGNMENT

1. Joan Kelly borrowed $1,250 at $6\frac{1}{2}$% simple interest for two years in order to finish her education. How much interest will she pay? *(5 points)*

2. A family, whose home was damaged by flooding, financed repairs by borrowing $1,750 for 9 months at 6% interest from a federal agency. How much interest will they pay? *(5 points)*

3. A lending institution charges borrowers different rates of ordinary interest based on credit rating. Compute the interest charges on the following loans:

Borrower	Amount	Term	Rate	Charge
Acme Appliances	$1,150	6 months	10.5%	$________
Corona and Son	745	45 days	9%	$________
E & S Hardware	2,400	75 days	$8\frac{1}{4}$%	$________
Garments Galore	950	85 days	12%	$________
Rings and Things	1,850	135 days	$9\frac{1}{2}$%	$________

(15 points)

4. a. What is the ordinary interest on a loan of $2,750 for a term of 147 days at a rate of 8.9%?

b. What is the ordinary interest on a loan of $2,750 for a term of 157 days at a rate of 8.9%? *(10 points)*

5. a. What is the ordinary interest on a loan of $1,400 for a term of 288 days at a rate of $8\frac{1}{4}$%?

b. What is the exact interest on a loan of $1,400 for a term of 288 days at a rate of $8\frac{1}{4}$%? *(10 points)*

6. Richard Levine borrowed $5,500 to expand his appliance store. He repaid the loan in 300 days with $10\frac{1}{2}$% exact interest.

a. How much interest did he pay?

b. What was his total repayment? *(10 points)*

7. On July 10, Jim Morgan borrowed $1,965 and agreed to pay the debt in 120 days with ordinary interest at $7\frac{3}{4}$%.

a. How much interest will Jim have to pay?

b. How much money will Jim have to repay? *(10 points)*

8. Ajamian Enterprises negotiated a loan of $10,750 to be repaid in 21 months with simple interest of 9.9%.

a. How much interest must be paid?

b. What is the total amount to be repaid? *(10 points)*

9. An investor borrowed $4,775 at 7.65% exact interest to finance a stock purchase. If the loan was repaid 279 days later, what was the total amount of repayment? *(5 points)*

10. Dorfman's Ice Cream Parlor borrowed $2,350 to renovate the seating area. The loan was negotiated for three months at 7.2% interest. How much interest will be charged? *(5 points)*

11. Brad Hartman is unable to pay his federal income tax liability of $719 by April 15. He will pay the full amount plus 12% exact interest 60 days later. How much interest must he pay? *(5 points)*

12. Pam Stahl borrowed $814 on December 18; she repaid the loan plus 9.15% ordinary interest 193 days later. What was the total amount of repayment? *(5 points)*

13. A service station borrowed $3,075 to cover the cost of new gas pumps and some minor repairs. The loan was repaid four months later at $8\frac{1}{2}$% interest. How much was repaid? *(5 points)*

14. Exact interest on a loan of $592.75 was 8.5%. If the loan was negotiated on March 21 and repaid 40 days later, how much interest was charged? *(5 points)*

15. Payment of a $735 invoice dated June 15 is due in one month, after which ordinary interest of 8.25% will be charged. If the invoice is settled on July 30, how much must be repaid? *(5 points)*

Chapter 12 Interest Variables

OBJECTIVE

- Evaluate each of the variables in the simple interest equation

SECTIONS

INTRODUCTION

How much money must you deposit in an account earning eight percent interest so that you could withdraw $1,200 each year without diminishing the principal? In some situations you need to determine one of the other variables rather than the interest.

Section 12.1 Finding the Principal

Similar to the percentage equation in chapter 7, the fundamental simple interest equation $I = PRT$ forms the core for three related equations that allow you to find principal, rate, or time.

You may want to memorize the following equations:

1. **Find I by covering I.** $\quad \frac{I}{PRT} \quad I = PRT$ — This is the fundamental equation.
2. **Find P by covering P.** $\quad \frac{I}{PRT} \quad P = \frac{I}{RT}$
3. **Find R by covering R.** $\quad \frac{I}{PRT} \quad R = \frac{I}{PT}$
4. **Find T by covering T.** $\quad \frac{I}{PRT} \quad T = \frac{I}{PR}$

In the three related equations, interest is the numerator of a fraction.

Other students may prefer to develop the related equations from the fundamental equation.

When deriving the principal, isolate the unknown P on the left side of the equation.

Fundamental equation:	$I = P \times R \times T$
Reverse the two sides:	$P \times R \times T = I$
Consider the product RT as one number:	$P \times (RT) = I$
Divide each side by RT:	$P \times (RT) \div (RT) = I \div (RT)$
Simplify:	$P = I \div (RT)$
Write the division as a fraction:	$P = I/RT$

When deriving the interest rate, isolate the unknown R on the left side of the equation.

Fundamental equation:	$I = P \times R \times T$
Reverse the two sides:	$P \times R \times T = I$
Change the order of multiplication:	$R \times P \times T = I$
Consider the product PT as one number:	$R \times (PT) = I$
Divide each side by PT:	$R \times (PT) \div (PT) = I \div (PT)$
Simplify:	$R = I \div (PT)$
Write the division as a fraction:	$R = I/PT$

When deriving the time, isolate the unknown T on the left side of the equation.

Fundamental equation:	$I = P \times R \times T$
Reverse the two sides:	$P \times R \times T = I$
Change the order of multiplication:	$T \times P \times R = I$
Consider the product PR as one number:	$T \times (PR) = I$
Divide each side by PR:	$T \times (PR) \div (PR) = I \div (PR)$
Simplify:	$T = I \div (PR)$
Write the division as a fraction:	$T = I/PR$

You can find the principal, given the interest, rate, and time.

$$\text{Principal} = \frac{\text{interest}}{\text{rate} \times \text{time}}$$

$$P = \frac{I}{RT}$$

Example 1 How large is an investment that yields interest of \$82.35 in one quarter of a year at $6\frac{1}{2}\%$?

$$T = 0.25$$

$$P = \frac{I}{RT} = \frac{82.35}{0.065 \times 0.25} = 5067.6923$$

Perform the operations in the denominator prior to division. Calculators with parenthesis keys alter the order of operations by performing operations inside parentheses first. If parenthesis keys are not available, compute the denominator first and store it in M-memory.

key w/paren:	82.35 [÷] [(] .065 [×] .25 [)] [=]
key w/o paren:	.065 [×] .25 [=] [M+] 82.35 [÷] [MR] [=]
display:	5067.6923
edited:	\$5,067.69

Example 2 How large is an investment that yields interest of \$82.35 in five months at $6\frac{1}{4}\%$?

$$T = 5/12$$

$$P = \frac{I}{RT} = \frac{82.35}{0.0625 \times 5/12} = 3162.24$$

key w/paren:	82.35 [÷] [(] .0625 [×] 5 [÷] 12 [)] [=]
key w/o paren:	.0625 [×] 5 [÷] 12 [=] [M+] 82.35 [÷] [MR] [=]
display:	3162.24
edited:	\$3,162.24

Example 3 What is the size of a loan requiring interest of $62.23 at 6.11 percent for 97 days? Assume that exact interest is paid.

$$T = 97/365$$

$$P = \frac{I}{RT} = \frac{62.23}{0.0611 \times 97/365} = 3832.4784$$

key w/paren: 62.23 [÷] [(] .0611 [×] 97 [÷] 365 [)] [=]
key w/o paren: .0611 × 97 [÷] 365 [=] [M+] 62.23 [÷] [MR] [=]
display: 3832.4784
edited: $3,832.48

PRACTICE ASSIGNMENT

Plan: Find the principal, given the interest, rate, and time. The formula is $P = \frac{I}{RT}$. T can be years, months/12, days/360, or days/365. If time is in days, assume exact interest unless otherwise indicated.

1. Find the principal in each of the following loans:

	Rate	Time	Interest	Principal
a.	5%	85 days	$ 12.00 (ordinary)	$________
b.	8.75%	4 months	150.00	________
c.	9.2%	$2\frac{1}{2}$ yrs	547.75	________
d.	7.9%	94 days	71.08	________
e.	$8\frac{1}{4}$%	27 months	210.55	________
f.	6.85%	295 days	91.70	________

2. If $500 interest is charged on a three-month loan at 10% interest, what is the size of the loan?

3. How much must someone invest at $6\frac{1}{4}$% in order to gain $250 seven months from today?

4. Timothy's Sandwich Shoppe borrowed money to pay for new kitchen equipment. If the loan plus $133.78, representing 9.1% ordinary interest, was repaid in 183 days, how much was borrowed?

5. A loan at 5% interest will be repaid with principal plus $50 on June 12. What is the size of the loan if it was negotiated on March 12?

Section 12.2 Finding the Interest Rate

Suppose you had $15,000 to invest. What interest rate would you need to gain $500 in six months?

You can find the interest rate, given the interest, principal, and time. Express the interest rate in percent form.

$$\text{Rate} = \frac{\text{interest}}{\text{principal} \times \text{time}}$$

$$R = \frac{I}{PT}$$

Example 1 If a simple interest loan of $1,500 for $1\frac{1}{2}$ years accrues $45 in interest, what is the interest rate?

$$T = 1.5$$

$$R = \frac{I}{PT} = \frac{45}{1500 \times 1.5} = 0.02$$

key w/paren: 45 [÷] [(] 1500 [×] 1.5 [)] [=]
key w/o paren: 1500 [×] 1.5 [=] [M+]
45 [÷] [MR] [=]
display: 0.02
edited: 2%

When numbers are simple, you can find the interest rate manually by manipulating the fraction:

$$\frac{45}{1500 \times 1.5} = \frac{\overset{30}{\cancel{450}}}{1500 \times \underset{1}{\cancel{15}}} = \frac{30}{1500} = \frac{2}{100} = 2\%$$

Example 2 What is the annual rate of interest on a loan of $2,500 that earns $68.33 in interest during five months?

$$T = 5/12$$

$$R = \frac{I}{PT} = \frac{68.33}{2500 \times 5/12} = 0.0655968$$

key w/paren: 68.33 [÷] [(] 2500 [×] 5 [÷] 12 [)] [=]
key w/o paren: 2500 [×] 5 [÷] 12 [=] [M+]
68.33 [÷] [MR] [=]
display: 0.0655968
edited: 6.56%

Example 3 At what interest rate will $15 be charged on a loan of $210.50 for 112 days? Assume exact interest.

$$T = 112/365$$

$$R = \frac{I}{PT} = \frac{15}{210.5 \times 112/365} = 0.2322276$$

key w/paren: 15 [÷] [(] 210.5 [×] 112 [÷] 365 [)] [=]
key w/o paren: 210.5 [×] 112 [÷] 365 [=] [M+]
15 [÷] [MR] [=]
display: 0.2322276
edited: 23.22%

PRACTICE ASSIGNMENT

Plan: Find the interest rate, given the interest, principal, and time. The formula is

$$R = \frac{I}{PT}.$$

T can be years, months/12, days/360, or days/365. Convert the interest rate to percent and round it to the nearest one-hundredth of one percent. If the time is in days, assume exact interest unless otherwise indicated.

1. Find the interest rate in each of the following loans:

	Principal	Time	Interest	Rate
a.	$ 1,870.00	145 days	$ 45.19 (ordinary)	____%
b.	18,500.00	10 months	816.46	____%
c.	6,532.00	$1\frac{3}{4}$ years	943.06	____%
d.	423.25	44 days	3.47	____%
e.	924.50	15 months	129.43	____%
f.	2,625.00	218 days	137.18	____%

2. At what interest rate will $10 be charged on a loan of $200 for 90 days assuming ordinary interest is paid?

3. A shopper borrowed $489.50 to finance a new wardrobe. He paid the principal plus $10.60 interest after 85 days. What rate of interest did he pay?

4. At what interest rate will \$450 yield \$78.75 interest in $3\frac{1}{2}$ years?

5. A payment of \$1,542.32 was rendered on a five-month loan of \$1,500. What interest rate was charged?

Section 12.3 Finding the Time

How long will it take for \$15,000 to increase to \$18,000 at eight percent simple interest? You can compute the time, given the interest, principal, and interest rate. The result is the number of years. If the result is fractional, edit it to a number of months or days. Multiply by 12 to obtain months; multiply by 365 or 360 to obtain days.

$$\text{Time} = \frac{\text{interest}}{\text{principal} \times \text{rate}}$$

$$T = \frac{I}{PR}$$

Example 1 Find the term if \$120 is charged for a loan of \$8,000 at an annual interest rate of six percent.

$$T = \frac{I}{PR} = \frac{120}{8000 \times 0.06} = 0.25$$

key w/paren: 120 [÷] [(] 8000 [×] .06 [)] [=]
key w/o paren: 8000 [×] .06 [=] [M+]
120 [÷] [MR] [=]
display: 0.25
edited: quarter of a year or three months

Example 2 What is the term of a loan of \$3,219 at 7.2 percent that yields \$135.20 in interest?

$$T = \frac{I}{PR} = \frac{135.20}{3219 \times 0.072} = 0.5833419$$

key w/paren: 135.2 [÷] [(] 3219 [×] .072 [)] [=]
key w/o paren: 3219 [×] .072 [=] [M+] 135.2 [÷] [MR] [=]
display: 0.5833419 How should this be edited? Try to find the number of months.

key: [×] 12 [=]
display: 7.0001036 (This number is very close to 7.)
edited: 7 months

Example 3 If someone borrows $2,400 at nine percent and pays $150 in interest, what is the term of the loan?

$$T = \frac{I}{PR} = \frac{150}{2400 \times 0.09} = 0.69444\ldots$$

key w/paren: 150 [÷] [(] 2400 [×] .09 [)] [=]
key w/o paren: 2400 [×] .09 [=] [M+]
150 [÷] [MR] [=]
display: 0.6944444 How should this be edited? Try to find the number of months first.

key: [×] 12 [=]
display: 8.3333333
edited: $8\frac{1}{3}$ months The number of days is a more meaningful answer. Assume ordinary interest.

key: .6944444 [×] 360 [=]
display: 249.99998
edited: 250 days

PRACTICE ASSIGNMENT

Plan: Find the time, given the interest, principal, and rate. The formula is $T = \frac{I}{PR}$. If T is a fractional number of years, you may need to convert it to an equivalent number of months or days. Assume exact interest unless otherwise indicated.

1. Find the time in each of the following loans.

	Principal	Rate	Interest	Time
a.	$ 750.00	5%	$ 7.50 (ordinary)	________
b.	6,621.00	8%	220.70	________
c.	589.50	5.5%	11.99	________
d.	1,498.00	9.1%	19.79	________
e.	28,000.00	6.8%	1,110.67	________
f.	8,370.00	7.2%	903.96	________

2. Find the time if $128 is charged for a loan of $8,000 at 16% ordinary interest.

3. How long will it take $1,000 to accrue $50 in interest at $6\frac{1}{2}$%?

4. Tony Williamson paid $112.75 interest on a loan of $1,650 with an interest rate of $10\frac{1}{4}$%. How long did it take him to repay the loan?

5. How many days are necessary to increase $380 to $389.50 at 5.75%?

Challenger

When Debra Manus received a sizable inheritance, she looked for safe, high-yield investments. One savings institution offered 7.8 percent interest. A money market fund guaranteed a gain of $1,020 on an investment of $10,000 after 15 months. Another firm had a plan by which $18,000 would be worth $19,080 after 245 days. Assume exact interest.

a. Which of the three plans offers the highest rate?

b. How much would Debra Manus need to invest (to the nearest $100) in order to earn interest of $10,000 in six months with the rate provided by the best offer?

CHAPTER 12 SUMMARY

Concept	*Example*	*Procedure*	*Formula*
SECTION 12.1 Finding the Principal			
Finding principal, given interest, rate, and time	What size investment yields interest of $100 at $6\frac{1}{2}$% for 100 days?	Divide interest by rate × time.	$P = \frac{I}{RT}$ R[×]T[=][M+]I[÷][MR][=] or I[÷][(]R[×]T[)][=] $I/(R*T)$
	$\frac{100}{0.065 \times 100/365}$ = 5615.3846 = 5615.38	If time is not a whole number of years, convert to a fractional year.	$P = \frac{I}{R\ m/12}$ R[×]m[÷]12[=][M+]I[÷][MR][=] or I[÷][(]R[×]m[÷]12[)][=] $I/(R*m/12)$
			$P = \frac{I}{R\ d/365}$ R[×]d[÷]365[=][M+]I[÷][MR][=] or I[÷][(]R[×]d[÷]365[)][=] $I/(R*d/365)$
			$P = \frac{I}{R\ d/360}$ R[×]d[÷]360[=][M+]I[÷][MR][=] or I[÷][(]R[×]d[÷]360[)][=] $I/(R*d/360)$

CHAPTER 12 SUMMARY—(*Continued*)

Concept	*Example*	*Procedure*	*Formula*
SECTION 12.2 Finding the Interest Rate			
Finding rate, given interest, principal, and time	What is the annual interest rate for a 100-day loan of $850 costing $16.30 in interest?	Divide interest by principal × time.	$R = \frac{I}{PT}$ P[×]T[=][M+]I[÷][MR][=] or I[÷][(]P[×]T[)][=] I/(P*T)
	$\frac{16.30}{850 \times 100/365}$ $= 0.0699941$ $= 7\%$	If time is not a whole number of years, convert to a fractional year.	$R = \frac{I}{P\,m/12}$ P[×]m[÷]12[=][M+]I[÷][MR][=] or I[÷][(]P[×]m[÷]12[)][=] I/(P*m/12) $R = \frac{I}{P\,d/365}$ P[×]d[÷]365[=][M+]I[÷][MR][=] or I[÷][(]P[×]d[÷]365[)][=] I/(P*d/365) $R = \frac{I}{P\,d/360}$ P[×]d[÷]360[=][M+]I[÷][MR][=] or I[÷][(]P[×]d[÷]360[)][=] I/(P*d/360)
SECTION 12.3 Finding the Time			
Finding time, given interest, principal, and rate	What is the term of a loan of $7,500 at 8% if $138.08 is charged? $\frac{138.08}{7500 \times 0.08} = 0.2301333$ 0.2301333×365 $= 83.998655$ $= 84$ days	Divide interest by principal × rate. T (time) is a whole or fractional number of years. If T is fractional, convert it to the appropriate number of months or days.	$T = \frac{I}{PR}$ P[×]R[=][M+]I[÷][MR][=] or I[÷][(]P[×]R[)][=] I/(P*R) $m = 12T$ T[×]12[=] 12*I/(P*R) $d = 365T$ T[×]365[=] 365*I/(P*R) $d = 360T$ T[×]360[=] 360*I/(P*R)

Chapter 12 **Spreadsheet Exercise**

Interest Variable Quiz

The chapter 12 template tests your ability to choose the correct interest variable formula and to translate it into a spreadsheet formula. Bring your chapter 12 template onto the screen. It will appear as follows:

A1: PR [W7] READY

INTEREST VARIABLE QUIZ

problem	principal	interest rate	interest	years	months	days		missing value
1		5.40%	100.00	1.50				principal
2	1,250.00		18.52		2			rate
3		5.50%	315.00		9			principal
4	1,000.00	7.55%	50.33					time in months
5	1,100.00		10.00		1			rate
6	1,200.00	6.81%	40.86					time in months
7	3,400.00		52.81			120	ord	rate

The odd-numbered rows 7, 9, 11, . . . , 29 contain twelve problems. If the time is in days, column H contains *ord* or *ex*.

Compose the spreadsheet formulas that compute the missing values. The formulas should be general; they should contain the proper cell addresses, and the only constants that should appear are 12, 360, 365.

Fractional spreadsheet formulas require parentheses to denote the order of operations. If a formula begins with a cell address, it must be preceded by a sign (+).

Spreadsheet Exercise Steps

1. Complete the following chart.
   Example: Suppose row 5 contains the following information:

Principal	Interest Rate	Interest	Time: Years	Time: Months	Time: Days	Missing Value
1,500.00	8.50%	29.45			ord	time in days

Mathematical Formula	Spreadsheet Formula	Storage Cell for Formula
T = I/PR × 360	+D5/(B5*C5)*360	G5

Problem	Mathematical Formula	Spreadsheet Formula	Storage Cell for Formula
1	________	________	____
2	________	________	____
3	________	________	____
4	________	________	____
5	________	________	____
6	________	________	____
7	________	________	____
8	________	________	____
9	________	________	____
10	________	________	____
11	________	________	____
12	________	________	____

2. Enter each formula. Do not copy formulas from one location to another.

3. For each problem, use your calculator to compute the interest ($I = PRT$). Compare your results with the interest displayed in column D. If your computed interest does not equal the interest in column D (rounded to the nearest cent) you have either computed incorrectly or you have stored an incorrect formula.

Problem	Interest Computed by Calculator $I = PRT$
1	________
2	________
3	________
4	________
5	________
6	________
7	________
8	________
9	________
10	________
11	________
12	________

NAME ______________________

CLASS SECTION ______________ DATE ______________

SCORE __________ MAXIMUM 75 PERCENT __________

CHAPTER 12 ASSIGNMENT

If time is in days, assume exact interest unless otherwise indicated. Round interest rates to two decimal places.

1. An investor earned $95.15 interest in three months on an investment that paid interest at the rate of 5.8%. How much was invested? *(5 points)*

2. The interest due on a 60-day loan of $320 was $5.60. Assuming ordinary interest, what was the interest rate? *(5 points)*

3. Find the time if $441 is charged for a loan of $6,000 at 14% ordinary interest. *(5 points)*

4. On May 17, Howard borrowed $185 from his friend Steve and promised to repay the loan with $12 interest. If Howard repaid the loan on October 17, what interest rate did he pay? *(5 points)*

5. A small grocery store accepted an *IOU* from one of its customers. The customer owed $75.32 and agreed to repay the debt plus $3 in 30 days. What was the interest rate, assuming ordinary interest? *(5 points)*

6. Darryl Lamont paid $8.20, or $8\frac{3}{4}$% ordinary interest, on a 45-day loan. What was the amount of the loan? *(5 points)*

7. How many months are required for $260 to yield $3.90 in interest at 6%? *(5 points)*

8. Lisa Arias repaid her used car loan after 10 months. If she paid $195.22 in interest on a 6.2% loan, what was the price of the car? *(5 points)*

9. The Sail-Away boat store financed its new stock with a 4.85% interest loan. If the loan was repaid 250 days later with interest of $805.59, what was the principal? *(5 points)*

10. If a loan of $7,500 carries an interest rate of 8.7% and a maturity value of $7,717.50, how long will it take for the loan to mature? *(5 points)*

11. At the end of 72 days, an individual who had borrowed $1,350 paid a total of $1,381.22. What was the interest rate? *(5 points)*

12. When Ted Broder finally paid his insurance bill of $350, he discovered that he had to pay an additional $17.50 in interest charges. If the interest rate was 12%, how long was the bill past due? *(5 points)*

13. On August 5 the Caribou Club took out a loan at 10.5% interest to renovate its club house. Eighty-seven days later the club repaid the loan including $73.83 interest. What was the amount of the loan? *(5 points)*

14. Sue Palumbo borrowed $2,125 at 8.9% interest. She repaid a total of $2,186.14 including interest. How long did she take to repay the loan? *(5 points)*

15. Leslie Mitsui borrowed $1,275 on April 14 and repaid the loan plus $18.34 interest on June 14. What rate of interest did she pay? *(5 points)*

Chapter 13 Promissory Notes, Present Value

OBJECTIVE

Determine present and future values of promissory notes

SECTIONS

INTRODUCTION

People who borrow money sometimes sign a form called a promissory note on which they promise to repay the loan amount at a specific rate of interest by a particular date. The note is the legal documentation of the loan.

Section 13.1 The Number of Days in a Term Defined by Two Dates

KEY TERMS **Due date** **Maturity date**

A promissory note (note) provides a record of the borrower (maker or payer), the lender or creditor (payee), the amount, the interest rate, and the term of a loan.

If the term is not a whole number of months, calculate the number of days between the starting and ending dates on the note in order to determine the time. If the two dates are not far apart, obtain the number of days by counting the days in each month of the term.

Years that are evenly divisible by four (with exceptions that will not arise until the year 2100) are leap years and contain an extra day in the month of February (February 29). 1992, 1996, 2000, . . . are leap years.

Month	Days
Jan	31
Feb	28 (or 29)
Mar	31
Apr	30
May	31
June	30
July	31
Aug	31
Sept	30
Oct	31
Nov	30
Dec	31

Example 1 Find the number of days between April 10 and July 25.
Do not count the first day of the loan.
April has 30 days. Count the period from April 10 to April 30 as 20 days.
May has 31 days. June has 30 days.
The period from July 1 to July 25 contains 25 days.
Count the last day of the loan.

Month	Days
Apr	20
May	31
June	30
July	25
Total	106

PRACTICE ASSIGNMENT 1

Plan: Compute the number of days in each of the following terms by counting the number of days in each month. Remember that in leap years February has 29 days.

1. May 12 to August 4:

May ______ days
June ______ days
July ______ days
Aug ______ days
Total ______ days

2. April 14 to July 9:

Apr ______ days
May ______ days
June ______ days
July ______ days
Total ______ days

3. December 8, 1991, to May 4, 1992 (1992 is a leap year):

Dec ______ days
Jan ______ days
Feb ______ days
Mar ______ days
Apr ______ days
May ______ days
Total ______ days

4. May 25, 1992, to August 15, 1993:

May 25, 1992, to May 25, 1993		______	days
Remainder of	May 1993	______	days
	June 1993	______	days
	July 1993	______	days
	Aug 1993	______	days
	Total	______	days

The date on which a loan must be repaid is called the **due date** or **maturity date.** Sometimes you will know the starting date and the term in days; then you must find the due date.

Example 2 On what date will a 100-day loan dated June 1 become due?

June 1 + 100 =	June	101	Improper date; June has 30 days.
		− 30	
	July	71	Improper date; July has 31 days.
		− 31	
	Aug	40	Improper date; August has 31 days.
		− 31	
	Sept	9	Proper date; it is the due date.

If a term is given in months or years, advance the date by that number of months or years; the day of the due date is the same as that of the beginning date. However, if that day results in a nonexistent date, the last day of the month becomes the due date.

Date of loan:	May 25
Term:	4 months
Due date:	September 25
Date of loan:	May 31
Term:	4 months
Due date:	September 30

PRACTICE ASSIGNMENT 2

Plan: Given the beginning date and the number of days in the term, determine the due date by adding the number of days to the opening date. If the resulting date is not a proper date, subtract the number of days in the month from the improper date and advance to the next month. Proceed in this manner until the resulting date is a proper date. If the term is given in months, advance the date by the number of months.

1. 35 days from June 20:

June 20 + 35 = June ______
= July ______

2. 72 days from July 11:

July 11 + 72 = July ______
= Aug ______
= Sept ______

3. 60 days from Oct 6:

October 6 + 60 = Oct ______
= Nov ______
= Dec ______

4. 100 days from June 23:

June 23 + 100 = June ______
= July ______
= Aug ______
= Sept ______
= Oct ______

5. 3 months from Mar 15: __________

6. 4 months from July 31: __________

Section 13.2 Date Serial Numbers

KEY TERM **Date serial number**

Tables that assign a consecutive number, a **date serial number,** to each day help simplify date computations. The date serial number in a table represents the number of days since the previous year expired. January 1 is day 1, January 2 is day 2, February 1 is day 32, . . . , December 31 is day 365 or day 366. Find the number of days between two dates by subtraction, remembering to adjust for leap years. These tables are most convenient for calculating dates within the same year.

Table 13.1 Date Serial Numbers

DAY	*Jan*	*Feb*	*Mar*	*Apr*	*May*	*Jun*	*Jul*	*Aug*	*Sep*	*Oct*	*Nov*	*Dec*
1	1	32	60	91	121	152	182	213	244	274	305	335
2	2	33	61	92	122	153	183	214	245	275	306	336
3	3	34	62	93	123	154	184	215	246	276	307	337
4	4	35	63	94	124	155	185	216	247	277	308	338
5	5	36	64	95	125	156	186	217	248	278	309	339
6	6	37	65	96	126	157	187	218	249	279	310	340
7	7	38	66	97	127	158	188	219	250	280	311	341
8	8	39	67	98	128	159	189	220	251	281	312	342
9	9	40	68	99	129	160	190	221	252	282	313	343
10	10	41	69	100	130	161	191	222	253	283	314	344
11	11	42	70	101	131	162	192	223	254	284	315	345
12	12	43	71	102	132	163	193	224	255	285	316	346
13	13	44	72	103	133	164	194	225	256	286	317	347
14	14	45	73	104	134	165	195	226	257	287	318	348
15	15	46	74	105	135	166	196	227	258	288	319	349
16	16	47	75	106	136	167	197	228	259	289	320	350
17	17	48	76	107	137	168	198	229	260	290	321	351
18	18	49	77	108	138	169	199	230	261	291	322	352
19	19	50	78	109	139	170	200	231	262	292	323	353
20	20	51	79	110	140	171	201	232	263	293	324	354
21	21	52	80	111	141	172	202	233	264	294	325	355
22	22	53	81	112	142	173	203	234	265	295	326	356
23	23	54	82	113	143	174	204	235	266	296	327	357
24	24	55	83	114	144	175	205	236	267	297	328	358
25	25	56	84	115	145	176	206	237	268	298	329	359
26	26	57	85	116	146	177	207	238	269	299	330	360
27	27	58	86	117	147	178	208	239	270	300	331	361
28	28	59	87	118	148	179	209	240	271	301	332	362
29	29		88	119	149	180	210	241	272	302	333	363
30	30		89	120	150	181	211	242	273	303	334	364
31	31		90		151		212	243		304		365

In leap years, February 29 is day 60. Add 1 to each date serial number after February 29.

Example 1 Find the number of days between April 10 and July 25 of the same year using date serial numbers.

Date	Date Serial Number
July 25	206
Apr 10	− 100
Number of Days	= 106

If two dates span the end of a year, add 365 or 366 to the serial number of the later date before subtracting (depending on whether the starting date is in a leap year or not).

Example 2 Find the number of days between November 15, 1991, and February 10, 1992. (1991 is not a leap year.)

Date	Date Serial Number
Feb 10, 1992	41 + 365 = 406
Nov 15, 1991	− 319
Number of Days	= 87

PRACTICE ASSIGNMENT 1

Plan: Use table 13.1 to compute the number of days in each of the following terms. Find the date serial number for each of the two dates. Assume both dates are within the same year unless the dates span the end of the year, in which case add 365 to the serial number of the later date. Remember to add 1 to the date serial number after February 29 in a leap year. Subtract the serial number corresponding to the earlier date from the serial number corresponding to the later date.

1. Mar 5 to Oct 29:

Date	Date Serial Number
Oct 29	______
Mar 5	− ______
Number of Days	= ______

2. Feb 5 to Oct 10: (in a leap year)

Date	Date Serial Number
Oct 10	______
Feb 5	− ______
Number of Days	= ______

3. Oct 13 to Jan 28:

Date	Date Serial Number
Jan 28	______ + ______ = ______
Oct 13	− ______
Number of Days	= ______

4. Nov 17 to Apr 21:

Date	Date Serial Number
Apr 21	______ + ______ = ______
Nov 17	− ______
Number of Days	= ______

To find the due date, given the beginning date and the number of days in a term, add the number of days to the serial number that corresponds to the beginning date. Use table 13.1 in reverse to translate the resulting serial number back to the calendar date.

Mar 5 + 50 days = ?
Mar 5 = day 64
Day 64 + 50 = day 114
Day 114 = Apr 24

If the resulting date serial number is over 365 (or over 366 if it spans a leap year), the due date is in the following year. Subtract 365 (or 366) prior to looking up the calendar date, and advance the year by 1.

Nov 15, 1991 + 170 days = ?

Date		Date Serial Number
Nov 15, 1991		319
		+ 170
	Day	489 of 1991
		− 365
May 3, 1992	Day	124 of 1992

(Note: 1992 is a leap year.)

PRACTICE ASSIGNMENT 2

Plan: Determine the due date by adding the number of days in the term to the serial number that represents the beginning date. Translate the date serial number back to the calendar date. If the resulting date serial number is over 365, the due date is in the following year. Assume no leap year unless indicated. Add 1 to the date serial number after February 29 in a leap year.

1. 141 days from May 8:

Date	Date Serial Number
May 8	______
	+ ______
Sept 26	______

2. 119 days from Feb 18 (in a leap year):

Date	Date Serial Number
Feb 18	______
	+ ______
June 16	______

3. 96 days from Dec 5, 1990:

Date	Date Serial Number
Dec 5, 1990	______
	+ ______
	day ______ of 1990
	− ______
Mar 11	day ______ of 1991

4. 104 days from November 19, 1991 (1992 is a leap year):

Date	Date Serial Number
Nov 19, 1991	______
	+ ______
	day ______ of 1991
	− ______
Mar 2	day ______ of 1992

Section **13.3 Promissory Notes at Simple Interest**

KEY TERMS **Non-interest-bearing note** **Face value** **Maturity value** **Interest-bearing note** **Amount due** **Amount**

A promissory note is the tangible evidence of a financial obligation and as such is a negotiable item since some lending institutions buy and sell notes prior to maturity. When an institution purchases a note, the payment of that note transfers to the purchaser.

The simple interest note is seldom used now except between individuals or small businesses.

A note on which no interest is stipulated is a **non-interest-bearing note,** and the principal, or **face value,** is the **maturity value** (the amount due at maturity). A note on which interest is stated is an **interest-bearing note** whose maturity value equals face value plus interest. The interest rate is always specified on an interest-bearing note.

Promissory Note (non-interest-bearing)

$800.00 Los Angeles, CA Aug. 17, 199X

Three months *after date* I *promise to pay*

to the order of Jim Williams, C.P.A.

Eight hundred and 00/100 *Dollars*

for value received.

Due Nov. 17, 199X D. D. Clark

Promissory Note (interest-bearing)

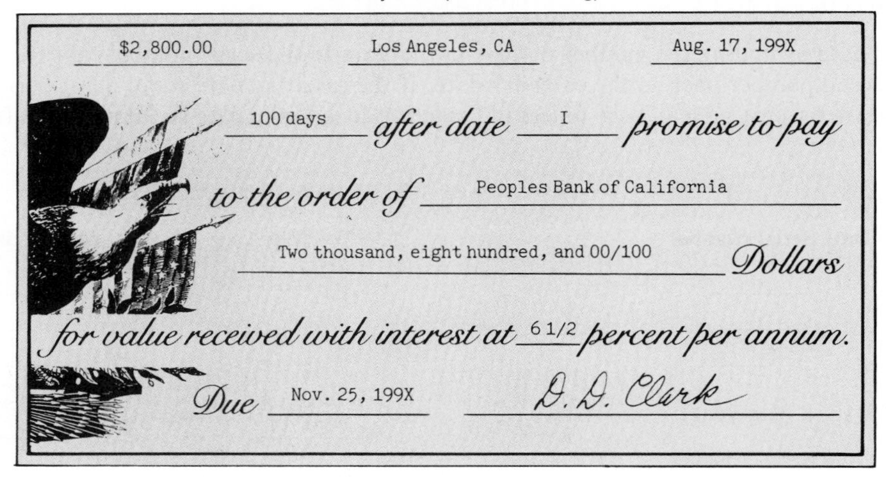

$2,800.00 Los Angeles, CA Aug. 17, 199X

100 days after date I promise to pay

to the order of Peoples Bank of California

Two thousand, eight hundred, and 00/100 Dollars

for value received with interest at 6 1/2 percent per annum.

Due Nov. 25, 199X D. D. Clark

The **amount due** upon a loan's maturity, the maturity value, is sometimes called the **amount** and its symbol is A. You can calculate the amount A with a special formula that is often more convenient than adding interest to face value.

$$\text{Amount due} = \text{face value} + \text{interest}$$
$$A = P + I$$
Since $I = PRT$,
$$A = P + PRT$$
$$A = P(1 + RT)$$

Example 1 Wayne Butler is the maker of a five-month note with face value of $6,000 at 6.5 percent. What is the maturity value?

Compute the interest and add it to the principal:

$$I = PRT = 6000 \times 0.065 \times 5/12 = 162.50$$

$$\text{Principal} + \text{interest} = 6000 + 162.50 = 6162.50$$

key: 6000 [×] .065 [×] 5 [÷] 12 [=]
display: 162.5

Or compute the maturity value directly:

$$A = P(1 + RT) = 6000 \times (1 + 0.065 \times 5/12) = 6162.50$$

Perform the multiplication and division inside the parentheses before adding 1. If a calculator features parenthesis keys, it usually does this for you. Use parenthesis keys if available. On an algebraic calculator without parenthesis keys (but with the M-memory) or on an arithmetic calculator, use M-memory to produce the correct result.

key w/paren: 6000 [×] [(] 1 [+] .065 [×] 5 [÷] 12 [)] [=]
key w/o paren: 1 [M+] .065 [×] 5 [÷] 12 [=] [M+] 6000 [×] [MR] [=]
display: 6162.5 (6162.4998 on some calculators)
edited: $6,162.50

PRACTICE ASSIGNMENT

Plan: First find the number of days by using table 13.1 or counting the days in the months. Then find the maturity value directly (without computing the interest first) using the amount formula $A = P(1 + RT)$. From this point on, when time is in days, assume exact interest unless otherwise stated. Assume no leap years unless indicated.

1. Compute the time and maturity value.

	Face Value	Date of Note	Due Date	Time in Days	Interest Rate	Maturity Value
a.	$ 760.00	Sept 30	Dec 22	______	8%	$ ______
b.	1,115.50	Mar 2	Apr 30	______	8.5%	______
c.	872.00	Dec 20	Feb 1	______	9.75%	______
d.	1,417.75	Mar 28	June 10	______	7.3%	______
e.	368.42	Sept 17	Nov 17	______	6.25%	______
f.	782.73	Oct 17	Feb 28	______	5.8%	______

2. A window cleaning service took out a loan of $3,500 at 7.5% interest. The note was repaid in 8 months. What was its maturity value?

3. On June 17 Matthew Falk received $1,250, the face value of a note bearing interest of 8.25%. If the note was settled on October 21, what was the total payment?

4. The First State Bank loaned George's Convenience Store $7,000 on January 31 in return for a ten-month promissory note at 8.5%.

a. What is the date of repayment?

b. What is the maturity value of the note?

5. Pam Hruska is the maker of a 6.8% promissory note dated December 5, 1991, with a face value of $2,700. She repaid the note on April 30, 1992.

a. For how many days did she have the use of this money?

b. What was her total payment?

Section 13.4 Present and Future Values at Simple Interest

KEY TERMS **Present value** **Future value**

If the maturity value of a six-month note at nine percent is $3,135, what is the face value? The value of a loan or investment at the beginning of the term, the principal or face value, is called the **present value.**

The **future value** is the value at the end of the term. It is the amount due or maturity value.

Given the present value (principal) of a loan or investment, you can find future value (amount). Or, given the future value (amount), you can find the present value (principal). Future value is greater than present value because as an investment acquires interest over time, its value grows.

What is the maturity value if a principal of $4,000 is invested at 8% for 100 days?

$$A = P(1 + RT)$$
$$= 4000 \times (1 + 0.08 \times 100/365)$$
$$= 4087.67$$

What principal is needed to obtain $5,000 in 100 days if money earns 8%?

$$P = \frac{A}{1 + RT} = \frac{5000}{1 + 0.08 \times 100/365}$$
$$= 4892.76$$

Present and Future Value
at Simple Interest

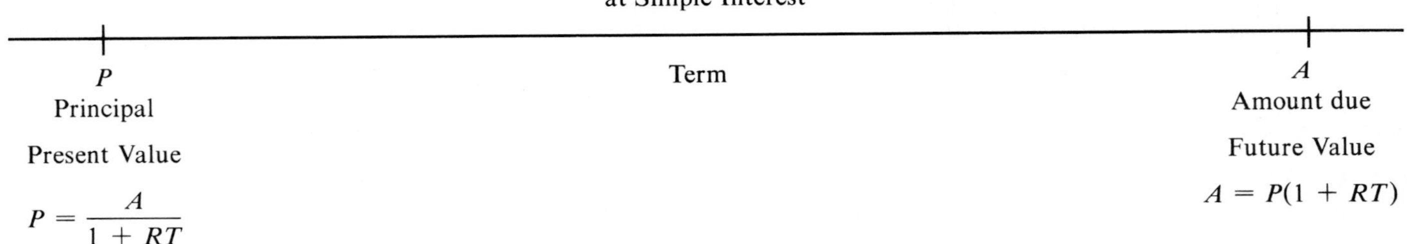

Example 1 How much money is needed to accumulate \$10,000 in 20 months if the principal is invested at 8.5 percent?

Find the present value of 10,000 at 8.5 percent over 20 months.

$$P = \frac{A}{1 + RT} = \frac{10000}{1 + 0.085 \times 20/12} = 8759.12$$

key w/paren:	10000 [÷] [(] 1 [+] .085 [×] 20 [÷] 12 [)] [=]
key w/o paren:	1 [M+] .085 [×] 20 [÷] 12 [=] [M+] 10000 [÷] [MR] [=]
display:	8759.1241
edited:	\$8,759.12

PRACTICE ASSIGNMENT

Plan: In all of these problems, find the present value. Assume exact interest unless ordinary interest is specified.

1.

	Principal (Present Value)	Interest Rate	Time	Maturity Value (Future Value)
a.	\$__________	7%	1 month	\$4,828.00
b.	__________	$8\frac{1}{4}$%	10 months	5,771.25
c.	__________	8%	155 days	6,203.84
d.	__________	$7\frac{1}{2}$%	5 months	3,326.30
e.	__________	8.5%	82 days	1,197.44
f.	__________	5.9%	170 days	770.61

2. The maturity value of a four-month loan at 6% was \$4,003.50. What was the principal?

3. The maturity value of an eight-month note at 7% is \$1,522.90. What is its face value?

4. Alicia Burke took 97 days to repay \$88.20. If the payment included 3% interest, how much did Alicia borrow?

5. A bookstore offers credit to students at $7\frac{1}{2}$% interest on any unpaid balance after 30 days. A student pays \$532.15 75 days after purchasing some books. How much did the books cost originally?

Challenger

Andrea Hopkins bought a used car with a down payment of 15 percent. She made no further payments until February 9, 1991, 185 days later, when she paid the remaining balance of \$2,616.72, including a $5\frac{1}{2}$ percent finance charge. On what date did she purchase the car, and what was its price?

CHAPTER 13 SUMMARY

Concept	*Example*	*Procedure*	*Formula*
SECTION 13.1 The Number of Days in a Term Defined by Two Dates			
Calculating the number of days in a term defined by two dates	June 5–Sept 15 June 5–June 30 25 July 31 Aug 31 Sept 15 102	Learn the number of days in each month. Count the days in each month for the term of the loan to calculate total number of days in the term. Do not count the first day of the loan.	
Finding the due date, given the starting date and number of days in a term	Starting date: Aug 10 Term: 80 days Aug 10 + 80 = Aug 90 improper date − 31 Sept 59 improper date − 30 Oct 29 proper date	Add the number of days to the starting date. If the resulting date is not proper, subtract the number of days in the month and advance to the next month. Proceed in this manner until the resulting date is proper.	
SECTION 13.2 Date Serial Numbers			
Using date serial numbers to find the number of days in a term	June 5–Sept 15 **Date** **Serial number** Sept 15 258 June 5 −156 102	Find the serial number that corresponds to each date. Subtract the serial number of the earlier date from that of the later date.	DATEVALUE(date2)–DATEVALUE(date1)
Using date serial numbers to find the due date	Starting date: Aug 10 Term: 80 days **Date** **Serial number** Aug 10 222 + 80 Oct 29 302	Add the number of days to the serial number corresponding to the starting date. Translate the resulting date serial number to the calendar date.	DATEVALUE(date1)+days
SECTION 13.3 Promissory Notes at Simple Interest			
Finding maturity value	What is the maturity value of a note dated May 10 which matures on July 31 if its face value is $1,285 and it specifies 7% interest? T = 82 days Amount due = 1285 × (1 + 0.07 × 82/365) = 1305.2079 = 1305.21	If the note is non-interest-bearing, maturity value = face value. If the note is interest-bearing, find maturity value by applying the amount formula.	$A = P(1 + RT)$ 1[M+]R[×]T[=][M+]P[×][MR][=] or P[×][(]1[+]R[×]T[)][=] P*(1+R*T)
			$A = P(1 + R\,m/12)$ 1[M+]R[×]m[÷]12[=][M+]P[×][MR][=] or P[×][(]1[+]R[×]m[÷]12[)][=] P*(1+R*m/12)
			$A = P(1 + R\,d/365)$ 1[M+]R[×]d[÷]365[=][M+]P[×][MR][=] or P[×][(]1[+]R[×]d[÷]365[)][=] P*(1+R*d/365)
			$A = P(1 + R\,d/360)$ 1[M+]R[×]d[÷]360[=][M+]P[×][MR][=] or P[×][(]1[+]R[×]d[÷]360[)][=] P*(1+R*d/360)

CHAPTER 13 SUMMARY—(*Continued*)

Concept	*Example*	*Procedure*	*Formula*
SECTION 13.4 Present and Future Values at Simple Interest			
Present value	What principal will yield an amount of $1,000 in 5 months at 8% interest? $P = \frac{1000}{1 + 0.08 \times 5/12} = 967.74194 = 967.74$	Present value is the value of an investment at the beginning of a term. Present value = principal	$P = \frac{A}{1 + RT}$ 1[M+]*R*[×]*T*[=][M+] *A*[÷][MR][=] or *A*[÷][(]1[+]*R*[×]*T*[)][=] *A*/(1+*R***T*)
			$P = \frac{A}{1 + m/12}$ 1[M+]*R*[×]*m*[÷]12[=][M+] *A*[÷][MR][=] or *A*[÷][(]1[+]*R*[×]*m*[÷]12[)][=] *A*/(1+*R***m*/12)
			$P = \frac{A}{1 + d/365}$ 1[M+]*R*[×]*d*[÷]365[=][M+] *A*[÷][MR][=] or *A*[÷][(]1[+]*R*[×]*d*[÷]365[)][=] *A*/(1+*R***d*/365)
			$P = \frac{A}{1 + d/360}$ 1[M+]*R*[×]*d*[÷]360[=][M+] *A*[÷][MR][=] or *A*[÷][(]1[+]*R*[×]*d*[÷]360[)][=] *A*/(1+*R***d*/360)
Future value	Future value is the same as maturity value.	Future value is the value of an investment at the end of a term. Future value = amount due	See section 13.3.

Chapter 13 Spreadsheet Exercise

Computing Present or Future Values

Use your chapter 13 template to compute the present or future value of interest-bearing notes, given the starting and ending date and the interest rate.

The date value function now becomes more meaningful. It produces a date serial number which is the number of elapsed days since December 31, 1899, thereby eliminating complications that arise when a term spans two or more years. The program also resolves all leap year difficulties. Bring your chapter 13 template onto the screen. It will appear as follows:

```
A1: PR [W15]                                                                  READY

              A            B         C          D        E          F        G       H
1                             PRESENT OR FUTURE VALUES OF NOTES
2
3                                                                                  present
4                          face   maturity              date of     due            or future
5            payer         value  value         rate    note        date     days   value
6    ---------------------------------------------------------------------------------------
7    Spunky Hunt         760.00                7.20%   09/30/90   12/22/90
8    Recycling Ltd.     1115.50                8.75%   02/25/90   04/30/91
9    US Tennis Fed.      418.00                5.00%   05/12/90   06/12/90
10   Arizona Bob         872.00                6.75%   12/20/90   03/01/91
11   Slimmetrics        1417.75                6.82%   03/28/90   06/10/90
12   Smith's Signs                  995.62     5.50%   07/24/90   08/31/90
13   Racks & Jacks                  368.42     6.25%   09/17/90   11/17/90
14   Inabinet & Co.                6312.13    10.15%   04/06/90   08/10/91
15   Yarley & Co.                   782.73     9.75%   10/17/90   02/29/92
16   Farmland Trust                4284.91     7.80%   03/21/90   07/01/90
17
18
19
20
```

Column A contains the names of ten payers in rows 7–16.
Column B contains face values in rows 7–11.
Column C contains maturity values in rows 12–16.
Column D contains the interest rates.
Column E contains the dates of the notes.
Column F contains the due dates.

Spreadsheet Exercise Steps

1. Column G: Number of Days in the Term
   The number of days in the term equals
   date value of due date − date value of note date.
   The due date is in F7 and the note date in E7.
   Complete the spreadsheet formula as shown:
   @DATEVALUE (______) −
   @DATEVALUE (______)
   Enter the formula into G7 and copy it from G7 to G8..G16.

2. Column H, rows 7–11: Future Values
   The face value is in B7.
   The rate is in D7 and the time in days will be in G7.
   Maturity value = $P \times (1 + R \times d/365)$.
   When a spreadsheet formula begins with a cell address, it must be preceded by a sign (+).
   Parentheses indicate the order of operations.
   The spreadsheet formula is
   \+ ______ * (_____ + ______ *
   ______ / _______).
   Complete the formula, store it in H7, and copy it from H7 to H8..H11.

3. Column H, rows 12–16: Present Values
   The maturity value is in C12.
   The rate is in D12, and the time in days will be in G12.
   Present value = $A/(1 + R \times d/365)$.
   When a spreadsheet formula begins with a cell address, it must be preceded by a sign (+).
   Parentheses indicate the order of operations.
   The spreadsheet formula is
   \+ _______ / (_____ + _______ *
   _______ / _______).
   Complete the formula, store it in H12, and copy it from H12 to H13..H16.

4. Computer software offers many advantages for obtaining present or future values, but you are responsible for correct programming. Computers usually do not make mistakes; people do.
   First, check for reasonable results.
   Are all future values greater than face values?
   Are all present values smaller than maturity values?

5. Verify one future and one present value with your calculator. (Use any method to find the number of days.)

Row		Number of Days in the Term	Mathematical Expression (with inserted numbers)	Result
7	future value	_______	_______________________________	__________
15	present value	_______	_______________________________	__________

NAME ____________________

CLASS SECTION ____________ DATE ____________

SCORE ________ MAXIMUM __110__ PERCENT ________

CHAPTER 13 ASSIGNMENT

1. A customer purchased merchandise from a department store on March 19 and rendered payment on June 8. How many days did it take this customer to pay? *(5 points)*

2. An appliance, purchased on April 28, was paid for 68 days later. On what date was payment rendered? *(5 points)*

3. On April 12 Donna Solari borrowed $100 from a friend. She repaid the loan on July 19. How many days did she take to repay the loan? *(5 points)*

4. Rocco's Appliances borrowed $3,750 and promised to repay the loan with 8.7% interest in eight months. How much will be repaid? *(5 points)*

5. A six-month loan at 5% was repaid with $779. What was the face value of the loan? *(5 points)*

6. A merchant borrowed $910.50 on June 15 and agreed to repay the note with interest of 7.4% in 66 days. What is the maturity value of the note? *(5 points)*

7. A family wants $2,000 for a pay-as-you-go vacation. How much money must they invest at $10\frac{1}{2}\%$ to produce the desired amount in one year? *(5 points)*

8. A note, issued on October 10, 1990, had a face value of $6,400. It was repaid on May 25, 1991, with 9% interest.

a. In how many days was the loan repaid?

b. What was the maturity value?

(10 points)

9. A used car dealer borrowed $6,900 by signing a ten-month promissory note bearing 8.9% interest. How much must be repaid at maturity? *(5 points)*

10. Statewide Credit charges 18% ordinary interest. If a borrower pays $724.10 to settle a debt incurred 228 days earlier, how much was borrowed? *(5 points)*

11. Twin Lakes Golf Club repaid $7,343.85 on October 19 for a debt incurred on August 10. If the interest rate was $6\frac{3}{4}\%$, what was the face value of the loan? *(10 points)*

12. A wholesaler received a promissory note in settlement of a customer's obligation of $744. The note, dated August 20, carried an interest rate of 6.5%. If the maturity date is November 3, how much will be paid at maturity? *(10 points)*

13. Liz's Boutique promised to repay a loan made on April 20 with a payment of $7,325.48 on September 6. The interest rate was $8\frac{1}{4}$%. What was the face value?
(10 points)

14. A shoe store borrowed $2,225 on March 5 at $8\frac{1}{4}$% ordinary interest. If the note's due date is July 18, how much must be repaid? *(10 points)*

15. Dan Cooper's investment counselor assured him that the principal he invested on March 12 at $9\frac{1}{2}$% would yield $7,500 in 175 days.

a. When will he receive the $7,500?
Maturity date:

b. How much did he invest? Investment:

(15 points)

Chapter 14 Bank Discount

OBJECTIVES

- Find bank discount proceeds
- Determine face value when certain proceeds are required
- Convert the discount rate to the true interest rate and the reverse
- Determine the sales price of a note prior to maturity by the discount method

SECTIONS

INTRODUCTION

For many years banks collected finance charges on loans at the beginning of the term. This finance charge is called bank discount. The quoted rate, or discount rate, is not the true interest rate. The United States Congress, therefore, passed legislation requiring lending institutions to disclose the true interest rate, and most banks no longer collect interest in advance. The method for determining bank discount, however, is still relevant to some loans and investments.

Section 14.1 Collecting Interest in Advance

KEY TERMS **Bank discount** **Proceeds** **Discount rate**

The finance charge for a three-month loan of $1,000 at eight percent interest equals

$$1000 \times 0.08 \times \frac{1}{4} = 20.$$

Under the **bank discount** method, the borrower receives the **proceeds** of $980 (face value of $1,000 − bank discount of $20) and repays the maturity value of $1,000 three months later.

The maturity value of a simple interest loan is the principal plus interest.

Simple Interest Loan of $1,000

Term = 3 months
Rate = 6%
Interest = $1000 \times 0.06 \times \frac{1}{4} = 15$
Principal = $1,000
Maturity value = $1,015

The bank discount method applies the rate, called the **discount rate,** to the face value of a loan. The resulting charge (discount) is subtracted from the face value first. Then the remainder (proceeds) is given to the borrower. At the end of the term, the borrower pays the face value (maturity value).

Bank Discount Loan of $1,000

Term = 3 months
Rate = 6%
Discount = $1000 \times 0.06 \times \frac{1}{4} = 15$
Proceeds = $985
Maturity value = $1,000

Example 1 A woman borrows $6,500 for 100 days from a lender who charges a discount rate of $7\frac{1}{2}$ percent. What is the discount, and how much does she receive?

The discount equals 6500 × 0.075 × 100/365 = 133.56.

The proceeds equal 6500.00 − 133.56 = 6366.44.

key:	6500 [×] .075 [×] 100 [÷] 365 [=]
display:	133.56164
rounded:	133.56 **discount**
key alg:	6500 [−] 133.56 [=]
key arith:	6500 [+] 133.56 [−]
display:	6366.44
edited:	$6,366.44 **proceeds**

A borrower in need of a specific amount of money can determine the face value of the loan that will result in the desired proceeds.

$$\text{Face value} = \frac{\text{proceeds}}{1 - \text{rate} \times \text{time}}$$

Example 2 Suppose the woman in example 1 needs the entire $6,500. What size loan should she apply for?

$$\text{Face value} = \frac{\text{proceeds}}{1 - \text{rate} \times \text{time}} = \frac{6500}{1 - 0.075 \times 100/365} = 6636.3636\ldots$$

key w/paren:	6500[÷] [(] 1 [−] .075 [×] 100 [÷] 365 [)] [=]
key w/o paren:	1 [M+] .075 [×] 100 [÷] 365 [=] [M−] 6500 [÷] [MR] [=]
display:	6636.3636
edited:	$6,636.36

(To verify the result, calculate the $7\frac{1}{2}$ percent discount on $6,636.36 for 100 days. It should be $136.36. The proceeds will be

6636.36 − 136.36 = 6500.00.)

PRACTICE ASSIGNMENT

Plan: All of these problems pertain to bank discount. As with interest, determine the charge given the rate and time. Remember to subtract the charge (discount) from the face value. The borrower then receives the difference, called the proceeds. The loan's face value is the maturity value. If the proceeds must be a particular figure, use the formula shown in this section to obtain face value.

1. Determine the discount and proceeds for each of the following loans:

	Face Value	Discount Rate	Time	Discount	Proceeds
a.	$1,200.00	9%	1 year	$______	$______
b.	727.50	$7\frac{1}{2}$%	1 month	______	______
c.	1,500.00	$8\frac{1}{2}$%	6 months	______	______
d.	3,700.00	9.85%	75 days	______	______
e.	2,800.00	$6\frac{3}{4}$%	10 months	______	______
f.	2,900.00	8.1%	218 days	______	______

2. Determine the face value of the following discounted loans:

	Proceeds	Discount Rate	Time	Face Value
a.	\$1,000	$6\frac{1}{2}\%$	1 year	\$________
b.	2,500	7.3%	3 months	________
c.	660	7%	100 days	________
d.	3,450	$9\frac{1}{4}\%$	10 months	________
e.	485	$7\frac{1}{2}\%$	83 days	________
f.	6,200	8.65%	297 days	________

3. Consider a seven-month loan of \$3,540 at a discount rate of $8\frac{1}{4}\%$.
   a. What is the discount?
   b. What are the proceeds?

4. The Eljay Company negotiated a 65-day loan of \$4,200 at a discount rate of 9.2%.
   a. What was the discount?
   b. What were the proceeds?

5. A borrower needs \$6,250 for five months. Determine the size of this loan given a discount rate of 10%.

6. George Dixon needs \$1,225 on April 10 and plans to repay the loan by July 31. What size loan should he request if the discount rate is 8.65%?

Section 14.2 The Relationship between Bank Discount and Simple Interest

What is the true interest rate on a bank discount loan?

The discount rate understates the true interest rate. A three-month loan of \$1,000 at six percent simple interest means that a borrower receives \$1,000 and repays \$1,015. In the case of bank discount, the borrower of \$1,000 at six percent receives \$985 and repays \$1,000; the finance charge of \$15 is imposed on \$985. Therefore, the true interest rate is somewhat higher than six percent.

Each discount rate corresponds to a slightly higher interest rate. For any specified term, the discount rate can be converted to the true interest rate.

$$R = \frac{\text{discount rate}}{1 - \text{discount rate} \times \text{time}}$$

Each interest rate corresponds to a slightly lower discount rate. For any specified term, the interest rate can be converted to the discount rate.

$$d = \frac{\text{interest rate}}{1 + \text{interest rate} \times \text{time}}$$

Example 1 Mary Rodino obtained a 100-day loan at a discount rate of $7\frac{1}{2}$ percent. What was the true interest rate?

$$R = \frac{\text{discount rate}}{1 - \text{discount rate} \times \text{time}}$$

$$= \frac{0.075}{1 - 0.075 \times 100/365} = 0.0765734$$

key w/paren: .075 [÷] [(] 1 [−] .075 [×] 100 [÷] 365 [)] [=]
key w/o paren: 1 [M+] .075 [×] 100 [÷] 365 [=] [M−] .075 [÷] [MR] [=]
display: 0.0765734
edited: 7.657%

Example 2 A lending institution charges interest of nine percent on three-month loans. What discount rate would provide the same gain?

$$d = \frac{\text{interest rate}}{1 + \text{interest rate} \times \text{time}}$$

$$= \frac{0.09}{1 + 0.09 \times 0.25} = 0.0880195$$

key w/paren: .09 [÷] [(] 1 [+] .09 [×] .25 [)] [=]
key w/o paren: 1 [M+] .09 [×] .25 [=] [M+] .09 [÷] [MR] [=]
display: 0.0880195
edited: 8.802%

PRACTICE ASSIGNMENT

Plan: In all of these problems, convert either a discount rate to a true interest rate or an interest rate to a discount rate. The conversion is correct only for the given term. Round all resulting rates to the nearest thousandth of a percent.

1. Compute true interest rates.

	Time	Discount Rate	Interest Rate
a.	1 year	11%	______%
b.	10 months	$6\frac{1}{2}$%	______%
c.	150 days	8.2%	______%
d.	200 days	8.2%	______%
e.	100 days	7%	______%
f.	200 days	7%	______%

2. Compute discount rates.

	Time	Interest Rate	Discount Rate
a.	1 year	6%	______%
b.	6 months	7.1%	______%
c.	100 days	$6\frac{3}{4}$%	______%
d.	200 days	$6\frac{3}{4}$%	______%
e.	75 days	7.85%	______%
f.	260 days	$6\frac{1}{4}$%	______%

3. A loan of $1,200 was negotiated on March 15 at a discount rate of $7\frac{3}{4}$%. If settlement was due on May 20, what was the true interest rate?

4. First Regional Credit granted a 250-day loan with a maturity value of $4,000 subject to a discount rate. If the true interest rate was 9.54%, what was the discount rate?

Section 14.3 Discounting Promissory Notes

KEY TERM **Discounting**

Discounting is a practice whereby a payee sells a promissory note to another party (a third party) prior to the note's maturity.

The method for setting the price of a note varies. Some banks may still compute discount proceeds on a note's maturity value, but many no longer do. Upon discounting a note, the payee receives proceeds calculated for the time from the discount date to the maturity date, the discounting period. Upon maturity, the holder (buyer) of the note collects the maturity value from the payer. (Unlike the relationship between discount and interest rates in Section 14.2, this discount rate is generally independent of the original interest rate of the note.)

Example 1 A 6-month note dated April 18 with a face value of $2,500 bears eight percent simple interest. What are the proceeds if the note is discounted at a bank four months prior to maturity at a nine percent discount rate?

Discounting an Interest-bearing Note

Origination Date 4/18	6 months at 8%	Maturity Date 10/18
2500		$A = 2500(1 + 0.08 \times 0.5)$ $= 2600$

4/18	Discounting Date 6/18		10/18
2500	2 months	4 months at 9% ($\frac{1}{3}$ of a year)	2600

$$2600 \times 0.09 \times 1/3 = 78$$

$$\text{Proceeds to payee} = 2600 - 78 = 2522$$

The discounter receives $2,522 four months before the maturity date. On the maturity date the bank collects the maturity value, $2,600, from the payer. In effect, the bank lends $2,600 to the payee for four months at a discount rate of nine percent.

PRACTICE ASSIGNMENT

Plan: A payee may sell a note to a third party prior to maturity at a discount rate. Calculate the maturity value of the note first. Next, calculate the time between date of transfer and maturity date. Then compute the discount on the maturity value at the discount rate for the calculated time period.

$$\text{Proceeds to the payee} = \text{maturity value} - \text{discount}.$$

You may use a shortcut for computing the proceeds from the maturity value A: Instead of computing $Pro = A - AdT$ (where d stands for discount rate), use

$$Pro = A(1 - dT).$$

1. Find the proceeds if the following notes are discounted on September 1, 1991.

a.

Date of Note	Term of Note	Face Value	Interest Rate	Discount Rate
June 11, 1991	4 months	$2,000	6%	7%

Maturity date: ____________
Discounting period: ____________
Maturity value: ____________
Proceeds: ____________

b.

Date of Note	Term of Note	Face Value	Interest Rate	Discount Rate
Dec 8, 1990	1 year	$1,775	$8\frac{1}{4}$%	8.85%

Maturity date: ____________
Discounting period: ____________
Maturity value: ____________
Proceeds: ____________

c.

Date of Note	Term of Note	Face Value	Interest Rate	Discount Rate
July 14, 1991	150 days	$1,850	6.8%	8%

Maturity date: ____________
Discounting period: ____________
Maturity value: ____________
Proceeds: ____________

2. A one-year note with a face value of $2,250 bears simple interest of $7\frac{1}{4}$%. If the note is discounted at 8% two months before maturity, what are the proceeds?

3. Rachel Clifton borrowed $4,800 by signing a 200-day note at $6\frac{1}{2}$% simple interest. Seventy-five days before maturity, the note was discounted at 7.45%. What were the proceeds?

4. On July 15 the holder of an $8\frac{1}{4}$% simple interest note with a face value of $1,600 discounted the note at a discount rate of 9%. The note was written on May 1, and the due date was September 30. What were the proceeds?

5. Chavez Contractors signed a 6-month note agreeing to pay Ridgemont Drywall $3,500 plus simple interest of $7\frac{3}{4}$%. Two months later Ridgemont sold the note to Fidelity Trust at a discount rate of $8\frac{1}{4}$%.

a. What was the maturity value?

b. What were the proceeds?

Challenger

On May 27, Merchants' Finance granted a loan to Radio Cottage at the customary discount rate. Radio Cottage received proceeds of $6,000; they paid the loan's maturity value of $6,222.07 on November 1. What was the discount rate? What was the true interest rate? If the maturity date had been December 1 with the same discount rate and maturity value, how much in proceeds would Radio Cottage have received on May 27?

CHAPTER 14 SUMMARY

SECTION 14.1 Collecting Interest in Advance

Concept	*Example*	*Procedure*	*Formula*
Bank discount	A loan of $1,500 for 6 months is contracted at 8% bank discount. $1500 \times 0.08 \times \frac{1}{2} = 60$ The discount (charge) equals $60, and the borrower receives the proceeds, $1500 - 60 = 1440$.	The discount rate is applied to the face value. The finance charge is collected in advance, and the borrower receives the remainder of the loan, called the proceeds. At the end of the term, the lender collects the full face value.	Discount $= AdT$ A[×]d[×]T[=] A*d*T Proceeds $= A - AdT$ A[M+][×]d[×][T][=][M−][MR] A−A*d*T
Computing the face value for specified proceeds	What face value is required for proceeds of $1,500? $\frac{1500}{1 - 0.08 \times 1/2} = 1562.50$		$A = \frac{\text{proceeds}}{1 - dT}$ 1[M+]d[×]T[=][M−]Pro[÷][MR][=]or Pro[÷][(]1[−]d[×]T[)][=] Pro/(1−dT)

SECTION 14.2 The Relationship between Bank Discount and Simple Interest

Concept	*Example*	*Procedure*	*Formula*
Relationship between bank discount and simple interest	What is the true interest rate corresponding to an 8% discount rate on a six-month loan? $\frac{0.08}{1 - 0.08 \times 1/2}$ $= 0.0833333 = 8\frac{1}{3}\%$	The true interest rate depends on the discount rate and the term of the loan	$R = \frac{d}{1 - dT}$ 1[M+]d[×]T[=][M−]d[÷][MR][=] or d[÷][(]1[−]d[×]T[)][=] d/(1−d*T)
	What discount rate secures the same gain as 10% simple interest on six-month loans? $\frac{0.1}{1 + 0.1 \times 1/2}$ $= 0.095238 = 9.524\%$		$d = \frac{R}{1 + RT}$ 1[M+]R[×]T[=][M+]R[÷][MR][=] or R[÷][(]1[+]R[×]T[)][=] R/(1+R*T)

SECTION 14.3 Discounting Promissory Notes

Concept	*Example*	*Procedure*	*Formula*
Discounting promissory notes	A 7% nine-month interest-bearing note with face value of $2,300 is discounted at 8% after three months. How much does the payee receive at this time? $2300 \times (1 + 0.07 \times \frac{3}{4})$ $= 2420.75$ = maturity value Discount $= 2420.75 \times 0.08 \times \frac{1}{2}$ $= 96.83$ Proceeds $= 2420.75 - 96.83$ $= 2323.92$	If a note is interest-bearing, determine its maturity value, using the interest rate. Then compute the discount for the remaining period, using the discount rate. The proceeds are computed by subtracting the discount from the maturity value. In effect, the bank lends the payee the maturity value at bank discount for the discounting period.	$A = P(1 + RT)$ 1[M+]R[×]T[=][M+]P[×][MR][=] or P[×][(]1[+]R[×]T[)][=] P*(1+R*T) Proceeds $= A - AdT$ A[M+][×]d[×][T][=][M−][MR] A−A*d*T or Proceeds $= A(1-dT)$ 1[M+]d[×]T[=][M−]A[×][MR][=] or A[×][(]1[−]d[×]T[)][=] A*(1−d*T)

Chapter 14 Spreadsheet Exercise

Value of Interest-Bearing Notes on Discount Date

The chapter 14 spreadsheet computes the proceeds of interest-bearing notes when they are discounted. The template will appear as follows:

A1: PR [W9] READY

	A	B	C	D	E	F	G	H	I
1		VALUE OF INTEREST-BEARING NOTES ON DISCOUNT DATE							
2									
3			term						
4	face	date of	in	maturity	int.	maturity	discount	discount	
5	value	note	days	date	rate	value	date	rate	proceeds
6									
7	1100.00	10/25/90	100		6.50%		11/30/90	7.00%	
8	1500.00	4/06/91	75		7.20%		5/25/91	8.00%	
9	2250.00	7/21/90	120		8.00%		9/01/90	8.75%	
10	750.00	11/01/90	90		6.75%		1/15/91	7.25%	
11	875.00	2/21/92	200		7.50%		6/01/92	7.75%	
12	645.50	12/15/91	60		5.80%		1/31/92	6.25%	
13	921.25	3/18/90	150		7.05%		5/15/90	7.45%	
14	1625.00	1/15/92	365		8.25%		6/01/92	9.00%	
15	3750.00	6/05/91	182		6.88%		8/20/91	7.29%	
16	8000.00	8/31/90	110		7.85%		10/06/90	8.47%	
17									
18									
19									
20									

To compute the value of an interest-bearing note at the time of transfer (sale), you must know the maturity date, the maturity value, and the number of days in the discounting period. You must know two rates and three dates; count days or subtract date serial numbers, and apply two formulas. If you program your computer properly, it will be highly efficient.

In rows 7–16, the template displays information for ten notes.

Columns	**Contents**
A	Face value (principal of note)
B	Date of note
C	Number of days in term
E	Interest rate
G	Discount date
H	Discount rate

Spreadsheet Exercise Steps

1. Column D: Maturity Date
Maturity date = date of note + number of days in term.
The date of the note is in cell B7. The number of days is in C7.
Complete the spreadsheet formula
@DATEVALUE (______) + ______.
Enter the formula into D7 and copy it from D7 to D8..D16.

2. Column F: Maturity Value
Maturity value = $P \times (1 + R \times d/365)$.
The principal (face value) is in A7; the rate is in E7; and the number of days is in C7.
The first cell address must be preceded by +.
The spreadsheet formula is
+ ______ * (______ + ______ * ______ / ______).
Complete the formula and store it in F7; copy it from F7 to F8..F16.

3. Column I: Proceeds
Proceeds = maturity value − maturity value × discount rate × discount time = $A - AdT$ = $A \times (1 - dT)$ where d stands for discount rate.
The maturity value will be in F7; the discount rate is in H7.
Discount time = (maturity date − discount date)/365.
The maturity date will be in D7; the discount date is in G7.
The first cell address must be preceded by +.
Complete the following spreadsheet formula:
+ ______ * (____ − ______ * (@DATEVALUE(______) − @DATEVALUE(______))/ ______).
Store the spreadsheet formula in I7 and copy it from I7 to I8..I16.

4. The programmer should understand enough mathematics to verify the results. Verify the results of row 11 with a calculator.

Face Value	Interest Rate	Days in Term	Interest	Maturity Value
______	______	______	______	______

Discount Rate	Maturity Date	Discount Date	Days in Discount Period	Discount	Proceeds
______	______	______	______	______	______

NAME ______________________________

CLASS SECTION ______________ DATE ______________

SCORE __________ MAXIMUM __80__ PERCENT __________

CHAPTER 14 ASSIGNMENT

Round computed rates to the nearest thousandth of a percent.

1. Determine the discount and the proceeds for a five-month loan of $1,850 at a discount rate of $7\frac{1}{4}\%$.
Discount: __________ Proceeds: __________
(5 points)

2. What is the true interest rate of a settlement due in two months at a discount rate of 8%?
(5 points)

3. Toffler's Notions received $7,500 in return for a 10-month note at an interest rate of 8.2%. The note was discounted at 8.9% four months after it had been negotiated. What were the proceeds?
(5 points)

4. National Savings and Loan granted Lori Pacelli a 200-day loan of $2,750 at a discount rate of 7.15%. Calculate the discount and proceeds.
Discount: __________ Proceeds: __________
(5 points)

5. On August 17, a bank negotiated a loan with a face value of $1,750 at 7.45% interest. If another institution charged equivalent discount rates, what was its discount rate for the same loan if it was paid on November 3? *(5 points)*

6. Cheryl's Crafts wants to borrow $7,500 to expand operations. If the bank charges a discount rate of $7\frac{3}{4}\%$, what size loan should Cheryl apply for if she plans to repay it in 10 months? *(5 points)*

7. A note bearing 7.9% simple interest with a face value of $4,000 and a term of 160 days was discounted 45 days after it was negotiated. If the discount rate was 9.3%, what were the proceeds? *(5 points)*

8. Iron Pump Sports Equipment received a 220-day note with a face value of $2,750 from an exercise gym. The note was dated October 10, 1991 and specified an interest rate of 7%. One hundred days before the note was due, it was discounted at $7\frac{3}{4}\%$. Calculate the maturity value and proceeds.
Maturity value: __________
Proceeds: __________ *(10 points)*

9. On August 17, First Manufacturers Bank granted a loan of $25,000 to Ritter Industries. The discount rate was $7\frac{1}{2}\%$, and the loan matured on December 3. Determine the bank discount and proceeds.
Discount: __________
Proceeds: __________ *(5 points)*

10. On May 17, a commercial finance company granted Sam Davies a loan at a discount rate of 6.85%. He repaid the loan on August 5. What was the true interest rate? *(5 points)*

11. Lasky's Furniture Town received proceeds of $5,000 from a loan negotiated on June 18. The lender charged a discount rate on all loans. Settlement was due on November 1. What was the discount rate if the true interest rate was $8\frac{3}{4}$%? *(5 points)*

12. Atkins' Plumbing charged the Old West Steak House $3,400 for major plumbing repairs. The restaurant manager gave a three-month non-interest-bearing note to Mr. Atkins. After one month, Mr. Atkins sold the note to a local bank at a discount rate of 10%.

a. How much will Old West Steak House pay? __________ To whom? ____________

b. How much did Mr. Atkins receive? ____________

c. How much will the bank receive upon maturity of the note? __________ *(20 points)*

Chapter 15 Installment Plans and Truth-in-Lending

OBJECTIVES

- Recognize various installment plans and their finance charges
- Find true interest rate
- Compute finance charge rebates in cases of early settlement

SECTIONS

INTRODUCTION

Installment loans have become an American tradition. Buy now–pay later plans permit people of even modest means to purchase everything from clothes to cars. If you meet a car dealer's credit criteria, you can make a small down payment, negotiate a series of payments to pay off the loan, and drive away in a brand-new car. Legislation enacted by the United States government has regulated lending practices and requires installment plans to disclose true interest rates, finance charges, payments, and terms.

Section 15.1 Add-on Interest

KEY TERMS **Installments** **Level payment plan** **Finance charge**

How are installment loans different from other loans?

A buyer may pay part of a purchase price with a down payment or a trade-in and the balance with a series of **installments.** The **level payment plan** with add-on interest is a simple arrangement that requires few calculations. The purchase price less the down payment or trade-in value is the amount financed.

The **finance charge** is interest that is added onto the amount financed for the duration of the plan. This plan, frequently used in the past, understates the interest rate.

> Amount financed
> = cash price − down payment
> or
> = cash price − trade-in value
>
> Periodic payment
> $= \dfrac{\text{amount financed} + \text{interest}}{\text{number of payments}}$
>
> Finance charge
> = add-on interest
> = amount financed × rate × time
> (time is duration of plan)

Example 1 A homeowner may purchase a shed for $1,176 cash, or pay 15% down, with add-on interest of 8.7% and equal monthly installment payments for one year.

a. What is the down payment?
15% of the purchase price = 1176 × 0.15 = 176.40.

b. What is the amount financed?
Purchase price − down payment
= 1176.00 − 176.40 = 999.60.

c. What is the finance charge?
Amount financed × add-on rate × time
= 999.60 × 0.087 × 1 = 86.97.

d. What are the payments?

$$\frac{\text{amount financed} + \text{finance charge}}{\text{number of months}} = \frac{999.60 + 86.97}{12} = 90.55$$

e. What is the total cost under the installment plan?
Purchase price + finance charge
= 1176 + 86.97 = 1262.97
or
down payment + number of payments × size of payments = 176.40 + 12 × 90.55 = 1263.00.
(Discrepancy of $0.03 due to rounding.)

Example 2 A college student purchased a used car priced at \$3,800 with \$500 down and 12 monthly payments of \$292 each.

a. What is the finance charge?

Finance charge = number of payments × size of payments − amount financed
= 12 × 292 − 3300
= 204

b. What is the add-on interest rate?

$$R = \frac{I}{PT} = \frac{\text{finance charge}}{\text{amount financed} \times \text{time}}$$

$$= \frac{204}{3300 \times 1} = 0.061818\ldots = 6.18\%$$

PRACTICE ASSIGNMENT

Plan: For these problems, assume level payment plans with add-on interest. The number of payments is given. The down payment may be a required percent of the purchase price. Given certain information, you will be able to find the missing items. You may have to find the following items:

Amount financed = Price − down payment or trade-in

Finance charge = Amount financed × rate × time

or

Number of payments × size of payments − amount financed

Size of payments = (amount financed + interest) ÷ number of payments

Interest rate = Interest ÷ (amount financed × time)

$$R = \frac{I}{PT}$$

Total cost of purchase on installment plan = Purchase price + finance charge

or

Down payment + number of payments × size of payments

(The two methods may produce slightly different results due to rounding.)

1. Compute the finance charge and the size of each payment.

	Purchase Price	Down Payment	Add-on Interest Rate	Number of Payments	Finance Charge	Payment
a.	\$1,000	\$100	9%	12	\$________	\$________
b.	725	75	$8\frac{1}{2}\%$	6	________	________
c.	320	25	7.6%	3	________	________
d.	3,000	450	11.3%	18	________	________
e.	437	50	$5\frac{3}{4}\%$	9	________	________
f.	850	85	6.9%	7	________	________

2. Compute the finance charge and the add-on interest rate to the nearest tenth of a percent.

	Purchase Price	Down Payment	Monthly Payment	Number of Payments	Finance Charge	Add-on Interest Rate
a.	\$3,000	\$300.00	\$245.25	12	\$________	______%
b.	1,600	125.00	161.14	10	________	______%
c.	845	84.50	89.57	9	________	______%
d.	5,700	750.00	318.31	18	________	______%
e.	2,400	500.00	330.52	6	________	______%
f.	148	40.00	27.66	4	________	______%

3. A refrigerator may be purchased for \$850 cash or \$100 down and \$70 per month for one year.

a. What is the amount financed?

b. What is the finance charge?

4. A customer may purchase a living room set costing $1,650 with 15% down, an 8% finance charge, and twelve monthly payments.

a. What are the monthly payments?

b. What is the total cost under the installment plan?

5. A bicycle priced at $250 may be purchased with monthly payments of $23.80 for one year.

a. What is the finance charge?

b. What is the add-on interest rate?

Section 15.2 APR and Truth-in-Lending

KEY TERMS **Annual Percentage Rate (APR)** **Effective Interest Rate**

Why Is the Add-On Interest Rate Not the True Annual Interest Rate?

Simple interest implies that principal and finance charge are paid in a lump sum at the end of a term. Under installment payment plans, however, parts of the principal and interest are paid throughout the term, and therefore the true interest rate is something higher than the stated interest rate.

What is Truth-in-Lending?

Under the federal Truth-in-Lending Act, installment plan contracts must reveal the true annual interest rate, which is called the **annual percentage rate** (APR) or **effective interest rate.** Because computing the APR can be complicated, the Federal Reserve Board furnishes APR tables.

To use the APR table knowing the number of payments, first compute the charge for each $100 financed.

$$\text{Finance charge per } \$100 = \frac{\text{finance charge} \times 100}{\text{amount financed}}$$

Table 15.1 Annual Percentage Rate (APR)
(finance charge per hundred dollars)

No. of Monthly Payments	*10.00%*	*10.25%*	*10.50%*	*10.75%*	*11.00%*	*11.25%*	*11.50%*	*11.75%*	*12.00%*	*12.25%*	*12.50%*	*12.75%*
1	0.83	0.85	0.87	0.90	0.92	0.94	0.96	0.98	1.00	1.02	1.04	1.06
2	1.25	1.28	1.31	1.35	1.38	1.41	1.44	1.47	1.50	1.53	1.57	1.60
3	1.67	1.71	1.76	1.80	1.84	1.88	1.92	1.96	2.01	2.05	2.09	2.13
4	2.09	2.14	2.20	2.25	2.30	2.35	2.41	2.46	2.51	2.57	2.62	2.67
5	2.51	2.58	2.64	2.70	2.77	2.83	2.89	2.96	3.02	3.08	3.15	3.21
6	2.94	3.01	3.08	3.16	3.23	3.31	3.38	3.45	3.53	3.60	3.68	3.75
7	3.36	3.45	3.53	3.62	3.70	3.78	3.87	3.95	4.04	4.12	4.21	4.29
8	3.79	3.88	3.98	4.07	4.17	4.26	4.36	4.46	4.55	4.65	4.74	4.84
9	4.21	4.32	4.43	4.53	4.64	4.75	4.85	4.96	5.07	5.17	5.28	5.39
10	4.64	4.76	4.88	4.99	5.11	5.23	5.35	5.46	5.58	5.70	5.82	5.94
11	5.07	5.20	5.33	5.45	5.58	5.71	5.84	5.97	6.10	6.23	6.36	6.49
12	5.50	5.64	5.78	5.92	6.06	6.20	6.34	6.48	6.62	6.76	6.90	7.04
18	8.10	8.31	8.52	8.73	8.93	9.14	9.35	9.56	9.77	9.98	10.19	10.40
24	10.75	11.02	11.30	11.58	11.86	12.14	12.42	12.70	12.98	13.26	13.54	13.82
36	16.16	16.58	17.01	17.43	17.86	18.29	18.71	19.14	19.57	20.00	20.43	20.87

Table 15.1—(*Continued*)

No. of Monthly Payments	*13.00%*	*13.25%*	*13.50%*	*13.75%*	*14.00%*	*14.25%*	*14.50%*	*14.75%*	*15.00%*	*15.25%*	*15.50%*	*15.75%*
1	1.08	1.10	1.12	1.15	1.17	1.19	1.21	1.23	1.25	1.27	1.29	1.31
2	1.63	1.66	1.69	1.72	1.75	1.78	1.82	1.85	1.88	1.91	1.94	1.97
3	2.17	2.22	2.26	2.30	2.34	2.38	2.43	2.47	2.51	2.55	2.59	2.64
4	2.72	2.78	2.83	2.88	2.93	2.99	3.04	3.09	3.14	3.20	3.25	3.30
5	3.27	3.34	3.40	3.46	3.53	3.59	3.65	3.72	3.78	3.84	3.91	3.97
6	3.83	3.90	3.97	4.05	4.12	4.20	4.27	4.35	4.42	4.49	4.57	4.64
7	4.38	4.47	4.55	4.64	4.72	4.81	4.89	4.98	5.06	5.15	5.23	5.32
8	4.94	5.03	5.13	5.22	5.32	5.42	5.51	5.61	5.71	5.80	5.90	6.00
9	5.49	5.60	5.71	5.82	5.92	6.03	6.14	6.25	6.35	6.46	6.57	6.68
10	6.05	6.17	6.29	6.41	6.53	6.65	6.77	6.88	7.00	7.12	7.24	7.36
11	6.62	6.75	6.88	7.01	7.14	7.27	7.40	7.53	7.66	7.79	7.92	8.05
12	7.18	7.32	7.46	7.60	7.74	7.89	8.03	8.17	8.31	8.45	8.59	8.74
18	10.61	10.82	11.03	11.24	11.45	11.66	11.87	12.08	12.29	12.50	12.72	12.93
24	14.10	14.38	14.66	14.95	15.23	15.51	15.80	16.08	16.37	16.65	16.94	17.22
36	21.30	21.73	22.17	22.60	23.04	23.48	23.92	24.35	24.80	25.24	25.68	26.12

No. of Monthly Payments	*16.00%*	*16.25%*	*16.50%*	*16.75%*	*17.00%*	*17.25%*	*17.50%*	*17.75%*	*18.00%*	*18.25%*	*18.50%*	*18.75%*
1	1.33	1.35	1.37	1.40	1.42	1.44	1.46	1.48	1.50	1.52	1.54	1.56
2	2.00	2.04	2.07	2.10	2.13	2.16	2.19	2.22	2.26	2.29	2.32	2.35
3	2.68	2.72	2.76	2.80	2.85	2.89	2.93	2.97	3.01	3.06	3.10	3.14
4	3.36	3.41	3.46	3.51	3.57	3.62	3.67	3.73	3.78	3.83	3.88	3.94
5	4.04	4.10	4.16	4.23	4.29	4.35	4.42	4.48	4.54	4.61	4.67	4.74
6	4.72	4.79	4.87	4.94	5.02	5.09	5.17	5.24	5.32	5.39	5.46	5.54
7	5.40	5.49	5.58	5.66	5.75	5.83	5.92	6.00	6.09	6.18	6.26	6.35
8	6.09	6.19	6.29	6.38	6.48	6.58	6.67	6.77	6.87	6.96	7.06	7.16
9	6.78	6.89	7.00	7.11	7.22	7.32	7.43	7.54	7.65	7.76	7.87	7.97
10	7.48	7.60	7.72	7.84	7.96	8.08	8.19	8.31	8.43	8.55	8.67	8.79
11	8.18	8.31	8.44	8.57	8.70	8.83	8.96	9.09	9.22	9.35	9.49	9.62
12	8.88	9.02	9.16	9.30	9.45	9.59	9.73	9.87	10.02	10.16	10.30	10.44
18	13.14	13.35	13.57	13.78	13.99	14.21	14.42	14.64	14.85	15.07	15.28	15.49
24	17.51	17.80	18.09	18.37	18.66	18.95	19.24	19.53	19.82	20.11	20.40	20.69
36	26.57	27.01	27.46	27.90	28.35	28.80	29.25	29.70	30.15	30.60	31.05	31.51

Example 1 The Winthrops purchased a piano priced at \$2,460 with twenty-four monthly payments of \$115. What is the true annual interest rate (APR)?

The total payments are: $115 \times 24 = 2760$.

Add-on interest is: $2760 - 2460 = 300$.

$$\text{Finance charge per \$100} = \frac{\text{finance charge} \times 100}{\text{amount financed}}$$

$$= \frac{300 \times 100}{2460} = 12.20$$

On the table, find the number closest to 12.20 in the rows for 24 monthly payments; 12.14 is closest, so the APR is approximately 11.25%.

PRACTICE ASSIGNMENT 1

Plan: In each of these problems, use table 15.1 to find the APR. Where necessary, compute the add-on interest first. Determine the finance charge per \$100; then refer to the table.

1. For the following installment plans, compute the interest and payments, and find the closest APR in table 15.1.

	Amount Financed	Number of Monthly Payments	Add-on Rate	Interest	Payment	APR
a.	$750.00	6	7.5%	$ ______	$ ______	______%
b.	1,000.00	6	8%	______	______	______%
c.	1,583.40	12	6%	______	______	______%
d.	1,750.00	12	7%	______	______	______%
e.	2,130.25	24	7%	______	______	______%
f.	950.00	10	9%	______	______	______%

2. Tim Folkers purchased a boat for $4,500. He paid $1,000 down and financed the balance with twelve monthly payments including 9% add-on interest. What was the true rate of interest?

3. A furniture store advertised a water bed: "$399! No money down! Six months to pay!" What is the true interest rate if the monthly payments are $70?

4. Laurie Bodansky bought a computer priced at $1,375 with a down payment of $175. Her 18 monthly installments included 9.5% add-on interest.

   a. What was the true interest rate?

   b. What was her total payment?

When APR tables are not available, apply the formula on the right to produce approximations. Although this formula is not absolutely correct according to financial mathematics, it produces reasonably good estimates in most cases and is not difficult to apply.

$$R = \frac{2mI}{P(n + 1)}$$

I = interest
P = principal (amount financed)
n = number of payments
m = number of payments per year (12 for monthly payments, even if term is less than a year)

Example 2 Apply the approximation formula to the Winthrops' piano purchase in example 1. The piano was priced at $2,460 with twenty-four monthly payments of $115. Compute the true annual interest rate.

The total payment is $115 \times 24 = 2760$.

Add-on interest is $2760 - 2460 = 300$.

$$\frac{2 \times 12 \times \text{interest}}{\text{principal} \times (\text{number of payments} + 1)}$$

$$= \frac{24 \times 300}{2460 \times 25}$$

$$= 0.1170731 = 11.71\%$$

Table 15.1 shows an APR of approximately 11.25%.

PRACTICE ASSIGNMENT 2

Plan: Recompute the APR to two decimal places for some of the problems in practice assignment 1. Compare results with APRs in table 15.1.

1.

	Amount Financed	Number of Monthly Payments	Add-on Rate	Interest	Computed APR	APR in Table 15.1
a.	$ 750.00	6	7.5%	$ ______	______%	______%
b.	1,000.00	6	8%	______	______%	______%

	Amount Financed	Number of Monthly Payments	Add-on Rate	Interest	Computed APR	APR in Table 15.1
c.	1,583.40	12	6%	______	______%	______%
d.	1,750.00	12	7%	______	______%	______%
e.	2,130.25	24	7%	______	______%	______%
f.	950.00	10	9%	______	______%	______%

2. Tim Folkers purchased a boat for $4,500. He paid $1,000 down and financed the balance with twelve monthly payments including 9% add-on interest. Compute the APR and compare your result with the APR in table 15.1.
Computed APR: ______%
APR from table: ______%

3. Jim Johnson purchased a motorcycle priced at $1,780 with a down payment of $250 and eight monthly payments of $200.50. Compute the APR and compare your result with the APR in table 15.1.
Computed APR: ______%
APR from table: ______%

4. A customer purchased a set of dinnerware costing $1,250 with a down payment of $100 plus nine monthly installments with add-on interest of 8%. Compute the APR and compare your result with the APR in table 15.1.
Computed APR: ______%
APR from table: ______%

Section **15.3 The Rule of 78**

KEY TERMS **Finance charge rebate** **Sum-of-the-digits method**

A borrower who pays off an installment plan earlier than scheduled is entitled to a partial **finance charge rebate,** since the charge was based on the original term. The rebate is usually deducted from the final payment. There are several accepted ways of determining the rebate; the Truth-in-Lending Act requires disclosure of the method of calculation.

An accepted method of calculating interest rebate is called the Rule of 78; the earlier the payoff, the larger the rebate.

The calculated rebate fraction generates a full rebate in the case of an immediate payoff (minus a minimum fee, if any), and progressively smaller rebates on payoffs that occur later. This procedure is called the **sum-of-the-digits method** because the numerator of the rebate fraction is the sum of the integers up to r (the number of remaining months), and the denominator is the sum of the integers up to t (the number of months in the term).

For example, for a term of 6 months ($t = 6$), the denominator is $1 + 2 + 3 + 4 + 5 + 6 = 21$. (For a term of 12 months, the sum-of-the-digits is 78. Thus, the Rule of 78.) It is easier, however, to work with the formula shown at the right, which produces the same rebate fractions.

Rebates for a 6-month Plan

r = number of remaining months
t = number of months in the term
I = finance charge (interest)

Time of Payoff	Rebate Fraction $\frac{r(r+1)}{t(t+1)} \times I$	Rebate
6 months early	$\frac{6 \times 7}{6 \times 7} \times I =$	$\frac{21}{21} \times I = I$
5 months early	$\frac{5 \times 6}{6 \times 7} \times I =$	$\frac{15}{21} \times I$
4 months early	$\frac{4 \times 5}{6 \times 7} \times I =$	$\frac{10}{21} \times I$
3 months early	$\frac{3 \times 4}{6 \times 7} \times I =$	$\frac{6}{21} \times I$
2 months early	$\frac{2 \times 3}{6 \times 7} \times I =$	$\frac{3}{21} \times I$
1 month early	$\frac{1 \times 2}{6 \times 7} \times I =$	$\frac{1}{21} \times I$

Example 1 William Taylor purchased an exercise machine priced at $425 on a twelve-month installment plan; his monthly payments were $38.50. He paid off the loan seven months ahead of schedule. How much should his last payment be?

Total payments for the entire term = 38.50 × 12 = 462.00.

Total finance charge = 462 − 425 = 37.

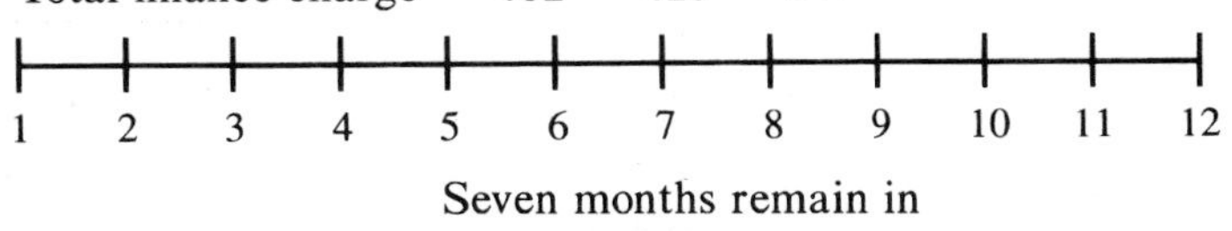

Seven months remain in the 12-month term.
$r = 7 \quad t = 12$

$$\frac{r(r+1)}{t(t+1)} = \frac{7(7+1)}{12(12+1)} = \frac{7 \times 8}{12 \times 13} = \frac{14}{39}$$

Rebate = rebate fraction × total finance charge
$= \frac{14}{39} \times 37 = 13.28$

Final payment = remaining number of payments × size of payment − rebate
$= 7 \times 38.50 - 13.28 = 269.50 - 13.28$
$= 256.22$

PRACTICE ASSIGNMENT

Plan: All problems relate to the Rule of 78 rebate method. The rebate is

$$\frac{r(r+1)}{t(t+1)} \times I$$

where r is the number of remaining months in the term at the time of payoff, t is the number of months in the term, and I is the finance charge.
Deduct the rebate from the final payment. This last payment should be:

Remaining number of payments × size of payment − rebate.

If an obligation is payed off immediately, the total finance charge is rebated; a small loan initiation fee, however, may be collected by the finance institution.

1. Compute all possible rebates on a loan carrying a finance charge of $50 for a term of 8 months assuming there is a minimum finance charge of $5.

Remaining Months r	Rebate Fraction in Lowest Terms	Rebate Fraction in Decimal Form to Four Places	Rebate
8	_____	________	$________
7	_____	________	________
6	_____	________	________
5	_____	________	________
4	_____	________	________
3	_____	________	________
2	_____	________	________
1	_____	________	________

2. Tom and Sally Peters financed a microwave oven with a ten-month installment loan. They paid off the loan six months early. What portion of the interest should be rebated?
   Fraction in lowest terms: _____
   Four-place decimal: ________

3. Millicent Jones purchased a snowblower priced at $850 with a twelve-month installment loan. The finance charge was $63.80. If she paid off the loan five months early, what was the final payment?

4. Leblanc Sport Shop sells Rossberg shotguns for $235, or 10% down and six monthly payments of $36.85. If a customer paid off the installment loan three months early, how much total interest did the customer pay?

Section 15.4 Interest Based on the Unpaid Balance

A mathematically correct way of determining a finance charge is to periodically compute the interest on the unpaid balance. In this method, the quoted annual rate is the APR, and the interest automatically increases or decreases with smaller or larger payments.

One approach is to pay an equal portion of the principal plus interest on the unpaid balance each month; however, this method creates unequal payments.

Example 1 Jan purchases a vacuum cleaner for $340 with $40 down and the $300 balance payable in four monthly installments. She will pay $75 per month on the principal and will be charged seven percent interest on the unpaid balance. For each monthly payment, interest = unpaid balance × 0.07 ÷ 12.

End of:	Unpaid Balance	Interest Computation	Interest	Payment	New Balance
Month 1	300	7% for 1 month on 300	1.75	76.75	225
Month 2	225	7% for 1 month on 225	1.31	76.31	150
Month 3	150	7% for 1 month on 150	0.88	75.88	75
Month 4	75	7% for 1 month on 75	0.44	75.44	0

Total finance charge = 4.38

You may also compute the total finance charge with the following formula:

$$\text{Finance charge} = \frac{(\text{no. of payments} + 1) \times \text{interest for first period}}{2}$$

$5 \times 1.75 \div 2 = 4.375 = 4.38$

The APR is the stated rate, 7%.

(An add-on interest rate of 7% with four monthly payments of $76.75 generates a finance charge of $7.00 and an APR of 11.25%.)

PRACTICE ASSIGNMENT

Plan: In this type of installment plan, an equal portion of the principal plus interest on the unpaid balance is paid each month.

$$\text{The payments on the principal} = \frac{\text{amount financed}}{\text{number of payments}}.$$

Complete the following schedule showing balances, interest, and payments. At the beginning, the amount financed is the unpaid balance. At the end of each month, compute the interest on the unpaid balance and add it to the payment. Reduce the unpaid balance by the payment on the principal. The total finance charge is the total interest, or you can use the formula to compute the total finance charge.

1. A stereo system priced at $875 is purchased with a down payment of $75 and eight monthly payments. The finance charge is 8% interest on the unpaid balance.

Monthly payment on principal = 800 ÷ 8 = 100
Monthly interest = unpaid balance × 0.08 ÷ 12
Complete the following payment schedule:

End of:	Unpaid Balance	Interest	Payment	New Balance
Month 1	$______	$______	$______	$______
Month 2	______	______	______	______
Month 3	______	______	______	______
Month 4	______	______	______	______
Month 5	______	______	______	______
Month 6	______	______	______	______
Month 7	______	______	______	______
Month 8	______	______	______	______

Finance Charge = $________

Challenger

The Whitsons purchased a set of encyclopedias priced at $1,925 seven months ago. They paid $250 down and agreed to finance the balance over 15 months with add-on interest of $7\frac{1}{2}$ percent.

a. What was the APR?

b. The Whitsons won a lottery prize of $2,500. How much money will they have on deposit if they pay off the debt (apply the Rule of 78) and invest the remainder in a certificate of deposit paying 8 percent simple interest for 300 days?

CHAPTER 15 SUMMARY

Concept	*Example*	*Procedure*	*Formula*
SECTION 15.1 Add-on Interest			
Add-on interest	An item costing $650 is purchased with $50 down and 6 payments that include 8% add-on interest. finance charge = $600 \times 0.08 \times \frac{1}{2} = 24.00$	The buyer takes immediate possession of an item, usually with a down payment (or trade-in) and a series of equal payments. The cash price less any down payment (or trade-in) is the amount financed (the principal).	$I = PRT$ P[×]R[×]T[=] P*R*T $I = PR\,n/12$ for monthly payments P[×]R[×]n[÷]12[=] P*R*n/12
Level payment plan	Refer to the preceding example. Each payment equals $\frac{600 + 24}{6} =$ 104.00		Payment $= \frac{P + I}{n}$ alg: P[+]I[=][÷]n[=] arith: P[+]I[+][÷]n[=] (P+I)/n

CHAPTER 15 SUMMARY—(*Continued*)

Concept | *Example* | *Procedure* | *Formula*

SECTION 15.2 APR and Truth-in-Lending

Annual percentage rate (APR)
Effective percentage rate

Refer again to the preceding example. The charge per $100 financed $= \frac{24}{600} \times 100 = 4.00$

In an APR table, the number closest to 4.00 for 6 monthly payments is 3.97. The APR is about 13.5%.

To find the true interest rate in an APR table, you must know the number of payments and the finance charge per $100.

$$\text{Charge per } \$100 = \frac{I}{P} \times 100$$

I[÷]P[×]100[=]

I/P*100

To estimate APR with the formula: $\frac{24 \times 24}{600 \times 7} = 0.1371428 = 13.71\%$

Estimate the APR with an approximation formula.

$$R = \frac{2mI}{P(n+1)} = \frac{24I}{P(n+1)} \text{ for monthly payments}$$

P[×]n+1[=][M+]24[×]
I[÷][MR][=]
or 24[×]I[÷][(]P[×]n+1[)][=]

24*I/(P*(n+1))

SECTION 15.3 The Rule of 78

Finance charge
Rebate
Rule of 78

Refer to the previous example. The 6-month installment plan is paid off halfway through the term.

$$\text{Rebate} = \frac{3 \times 4}{6 \times 7} \times 24 = \frac{12}{42} \times 24 = 6.86$$

Early payoff of an installment plan with interest based on the entire term entitles the borrower to a partial rebate of the finance charge.

$$\text{Rebate} = \frac{r(r+1)}{t(t+1)} \times I$$

t[×]t+1[=][M+]r[×]r+1
[÷][MR][×]I[=]
or r[×]r+1[÷][(]t[×]t+1[)]
[×]I[=]

r*(r+1)/(t*(t+1))*I

Sum-of-the-digits

$$\frac{3 \times 4}{6 \times 7} = \frac{12}{42} = \frac{6}{21}$$

The numerator of the rebate fraction is the sum of integers up to r (inclusive).

$$\frac{1+2+3}{1+2+3+4+5+6} = \frac{6}{21}$$

The denominator of the rebate fraction is the sum of integers up to t (inclusive).

Last payment $= 3 \times 104 - 6.86 = 305.14$

Deduct the rebate from the last payment.

Last payment $= r \times$ regular payment $-$ rebate

alg:
r[×]reg.payment[=][−]rebate[=]
arith:
r[×]reg.payment[=][+]rebate[−]

r*regpmt−rebate

SECTION 15.4 Interest Based on the Unpaid Balance

Interest charged on unpaid balance

Suppose 8% interest on the unpaid balance is charged in the previous example.

Interest is paid on the unpaid balance. The interest rate is the true rate.

M	Bal.	Paid on Bal.	New Bal.	Int.	Payment
1	600	100	500	4.00	104.00
2	500	100	400	3.33	103.33
3	400	100	300	2.67	102.67
4	300	100	200	2.00	102.00
5	200	100	100	1.33	101.33
6	100	100	0	0.67	100.67
			Finance Charge =	14.00	

Each monthly payment consists of an equal part of the principal plus monthly interest (unequal payments).

$$I = \text{bal} \times R \times 1/12$$

bal[×]R[÷]12[=]

bal*R/12

$$\text{Payment} = \frac{P}{n} + I$$

alg: P[÷]n[+]I[=]
arith: P[÷]n[=][+]I[+]

P/n+I

$$\text{Total charge} = \frac{7 \times 4}{2} = 14.00$$

The total finance charge, the sum of the monthly interest charges, can be found with a formula.

$$\text{Total charge} = \frac{(n+1)\,\text{first}I}{2}$$

n+1[×]firstI[÷]2[=]

(n+1)*firstI/2

Chapter 15 Spreadsheet Exercise

APR Approximations

If merchandise priced at $500 is purchased with six equal monthly payments and an add-on interest charge of 12 percent, a complete APR table shows a rate of 20.25 percent. The formula

$$\frac{24I}{P(n + 1)}$$

yields 20.57 percent.

A more accurate formula,

$$\frac{72I}{3P(n + 1) + I(n - 1)}$$

may seem formidable, but you should be able to program it. Your Chapter 15 template will appear as follows:

```
A1: PR [W11]                                                         READY

          A          B          C          D          E          F          G          H
1                          APR APPROXIMATIONS
2   ---------------------------------------------------------------------------
3      amount     number    add-on    add-on                 estimate
4     financed   of pmts     rate    interest   payments      of APR
5   ---------------------------------------------------------------------------
6       500.00         1    12.00%
7       500.00         2    12.00%
8       500.00         3    12.00%
9       500.00         4    12.00%
10      500.00         5    12.00%
11      500.00         6    12.00%
12      500.00         7    12.00%
13      500.00         8    12.00%
14      500.00         9    12.00%
15      500.00        10    12.00%
16      500.00        11    12.00%
17      500.00        12    12.00%
18
19
20
```

The template displays $500, the amount financed, in column A; the number of monthly payments (1–12) in column B; and the add-on interest rate of 12% in column C. The twelve APR computations are programmed into rows 6–17.

Spreadsheet Exercise Steps

1. Column D: Add-on Interest
   Compute the add-on interest using the simple interest formula from the amount financed, the add-on rate, and the number of payments. (The number of payments is the number of months in the term.)
   The amount financed is in A6; the add-on rate is in C6; and the number of payments is in B6.
   If a spreadsheet formula begins with a cell address, it must be preceded by +.
   The spreadsheet formula is
   + ______ * ______ * ______ / ______ .
   Complete the formula and store it in D6.

2. Column E: Payments
   Compute the size of the payments from the amount financed, the interest, and the number of payments. The amount financed is in A6; the interest will be in D6; and the number of payments is in B6.
   The spreadsheet formula is
   (______ + ______) / ______ .
   Complete the formula and store it in E6.

3. Column F: APR
When you program the formula

$$\frac{72I}{3P(n+1)+I(n-1)}$$

I stands for interest, in D6,
P stands for principal, the amount financed, in A6,
n stands for the number of payments, in B6.

The spreadsheet formula is
______ * ______ / (_____ * ______ * (______ + _____) + ______ * (______ − _____)).
Complete the formula and store it in F6.

4. Copy the formulas from D6..F6 to D7..F17.
5. Compare the results in row 8 with the APR from table 15.1.

Amount Financed	Number of Pmts.	Finance Charge	Add-on Rate	Computed APR	Charge per $100	APR from Table 15.1
$500.00	3	________	12.00%	________%	________	_____%

6. Produce the following APRs by varying the data on the spreadsheet, and compare the computed APRs with the APRs in table 15.1.
Enter the add-on rates in decimal form. For example, enter 8% as .08.

	Amount Financed	Number of Pmts.	Add-on Rate	Add-on Interest	Computed APR	Charge per $100	APR from Table
1.	$3000.00	8	8.00%	$________	________%	$________	________%
2.	700.00	2	13.00%	________	________%	________	________%
3.	1200.00	5	10.50%	________	________%	________	________%
4.	850.00	4	11.25%	________	________%	________	________%
5.	1850.00	6	9.50%	________	________%	________	________%
6.	831.50	4	9.25%	________	________%	________	________%

NAME ________________________

CLASS SECTION ____________ DATE ____________

SCORE ________ MAXIMUM 180 PERCENT ________

CHAPTER 15 ASSIGNMENT

1. An appliance store sells a food processor for $65 cash or 10% down, 10% interest, and six monthly payments.

a. What is the finance charge?

b. What are the monthly payments?

(10 points)

2. A rider lawn mower is priced at $780. On an installment plan, a 20% down payment is required, and the twelve monthly payments include a total finance charge of $53.04.

a. What are the monthly payments?

b. What is the add-on interest rate? *(10 points)*

3. A small boat may be purchased for $3,175 cash or $500 down and 10 monthly payments of $285. Use two methods to find the APR for this installment plan.
With APR table: ________
Without APR table: ________ *(10 points)*

4. Betsy Carvello repaid her 15-month installment loan seven months ahead of schedule. What portion of the interest must she pay in accordance with the Rule of 78?
Fraction in lowest terms: ________
Four-place decimal: ________ *(10 points)*

5. Lavonne Wills purchased a gold pendant priced at $495 with a down payment of $50. The balance, including a finance charge of $40.45, is to be paid in seven monthly installments. If she pays off the installment loan after three months, what is her interest rebate in accordance with the Rule of 78? *(5 points)*

6. The Browns agreed to purchase a TV priced at $260 with 15% down, a finance charge of $25, and 12 equal installments.

a. What are the payments?

b. What is the total cost on the installment loan?

c. What is the add-on interest rate?

(15 points)

7. Shear's offers $8\frac{1}{5}$% financing for its household appliances. If a dishwasher is priced at $395 with an installment plan of eight months, what are the monthly payments if there is a down payment of:

a. $50? ________

b. $75? ________ *(10 points)*

8. Mark Schneider purchased an electric organ priced at $1,750 with a down payment of $200. He arranged to pay the balance in 14 monthly installments of $131.80. If Mark paid off the loan after six months, what was his final payment in accordance with the Rule of 78? *(5 points)*

9. A home entertainment center priced at $1,299 can be purchased for 20% down and six monthly payments with 7% add-on interest. Use two methods to find the APR.
With APR table: ________
Without APR table: ________ *(10 points)*

10. A compact disc player costs $275 and it can be purchased for $35 down and 10 monthly payments of $31.60.

a. What is the total payment?

b. What is the finance charge?

c. What is the add-on interest rate? *(15 points)*

11. Clark Zimmer purchased an outboard motor with a down payment of $40 and an installment plan of nine monthly payments of $132.60 each. The motor was priced at $1,170, and the loan was paid off five months ahead of schedule. Use the Rule of 78.

a. What was the interest rebate?

b. What was the total interest paid?

c. What was the final payment? *(15 points)*

12. A young couple purchased a set of nursery furniture priced at $1,000 with a down payment of 10% and an eight-month installment plan. The interest is $7\frac{1}{2}$% of the unpaid balance. *(48 points)*

a. Complete the following schedule:

End of:	Unpaid Balance	Interest	Payment	New Balance
Month 1	$________	$________	$________	$________
Month 2	________	________	________	________
Month 3	________	________	________	________
Month 4	________	________	________	________
Month 5	________	________	________	________
Month 6	________	________	________	________
Month 7	________	________	________	________
Month 8	________	________	________	________

b. What is the unpaid balance after the third month? *(2 points)*

c. How much interest is charged for the third month? *(2 points)*

d. How much of the principal is unpaid after the fifth month? *(2 points)*

e. How much is the payment at the end of the sixth month? *(2 points)*

f. How much interest is charged for the first half of the installment loan? *(2 points)*

g. Add up all of the finance charges. *(2 points)*

h. Compute the total finance charge using the formula, and compare the result with the charge obtained in part *g* above. Show your calculations. *(5 points)*

PART

4 CONSUMER APPLICATIONS

Chapter 16 Revolving Credit

OBJECTIVES

- Understand charge accounts, credit cards, and financial packages
- Compute an average daily balance with or without a running balance

SECTIONS

INTRODUCTION

Revolving credit, an extension of the "buy now—pay later" plan, is offered to customers by many financial institutions and retailers. Revolving credit accounts allow customers to purchase goods and services month after month, year after year, without ever having to reach a zero balance, as long as they make the minimum monthly payments and do not exceed the credit limit set by the account issuer. Thus the account is said to "revolve."

Section 16.1 Charge Accounts

KEY TERMS **Open-end credit** **Minimum monthly payment** **Credit limit** **Revolving credit** **Monthly interest rate**

How Does Open-end or Revolving Credit Differ from Installment Credit?

Installment credit requires a fixed number of payments in a set period of time to clear an account balance. **Open-end credit** allows purchases and finance charges to accrue indefinitely on accounts that need never reach zero balance as long as the **minimum monthly payment** is made and the balance is within the **credit limit.** Open-end or **revolving credit** is offered by many major department stores and chains; some small retailers still offer installment contracts.

Example 1 Jan used to make installment purchases, but now she uses a department store charge account to buy on credit up to a limit of $2,000.

Her minimum monthly payment may be as small as $20, but she may make large payments if she desires.

The store charges a **monthly interest rate** of 1%, equivalent to an annual interest rate of 12%. Jan may pay off her balance at any time but prefers to pay $100 each month, unless she is short of cash. The interest is deducted from the payment first; the remainder of the payment reduces the balance.

Month	Previous Month's Balance	Interest	Payments and Returns	Balance Reduction	Purchases	New Balance
Apr	844.17	8.44	100.00	91.56	0.00	752.61
May	752.61	7.53	20.00	12.47	50.72	790.86
June	790.86	7.91	100.00	92.09	300.00	998.77
July	998.77	9.99	100.00	90.01	12.50	921.26
Aug	921.26	9.21	100.00	90.79	6.75	837.22
Sept	837.22	8.37				

PRACTICE ASSIGNMENT

Plan: The record of a charge account shows balances, interest, purchases, and payments or returns. At the end of each month, the finance charge–the interest on the previous month's ending balance–is computed. Instead of applying the monthly interest = balance × annual rate × $\frac{1}{12}$ formula, it is more convenient here and just as correct to compute:

monthly interest = balance × monthly rate.

Payments and returns minus monthly interest reduce the balance. Purchases are added to the balance:

Balance reduction = payments + returns − interest
New balance = old balance − balance reduction + purchases

If you are only interested in finding the new balance compute:

New balance = old balance + interest − payment − returns + purchases.

1. Complete the following charge account record. The monthly interest rate is 1.25%, equivalent to an annual interest rate of 15%. The customer alternates payment of the monthly minimum of \$25 with a payment of \$125. There were returns of \$75.50 in August and \$50 in October.

Month	Previous Month's Balance	Interest	Payments and Returns	Balance Reduction	Purchases	New Balance
June	1,765.40	______	125.00	______	132.55	______
July	______	______	25.00	______	0.00	______
Aug	______	______	125.00		216.99	
			75.50	______		______
Sept	______	______	25.00	______	23.51	______
Oct	______	______	125.00		95.90	
			50.00	______		______
Nov	______	______	25.00	______	12.50	______

What is the total finance charge for the 6-month period?

2. A monthly charge account record shows a previous balance of \$1,045.75 and no additional purchases. Monthly interest of 1% is charged. If \$225 is remitted, how much of this payment will be used to reduce the balance?

3. A customer of Hargrove's Department Store received a monthly charge account statement recording a previous balance of \$734.29 and additional purchases totaling \$87.50. If monthly interest is $1\frac{1}{2}$% and the customer remits \$130, by how much will the balance be reduced?

4. A monthly charge account statement lists a previous balance of \$519.49, purchases of \$97.60, and returns of \$35.50. The monthly interest rate is 1.25%. What is the new balance after a payment of \$150?

Section 16.2 Credit Cards

KEY TERMS **Annual fee** **Average daily balance**

The credit card represents a highly convenient form of revolving credit. National retail chains issue cards that may be used in any of their outlets. Credit companies issue cards that are honored by businesses nationally and internationally. Some financial institutions' plans allow consumers to make purchases without cash and then pay the balance within 30 days. Certain cards also entitle the holder to a package of financial benefits. Many people who travel use MasterCard or Visa instead of cash. While traveling in the United States and many other countries worldwide, they may charge most expenses including transportation, hotel/motel accommodations, and meals at restaurants. A cardholder may also withdraw cash at automated teller machines or secure cash advances at participating banks. Credit cards must be used with caution, however. People who spend "plastic money" too easily may find themselves deeply in debt.

A credit card transaction is recorded on a charge receipt, which the customer signs. The customer keeps one copy, and the merchant sends a copy to the credit card company. The credit card company pays the merchant after assessing a small percentage on the purchase price, and then bills the customer.

CUSTOMER COPY

2556874

1234 5678 9012 3456

3144 06/9X

George Adams

12 345 678

Centre Clinic Pharmacy

Centre City, KY

QUAN.	CLASS	DESCRIPTION	PRICE	AMOUNT
1	RX	398371		66 46

DATE 1/5/9X AUTHORIZATION 790573

REFERENCE NO. CLERK/DEPT.

SUB TOTAL

TAX

SALES SLIP

TOTAL 66 46

PURCHASER SIGN HERE

X George Adams

Cardholder acknowledges receipt of goods and/or services in the amount of the Total shown hereon and agrees to perform the obligations set forth in the Cardholder's agreement with the issuer.

IMPORTANT: RETAIN THIS COPY FOR YOUR RECORDS

Although the annual interest rate for the finance charge on most credit cards is high, most cardholders consider the convenience worth the cost. Some credit card companies charge the cardholder a small **annual fee** in addition to the monthly finance charge.

Historically, there has been more than one way to determine the balance of a credit card account. One way is to calculate the average of the daily ending balances.

The previous month's interest does not increase the average balance because that interest is not yet due. As with installment accounts, interest is deducted from a payment before the balance is reduced.

The fundamental formula for computing the **average daily balance** is:

$$\frac{\text{sum of (balance} \times \text{outstanding days)}}{\text{number of days in billing cycle}}$$

Because credit card companies do not show a running balance on monthly statements, they prefer to use a formula that computes the average daily balance without reference to running balances. To do so, one calculation is performed for balances and another for transactions.

$$\text{Average daily balance} = \text{balance average} + \text{transaction average}$$

$$\text{Balance average} = \frac{\text{sum of (balance} \times \text{number of outstanding days)}}{\text{number of days in billing cycle}}$$

Only two balances enter the numerator: the initial balance (after previous interest has been deducted) and the balance after payment (initial balance − payment on balance).

$$\text{Transaction average} = \frac{\text{sum of (amount} \times \text{remaining days)}}{\text{number of days in billing cycle}}$$

The balance-transaction method makes it possible for credit card companies to charge higher interest on cash advances. The cash average can then be computed separately in a similar manner. Example 1 shows a comparison of the two methods.

Example 1

Opening date: 10/21
Closing date: 11/21
Number of days in billing cycle: 31
Monthly interest rate: 1.3%

1. Running balance method:

Posting Date		Transaction	Amount	Running Balance	Outstanding Days	Balance × Days
10/21		Opening balance	946.38	946.38		
	−	Previous interest	14.23	932.15	3	2796.45
10/24		REVCO 4401	15.72	947.87	6	5687.22
10/30		True Value 523	38.49	986.36	3	2959.08
11/02		Payment	150.00			
	−	Previous interest	14.23			
			135.77	850.59	10	8505.90
11/12		Martin's	74.25	924.84	6	5549.04
11/18		MCI 4B609811	21.36	946.20	3	2838.60
						28336.29

Average daily balance = 28336.29 ÷ 31 = 914.07
Interest = 0.013 × 914.07 = 11.88

key:	932.15 [×] 3 [=] [M+]	
	947.87 [×] 6 [=] [M+]	
	986.36 [×] 3 [=] [M+]	
	850.59 [×] 10 [=] [M+]	
	924.84 [×] 6 [=] [M+]	
	946.20 [×] 3 [=] [M+]	**sum of products**
	[MR] [÷] 31 [=]	**average daily balance**
	[×] .013 [=]	**interest**
display:	11.88296	
edited:	$11.88	

2. Balance-transaction method:

Balance average:

10/21–11/02 = 12	932.15 × 12 = 11185.80	
11/02–11/21 = 19	796.38 × 19 = 15131.22	932.15 − 135.77 = 796.38
	26317.02 ÷ 31 = 848.94	

Transaction average:

10/24–11/21 = 28	15.72 × 28 =	440.16
10/30–11/21 = 22	38.49 × 22 =	846.78
11/12–11/21 = 9	74.25 × 9 =	668.25
11/18–11/21 = 3	21.36 × 3 =	64.08
		2019.27 ÷ 31 = 65.14
Average daily balance:		914.08

(The difference of $0.01 is due to rounding.)

PRACTICE ASSIGNMENT

Plan: When you use the running balance method to compute the average daily balance remember the following:

Number of outstanding days for initial balance = date of first transaction − previous statement date
Number of outstanding days for last balance = statement date − transaction date
Number of outstanding days for other balances = transaction date − previous transaction date
Multiply each running balance by the outstanding days.
Add these products, and divide the sum by the number of days in the billing cycle to obtain the average daily balance.

When you use the balance-transaction method remember the following:

The two balances in the balance average are the initial balance and the difference between the initial balance and the applied payment (after first deducting interest). Multiply these two balances by the outstanding days. Divide the sum of the two products by the number of days in the billing cycle.

The number of remaining days for a transaction is the number of days from the transaction to the end of the cycle. Multiply each transaction by the remaining days. Add these products, and divide the sum by the number of days in the billing cycle.

Add this transaction average to the previously obtained balance average and obtain the average daily balance.

1. Compute the average daily balance with the running balance method.

Last statement date: June 16 Statement date: July 16 Number of days in billing cycle: 30

Date		Transaction	Amount	Balance	Bal.	×	Days	=	
June 16		Opening balance	829.22	829.22					
	−	Prev. interest	10.09	819.13	______	×	______	=	______
June 21		S & J Odds 'n Ends	59.65	______	______	×	______	=	______
June 29		Payment	150.00						
	−	Prev. interest	10.09						
			139.91	______	______	×	______	=	______
July 5		Thrift Stores	38.49	______	______	×	______	=	______
July 10		Rick's Sports	74.25	______	______	×	______	=	______
July 14		Estelle's Salon	21.36	______	______	×	______	=	______

Average daily balance = ______ [÷] ______ [=] ______

2. Recompute the average daily balance with the balance-transaction method.

Date			Amount	Balance
June 16		Opening balance	829.22	829.22
	−	Prev. interest	10.09	819.13
June 21		S & J Odds 'n Ends	59.65	
June 29		Payment	150.00	
	−	Prev. interest	10.09	
			139.91	______
July 5		Thrift Stores	38.49	
July 10		Rick's Sports	74.25	
July 14		Estelle's Salon	21.36	

Balance average:
6/16–6/29 ______ × ______ = ______
6/29–7/16 ______ × ______ = ______
______ ÷ ______ = ______

Transaction average:
6/21–7/16 ______ × ______ = ______
7/05–7/16 ______ × ______ = ______
7/10–7/16 ______ × ______ = ______
7/14–7/16 ______ × ______ = ______
______ ÷ ______ = ______

Average daily balance: ______

Section 16.3 Financial Packages

KEY TERMS **Cash-back bonus** **Travel bonus**

What Is a Financial Package?

The financial services industry has become highly competitive, and in an effort to lure customers, many institutions offer attractive benefit packages that accompany their credit cards. Cards that carry these benefit packages, however, may carry a higher interest rate.

A financial benefit package plan may include the following:

No annual fee (except in states where APR is limited)
Cash-back bonuses
Grace period on interest
Savings accounts and certificates of deposit at high rates
Discount coupons offered by participating merchants
5% travel discount
Travel insurance
Car rental liability insurance

The **cash-back bonus** is a much-touted selling point of some packages. The bonus is calculated annually on the amount of purchases and is either sent to the cardholder in the form of a check or credited to the account.

Those who will benefit most from a package plan are people who will use most or all of the benefits offered. The additional cost of a benefit package that contains a sizable **travel bonus** may not be worthwhile for a person who does not travel much.*

$\frac{1}{4}$% on first $1,000
$\frac{1}{2}$% on second $1,000
$\frac{3}{4}$% on third $1,000
1% on purchases over $3,000
Example: purchases average $200 per month = $2,400 per year.

$\frac{1}{4}$% of 1000 = 0.0025 × 1000 = 2.50
+ $\frac{1}{2}$% of 1000 = 0.005 × 1000 = 5.00
+ $\frac{3}{4}$% of 400 = 0.0075 × 1000 = 3.00
Cash-back bonus for year = 10.50

Example 1 Each year a consumer charges travel expenses of $2,000 and other purchases of $2,000. The average daily balance is $1,000. The consumer has two credit cards. Which is less expensive?

Credit card 1

Monthly Rate	Annual Fee	Annual Cost
1.5%	20	?

1000 × 0.015 × 12 + 20 = 200

Credit card 2 (package plan)

Monthly Rate	Annual Fee	Cash Back	Travel Discount	Annual Cost
1.65%	0	25.00	100.00	?

Cash-back bonus:

$\frac{1}{4}$% of first $1000 = 2.50
$\frac{1}{2}$% of next $1000 = 5.00
$\frac{3}{4}$% of next $1000 = 7.50
1% of next $1000 = 10.00
25.00

Travel discount = 5% of 2000 = 100
Interest − cash back − travel bonus
= 1000 × 0.0165 × 12 − 25 − 100 = 73

key alg: 16.5 [×] 12 [−] 125 [=]
key arith: 16.5 [×] 12 [=] [+] 125 [−]
display: 73.
edited: $73.00

Credit card 2 (package plan) is less expensive.

*Although not always the case, assume that charged travel expenses count toward the cash-back bonus.

PRACTICE ASSIGNMENT

Plan: Compute package plan net costs deducting the travel discount and cash-back bonus as listed in example 1. Assume a monthly interest rate of 1.65%.

1. Luke Delaney charged $3,278.37 to his package plan credit card last year. What was his cash-back bonus?

2. In 1991 the Hoffmans charged $2,978.67 to their package plan credit card. Their average daily balance was $541.28. What was their net credit card expense for the year?
Interest:
Cash-back bonus:
Net expense:

3. Anne-Marie Reynard charged $2,174.83 to her credit card last year. Of this amount, she spent $744.80 on travel. Her average daily balance was $423.68. How much did she save on the travel discount and cash-back bonus? What was her net credit card expense for the year?
Interest:
Travel discount:
Cash-back bonus:
Net expense:

Challenger

A cardholder charges approximately $350 each month and has an average balance of $1,000 per month.

a. Compute an effective annual interest rate for various credit cards, prorating the annual fee over the whole year.

Card	Monthly Rate	Annual Fee	Effective Annual Rate
1	1.65%	0	______%
2	1.50%	20	______%
3	1.50%	25	______%

b. Assume card 1 offers the cash-back feature shown in example 1 of section 16.3 and recalculate the effective rates.

Card	Monthly Rate	Annual Fee	Cash Back	Effective Annual Rate
1	1.65%	0	________	______%
2	1.50%	20	0	______%
3	1.50%	25	0	______%

CHAPTER 16 SUMMARY

Concept | *Example* | *Procedure* | *Formula*

SECTION 16.1 Charge Accounts

Concept: Open-end credit; Revolving credit; Charge account

Example: Monthly interest rate is 1%.

M	Bal.	Int.	Pmt.	Purch.
1	634.56	6.35	100	16.85
2	557.76	5.58	100	0.00
3	463.34	4.63	100	76.80
4	444.77	4.45		

Procedure: Interest, payments, and purchases continue indefinitely. The balance need not be paid off, but must remain within credit limit.

Formula:

$I = \text{bal.} \times R \times 1/12$

bal[×]*R*[÷]12[=]

bal*R/12

$I = \text{bal.} \times \text{monthly rate}$

bal[×]monthrate[=]

bal*monthrate

SECTION 16.2 Credit Cards

Concept: Credit card

Example: Monthly interest rate is 1.5%.
Previous month's interest: $10.16
Date of statement: 4/21

Concept: Average daily balance

Example:

Date	Trans.	Amount	Balance
3/21			506.25
3/25	purchase 1	17.89	524.14
4/1	payment	75.00	
	−interest	10.16	459.30
4/6	purchase 2	104.50	563.80
4/10	purchase 3	25.43	589.23

Procedure: Calculate the average daily balance by multiplying each balance by the number of days outstanding.

Divide the sum of these products by the number of days in the billing cycle.

Formula:

$$\text{Average daily balance} = \frac{\text{sum of (bal.} \times \text{days)}}{\text{days in cycle}}$$

bal1[×]days1[=][M+]
bal2[×]days2[=][M+]
·
·
[MR][÷]cdays[=]

SUM(bal*days)/cdays

Concept: Running balance method

Example:

Average daily balance =
506.25 × 4 = 2025.00
524.14 × 7 = 3668.98
459.30 × 5 = 2296.50
563.80 × 4 = 2255.20
589.23 × 11 = 6481.53
16727.21 ÷ 31
= 539.59
Interest = 539.59 × 0.015 = 8.09

Procedure: The finance charge is average daily balance × monthly rate.

Formula: Finance charge = avg. daily bal. × monthly rate

Concept: Balance-transaction method

Example:

Average daily balance = balance avg. + transaction avg.

Balance average:
506.25 × 11 = 5568.75
441.41 × 20 = 8828.20
14396.95 ÷ 31 = 464.42

Transaction average:
17.89 × 27 = 483.03
104.50 × 15 = 1567.50
25.43 × 11 = 279.73
2330.26 ÷ 31 = 75.17

464.42 + 75.17 = 539.59

Procedure: You can calculate the average daily balance (without running balances) as balance average + transaction average. The balance average is the daily average of the initial balance and the balance after the payment. Compute the transaction average by multiplying each transaction amount by the number of days remaining in the cycle and dividing the sum of the products by the days in the cycle.

The finance charge is average daily balance × monthly rate.

Formula:

Balance average = balance 1 + balance 2

$$\text{Transaction average} = \frac{\text{sum of (trans.} \times \text{remaining days)}}{\text{days in cycle}}$$

trans1[×]remdays1[=][M+]
trans2[×]remdays2[=][M+]
·
·
[MR][÷]cdays[=]

SUM(trans*remdays)/cdays

Average daily balance = balance average + transaction average

Finance charge = average daily balance × monthly rate

SECTION 16.3 Financial Packages

Concept: Financial packages

Example:

Cash-back bonus:
$\frac{1}{4}$% on 1st $1,000 of purchases
$\frac{1}{2}$% on 2d $1,000 of purchases
$\frac{3}{4}$% on 3d $1,000 of purchases
1% on purchases above $3,000

Annual purchase = $2,000
Bonus = 2.50 + 5.00 = $7.50

Procedure: The credit cards of some institutions have additional benefits and services, such as cash-back bonuses, travel bonuses, discounts, and savings accounts.

Chapter 16 **Spreadsheet Exercise**

Two Methods of Computing Average Daily Balance

Essentially, average daily balance equals

$$\frac{\text{sum of (balance} \times \text{outstanding days)}}{\text{number of days in billing cycle}}.$$

With a second method, you can calculate the average daily balance as the sum of the balance average and the transaction average.

The balance average depends on the outstanding days of the balances, while the transaction average depends on the days remaining in the billing cycle after each transaction has been posted.

The running balance is not required for the second method, and the two computations appear to be quite different. This spreadsheet shows how the two methods surprisingly yield the same result.

The chapter 16 template appears as follows:

A1: PR [W13] READY

	A	B	C	D	E	F	G	H	I
1	TWO METHODS OF COMPUTING THE AVERAGE DAILY BALANCE								
2	----------								
3			running balance method				balance - trans. method		
4	----------								
5			trans.	running	outst.	bal.	balance	rem.	trans.
6	transaction	date	amount	balance	days	x days	products	days	products
7	----------								
8	opening bal.	3/18		521.14	2				
9	Valley Bout.	3/20	84.77	605.91	3			28	
10	Drake Hardw.	3/23	36.18	642.09	4			25	
11	Vic's Body Sh	3/27	491.69	1133.78	4			21	
12	Hi-Tone Styl.	3/31	17.00	1150.78	3			17	
13	payment	4/03	325.00	825.78	3				
14	Handy Shop	4/06	138.75	964.53	3			11	
15	L&L Furniture	4/09	318.04	1282.57	3			8	
16	Sim's Dept.St	4/12	73.82	1356.39	5			5	
17	closing bal.	4/17		1356.39					
18	----------								
19	sums of products								
20	average daily balance								

Column A shows credit card transactions, column B transaction dates, and column C transaction amounts in rows 8–17.

Running Balance Method

Column D: Running Balances

These have been computed and are displayed.

Column E: Outstanding Days

The number of days during which each balance is maintained is the difference between the next transaction date and the current one.

The outstanding days have been computed and are displayed.

Spreadsheet Exercise Steps

1. Column F: Balance × Days
   The balance is in D8 and the number of days in E8. The spreadsheet formula begins with a cell address and must be preceded by +.
   The formula is + _______ * _______ .
   Complete the formula, store it in F8, and copy it from F8 to F9..F16.

2. Cell F19: Sum of (Balance × Days)
   The balance × days products are in F8..F16.
   Complete
   @SUM(______ .. ______) and store it in F19.
3. Cell F20: Average Daily Balance

   $$\text{Average daily balance} = \frac{\text{sum of (balance} \times \text{days)}}{30}$$

   The sum is in F19. The spreadsheet formula must be preceded by +.
   Complete + ______ / ______ and store it in F20.

Balance-Transaction Method

4. Column G: Balance Products
   The opening balance is in D8 and is in effect for 16 days (from the first day to the day of the payment).
   The spreadsheet formula for 16 × opening balance is
   ______ * ______ .
   Complete the formula and store it in G8.
   For the remaining 14 days (30 − 16 = 14), the balance is reduced by the payment. The payment is in C13.
   The spreadsheet formula is
   ______ * (______ − ______).
   Complete the formula and store it in G13.

Column H: Remaining Days after a Transaction

The number of remaining days is the difference between the closing date and the transaction date.

The remaining days have been computed and are displayed.

5. Column I: Transaction Amount × Remaining Days
   The transaction amount is in C9 and the number of days in H9.
   The spreadsheet formula begins with a cell address and must be preceded by +.
   The formula is + ______ * ______ .
   Complete the formula, store it in I9, and copy it from I9 to I10..I16.
6. Cell G19: Sum of Balance Products
   The first balance product is in G8 and the second in G13.
   The spreadsheet formula begins with a cell address and must be preceded by +.
   The formula is + ______ + ______ . Complete the formula and store it in G19.
7. Cell I19: Sum of (Transaction Amount × Days)
   The transaction × days products are in I9..I16.
   Complete
   @ SUM(______ .. ______) and store it in I19.
8. Cell I20: Average Daily Balance

   $$\text{Average daily balance} = \frac{\text{sum of balance products} + \text{sum of transaction products}}{30}$$

   The sum of the balance products is in G19, and the sum of the transaction products is in I19.
   The formula is (______ + ______)/ ______.
   Complete the formula and store it in I20.
9. Do the two methods of computing the average daily balance produce the same result? If not, your program contains an error. If yes, average daily balance = ________
10. What is the interest charge for the month if the monthly rate is 1.5%? ________
11. What would the interest charge have been at 1.5% if the payment had been $500 rather than $325? ________

NAME ____________________

CLASS SECTION ____________ DATE ____________

SCORE ________ MAXIMUM 150 PERCENT ________

Chapter 16 Assignment

1. Compute the monthly interest for the following average daily balances if the monthly interest rate is 1.5%.

	Average Daily Balance	Finance Charge
a.	506.42	________
b.	1,213.04	________
c.	882.61	________
d.	478.26	________
e.	783.14	________

(5 points)

2. Complete the following charge account record. The monthly interest rate is 1.45%

Month	Previous Month's Balance	Interest	Payments and Returns	Balance Reduction	Purchases	New Balance
Feb	1,534.85	________	120.00	________	275.79	________
Mar	________	________	150.00	________	97.50	________
Apr	________	________	100.00	________	103.84	________
May	________	________	75.00	________	58.17	________
June	________	________	110.00	________	361.35	________
July	________	________	200.00	________	89.75	________
Aug	________	________	175.00	________	0.00	________
Sept	________	________	180.00	________	124.65	________

What is the total finance charge for the period? $203.77 *(32 points)*

3. Compute the interest for an average daily balance of $803.14.

	Monthly Interest Rate	Monthly Interest
a.	1.00%	________
b.	1.25%	________
c.	1.50%	________
d.	$1\frac{7}{12}\%$	________
e.	1.65%	________

(5 points)

4. During the month of September, Peter Schmidt charged $129.88 in purchases from his favorite clothing store. His charge account statement at the end of the month showed a previous balance of $375.45. If the store's monthly finance charge is 1% and Peter remits $75, what is the new balance? *(5 points)*

5. Morris-Chandler's charge account policy requires a minimum monthly payment of $20 or five percent of the old balance, whichever is larger. The monthly interest rate is 1.2%. If a customer receives a statement recording a previous balance of $323.50 and purchases of $75.05, what will the new balance be assuming the customer makes only the minimum payment? *(5 points)*

6. Complete the following credit card computation using the running balance method.

Last statement date: March 12

Statement date: April 12

Number of days in billing cycle: 31

Monthly interest rate: 1.5%

Date		Transaction	Amount	Balance	Bal.	×	Days	=	
		Opening balance		581.60					
Mar 12	–	Prev. interest	8.68	______	______	×	____	=	______
Mar 14		Step-Right Shoes	48.79	______	______	×	____	=	______
Mar 20		Lee's Nursery	30.14	______	______	×	____	=	______
Mar 23		Sharon's Shoppe	62.25	______	______	×	____	=	______
Apr 1		Payment 120.00							
	–	Interest 8.68	______	______	______	×	____	=	______
Apr 3		Power Gas	10.00	______	______	×	____	=	______
Apr 4		RoadRest	51.21	______	______	×	____	=	______
Apr 7		Candlelight Inn	38.52	______	______	×	____	=	______
									______

Average daily balance:

__________ [÷] ______ [=] __________

Interest: __________ [×] __________ [=] ________

(43 points)

7. Tanya Michaels charges merchandise from a store that requires $1\frac{1}{4}\%$ monthly interest. Last month's statement recorded a previous balance of $628.67 and purchases totaling $122.89. If Tanya wants to reduce her balance to $500, how much must she pay? *(10 points)*

For problems 8 through 10, refer to the following package plan.

Monthly Rate	Cash-back Bonus	Travel Discount
1.65%	$\frac{1}{4}\%$ on first $1,000	5%
	$\frac{1}{2}\%$ on second $1,000	
	$\frac{3}{4}\%$ on third $1,000	
	1% on purchases over $3,000	

8. A family used their credit card to finance airplane tickets costing $1,850. They charged an additional $2,436.65 to the card that year. How much did they save by using the travel discount and cash-back bonuses?

Travel discount: __________

Cash-back bonus: __________

Total saved: __________ *(15 points)*

9. The Sardis family is choosing a credit card. Card 1 charges 1.5% monthly interest and no annual fee, while card 2 offers the package plan. If this family annually charges purchases of $2,050 with no travel expenses, and has an average daily balance of $650, which card will be most economical for them?

Annual cost of card 1: __________

Annual cost of card 2: __________

Better choice: Card 1 (regular): _____

Card 2 (package): _____ (check one) *(15 points)*

10. Lisa and Richard Swensen are deciding which credit card to apply for. Card 1 is a regular credit card with an annual fee of $15 and monthly interest of 1.5%. Card 2 is the package plan. The Swensens charge average annual purchases of $2,200. They plan to charge an additional $1,000 in travel expenses. Their average daily balance is expected to be $400. Which card should they apply for?

Annual cost of card 1: __________

Annual cost of card 2: __________

Better choice: Card 1 (regular): _____

Card 2 (package): _____ (check one) *(15 points)*

Chapter 17 Checking Accounts

OBJECTIVES

- Determine the cost of checking
- Endorse checks
- Record checking account transactions
- Perform a checkbook reconciliation

SECTIONS

INTRODUCTION

Before the Federal government deregulated banking in the 1980s, bank services were fairly straightforward: savings accounts earned interest and checking accounts did not. Today distinctions may no longer be obvious with the availability of savings accounts that permit checking and checking accounts that earn interest.

Section 17.1 Checking; Automated Teller Machines

KEY TERMS **Service charge** **Transaction charge** **Minimum balance**

While notes are promises to pay, drafts are orders to pay. A person or organization issues a draft as an order for another person or organization to render a payment to a third party. The most commonly used draft is the check. Checking accounts provide a convenient way to pay bills. Money deposited into a checking account is paid out by the bank in the amount and to the party specified on the check. The recipient of a check may cash it, deposit it, or sign it over to another party.

In addition to personal and business checking accounts, banks offer other checking services such as cashier's checks, traveler's checks, and automated teller machines (ATMs).

A bank issues a cashier's check at the request of an account holder in cases where the amount of a check must be assured. The issuing bank guarantees the payee that the cashier's check is backed by sufficient funds.

Many banks issue American Express or Bank of America traveler's checks. Traveler's checks may be purchased in several denominations for a nominal fee and are considered as good as cash by most merchants worldwide. If traveler's checks are lost or stolen, the issuing company replaces them.

Automated teller machines are springing up in every conceivable location, from bank entrances to grocery stores, shopping malls, airports, and college campuses. National ATM networks form on-line computer systems that make bank transactions possible at any time of the day or night, on any day of the week, in practically any city in the world. It is as easy as inserting a plastic card and pressing a few keys. ATM transactions include cash withdrawals from savings or checking accounts, transfer of money between accounts, or deposits into accounts. Table 17.1 shows procedures for typical ATM transactions. Since ATMs are not standardized, procedures may vary.

Table 17.1 ATM Procedures

SELECT TRANSACTION	withdrawal deposit transfer other	[punch]
SELECT TYPE OF WITHDRAWAL	savings checking credit card	[punch]
ENTER AMOUNT: DENOMINATIONS OF TEN DOLLARS		[100]
PRESS IF CORRECT PRESS IF INCORRECT		[punch]
TRANSACTION COMPLETED PLEASE TAKE YOUR CASH		

SELECT TRANSACTION	withdrawal deposit transfer other	[punch]
SELECT TYPE OF DEPOSIT	savings checking	[punch]
ENTER AMOUNT:	$.	[150.00]
PRESS IF CORRECT PRESS IF INCORRECT		[punch]
PLACE DEPOSIT ENVELOPE INTO SLOT		
TRANSACTION BEING PROCESSED		
TRANSACTION COMPLETED		
THANK YOU!		

Electronic fund transfers of many kinds promise to become widely available; however, checks are still the most common way to pay bills and settle accounts.

Checking services vary from bank to bank, and many banks offer several choices. Table 17.2 outlines several common plans on the left and illustrates one bank's options on the right.

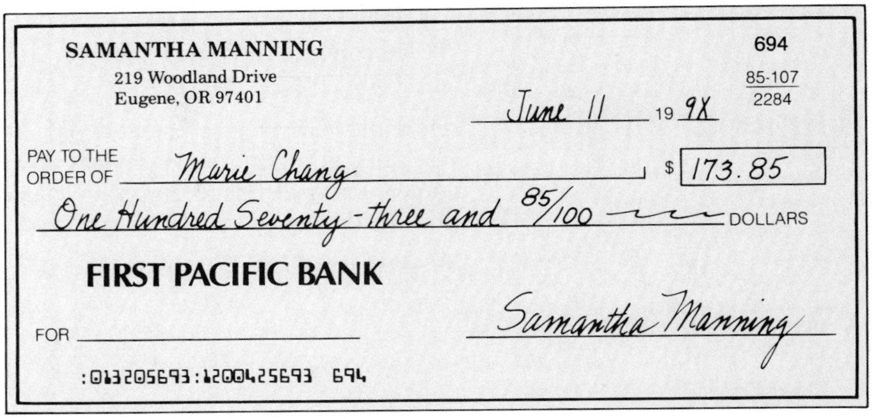

SAMANTHA MANNING 694
219 Woodland Drive
Eugene, OR 97401
85-107/2284
June 11 19 9X
PAY TO THE ORDER OF Marie Chang $ 173.85
One Hundred Seventy-three and 85/100 DOLLARS
FIRST PACIFIC BANK
FOR
Samantha Manning
:013205693:1200425693 694

Table 17.2 Checking Account Services

Common Checking Plans

Free Checking

No **service charge**
No **transaction charge**
Unlimited checking

Transaction Plan

Charge per transaction (the higher the balance, the lower the monthly charge)
Example:
$0.10 per check
$0.05 per deposit
credit of $0.15 for $100 average balance maintained

Activity Plan

Flat monthly fee and a charge per check
Example:
$2.00 per month and $0.10 per check

Minimum Balance Plan

The higher the maintained balance, the lower the service charge; no charge if **minimum balance** or above is maintained.

United Bank

Regular Checking Accounts
(personal or business)

Minimum Daily Balance	**Monthly Service Charge**
300.00 and higher	none
200.00–299.99	$3.00
100.00–199.99	$4.00
0.00– 99.99	$5.00

First Flight Club

Monthly service charge: $5.00
No minimum balance
Personalized checks
Free traveler's and cashier's checks
$10,000 accidental death insurance

Collegiate Club (for students only)

Monthly service charge: $2.00
No minimum balance
10 checks per month free
$0.25 for each check over 10

Cornerstone Club

5% interest paid, (rate may vary)
No service charge if minimum balance is at least $2,500
Monthly charge of $6 if balance falls below $2,500
Free traveler's and cashier's checks

Banks may also offer:

Free safe-deposit box
Lower rates on loans
Discounts at certain local businesses
Overdraft protection (If a check is written that drops the balance below zero, the bank will honor the check, usually for a small fee.)

Example 1 Comparison of Checking Account Costs
Jennifer Stone has a checking account at Citizens Bank and routinely writes 20 checks per month, deposits money twice each month, and maintains an average daily balance of \$450 with a minimum daily balance of \$400. Under a transaction plan, the bank charges \$0.12 for each check and \$0.06 for each deposit, and pays a credit of \$0.20 for each \$100 of average daily balance. What would the monthly charge be under this plan?

20 × 0.12 =	2.40	Checking charge
2 × 0.06 =	0.12	Deposit charge
	2.52	Transaction charge
4 × 0.20 =	0.80	Balance credit
	1.72	Monthly charge

Under an activity plan with a monthly service charge of \$2.00 and an activity charge of \$0.10 per check, the monthly charge would be 2.00 + 20 × 0.10 = 4.00.

Had she opted for the minimum balance plan at United Bank (table 17.2), Jennifer would have no monthly charge.

With the First Flight Club at United Bank, her service charge would be \$5.00.

If Jennifer were eligible for the Collegiate Club, her charge would be:

Service charge		= 2.00
Charge for checks above 10	= 10 × 0.25	= 2.50
Total monthly charge		4.50

Under the Cornerstone Club, Jennifer would pay a service charge of \$6, but she would also earn interest of five percent on the average daily balance.

Interest earned = 450 × 0.05 ÷ 12 = 1.88
Monthly charge = 6.00 − 1.88 = 4.12

PRACTICE ASSIGNMENT

Plan: Determine the estimated monthly charges for the checking plans described. Refer to table 17.2.

1. Toni Wisnewski has a regular checking account at United Bank under their minimum daily balance plan. What was her service charge last month if she maintained a balance of \$235?

2. A student who has the Collegiate Club checking account wrote 18 checks during October. What was the monthly charge?

3. John Leone has a Cornerstone Club checking account. During one month, he wrote 52 checks and maintained an average daily balance of \$900. The next month he wrote 41 checks and maintained an average daily balance of \$1,500. What were his monthly charges?
First month: ________
Second month: ______________

4. Tom Lovelace is opening a checking account at United Bank and must choose between regular checking, First Flight Club, or Cornerstone Club. He does not want to worry about minimum balances as he often lets the balance drop below \$100 in the last week of the month. His average daily balance, however, is about \$500 because his balance is quite high in the first half of the month. Which plan would be his best choice? (Disregard the savings from free traveler's and cashier's checks.)

Account	Monthly Charge
Regular	________
First Flight Club	________
Cornerstone Club	________
Least expensive plan	______________

Section 17.2 Endorsements

The recipient of a check may cash it, deposit it, or remit it to someone else who may cash it or deposit it. When a check is cashed, deposited, or transferred, it must be endorsed on the back. The endorsement must be contained in a specific space designated by the Federal Reserve Bank. Banks may refuse to honor a check if an endorsement extends beyond the designated space.

Marie Chang received a check in the amount of $214.85. She can endorse it with a blank, full, or restrictive endorsement.

Endorsements

Blank endorsement allows any bearer to use the check.

signature

Full endorsement names the person to whom the check is transferred.

pay to the order of

signature

Restrictive endorsements restrict the check to a single use.

Examples:

for deposit only signature

for deposit only into the account of

signature

pay to only

signature

ENDORSE HERE

x Marie Chang

DO NOT WRITE, STAMP OR SIGN BELOW THIS LINE
RESERVED FOR FINANCIAL INSTITUTION USE*

ENDORSE HERE

x Pay to the order of

James N. McGrath

Marie Chang

DO NOT WRITE, STAMP OR SIGN BELOW THIS LINE
RESERVED FOR FINANCIAL INSTITUTION USE*

ENDORSE HERE

x Pay to First Federal

Bank of Sacramento

only

Marie Chang

DO NOT WRITE, STAMP OR SIGN BELOW THIS LINE
RESERVED FOR FINANCIAL INSTITUTION USE*

PRACTICE ASSIGNMENT

1. Write an endorsement that will allow a check you have received to be used by anyone who owns it.

ENDORSE HERE

x

DO NOT WRITE, STAMP OR SIGN BELOW THIS LINE
RESERVED FOR FINANCIAL INSTITUTION USE*

2. Write an endorsement that will allow a check you have received to be used only by Mary Forsythe for any purpose.

ENDORSE HERE

x

DO NOT WRITE, STAMP OR SIGN BELOW THIS LINE
RESERVED FOR FINANCIAL INSTITUTION USE*

3. Write an endorsement that will allow a check you have received to be deposited only.

ENDORSE HERE

x

DO NOT WRITE, STAMP OR SIGN BELOW THIS LINE
RESERVED FOR FINANCIAL INSTITUTION USE*

Section 17.3 Records and Reconciliation

KEY TERMS **Checkbook** **Check stub** **Check register** **Reconciliation** **Cancelled checks** **Outstanding checks** **Deposits in transit**

Periodically, money must be deposited into a checking account to restore the balance. Generally, a bank will not honor a check for which there are insufficient funds, and the check is said to "bounce." Accurate record keeping can prevent you from writing "rubber checks."

Deposit Slip

DEPOSIT TICKET
Samantha Manning
219 Woodland Drive
Eugene, OR 97401

DATE *July 19* 19*9X*
DEPOSITS MAY NOT BE AVAILABLE FOR IMMEDIATE WITHDRAWAL

CASH	CURRENCY	75	00
	COIN		
LIST CHECKS SINGLY		278	62
		107	45
TOTAL FROM OTHER SIDE			
TOTAL		461	07
LESS CASH RECEIVED			
NET DEPOSIT		461	07

54-1/114

USE OTHER SIDE FOR ADDITIONAL LISTING

BE SURE EACH ITEM IS PROPERLY ENDORSED

First Pacific Bank
Eugene, OR 97401

⑆0 123456789 12 345 6⑈ 789

CHECKS AND OTHER ITEMS ARE RECEIVED FOR DEPOSIT SUBJECT TO THE PROVISIONS OF THE UNIFORM COMMERCIAL CODE OR ANY APPLICABLE COLLECTION AGREEMENT

It is important to record all transactions—checks written, deposits, service charges, adjustments, etc.—and to keep a running balance. A transaction in a business **checkbook** is recorded on a **check stub.** Transactions in a personal checking account are usually recorded in a **check register.**

Example 1 Hensley's Pizza purchased a battery from Gene's Car Repair with check no. 482 in the amount of $71.15. The transaction is recorded on the check stub, which remains in the checkbook.

no. *482* $ *71.15*
date *April 16* 19 *9X*
to *Gene's Car Repair*
for *battery*

prev. bal	587	42
deposit		
balance		
check	71	15
balance	516	27

Hensley's Pizza
147 Jasper Lane
Buffalo, NY 12645

482

31-207
1012

April 16 19 *9X*

Pay to the order of *Gene's Car Repair* $ *71.15*

Seventy-One and 15/100 DOLLARS

First Citizens Bank of Buffalo

for *battery* *Paul Hensley*

123456789 12 345 6 789 482

Each month, the bank issues a statement of the account listing the previous month's balance, checks paid, deposits posted, interest earned, service charges, and the new balance. The bank statement balance should agree with the checkbook balance, allowing for transactions not yet posted. The process of verifying that there are no errors and that the two balances agree is called **reconciliation** and can be done by two methods.

The items required to reconcile a bank statement and a checkbook are the cancelled checks, outstanding checks, deposits in transit, and unrecorded charges or interest.

Businesses use the two-part reconciliation method (method 1) because it conforms to formal accounting procedures and is designed to identify bank errors as quickly as accounting errors. Most personal account holders use the three-step shortcut method (method 2).

The items considered when reconciling a bank statement and a checkbook are the cancelled checks, outstanding checks, deposits in transit, and unrecorded charges or interest.

Cancelled Checks

Most banks send all **cancelled checks** (the checks that have been paid) to the account holder.

Outstanding Checks

Checks that do not reach the bank before the statement is issued are **outstanding checks.**

Deposits in Transit

Deposits not posted to an account by the bank before the statement is issued are **deposits in transit.**

Unrecorded Charges or Interest

The bank statement lists service charges or interest that an account holder may not have recorded in the checkbook.

Method 1 Two-Part Reconciliation for Business Accounts

1. Checkbook adjustment:
   Checkbook balance
   + Interest received
   − Charges
   = Adjusted checkbook balance
2. Bank record adjustment:
   Bank balance
   + Deposits in transit
   − Outstanding checks
   = Adjusted bank balance

The adjusted bank balance should equal the adjusted checkbook balance. If not, there is an error.

Method 2 Three-Step Shortcut for Personal Accounts

1. Mark each cancelled check in your checkbook. Those remaining are outstanding checks. Sum the amounts of outstanding checks.
2. Mark each deposit recorded on the bank statement. Remaining deposits are in transit. Sum the amounts of the deposits in transit.
3. Begin with the checkbook balance. Add any interest received. Subtract any unrecorded service charges. The result is the adjusted checkbook balance.
   Adjusted checkbook balance
   + Outstanding checks
   − Deposits in transit
   = Bank balance (if no errors)

This method is only recommended for individuals who update balances regularly and carefully. (A calculator helps.)

If the balances do not reconcile, there is either a calculation error or a recording error. Bank errors do occur, but only rarely.

The following are common errors that account holders commit:

1. Failure to record checks or deposits in the checkbook.
2. Failure to adjust the check register for voided checks or deposits.
3. Error in the running balance.
4. Failure to take into account *all* outstanding checks.

Example 2 Method 1: Two-part Reconciliation
Rosen and Scalese, Inc. has a checkbook balance of $957.91. The bank statement shows a balance of $723.45. Checks totaling $253.37 are outstanding, and a deposit of $480.00 is in transit. The bank's service charge was $12.00. Interest of $4.17 was credited to the account.

Checkbook Adjustment

Balance	957.91
+ Interest	4.17
	962.08
− Service charge	12.00
= Adjusted balance	950.08

Bank Record Adjustment

Balance	723.45
+ Deposits in transit	480.00
	1203.45
− Outstanding checks	253.37
= Adjusted balance	950.08

Example 3 Method 2: Three-Step Shortcut for Personal Accounts
Here is a portion of Peter Linden's check register.

RECORD ALL CHARGES OR CREDITS THAT AFFECT YOUR ACCOUNT

Number	Date	Description of Transaction	Payment/Debit (−)		✓ T	FEE IF ANY (−)	Deposit/Credit (+)		Balance	
									411	30
478	4/05	City Utilities	34	58	✓				376	72
479	4/07	Acme Home Supplies	67	39	✓				309	33
	4/10				✓		305	47	614	80
480	4/12	Amer. Cancer Society	15	00					599	80
481	4/15	Family Dollar	12	38	✓				587	42
482	4/16	Charles' Shoe Store	71	15	✓				516	27
483	4/19	TV Cable Service	20	00	✓				496	27
484	4/23	College Book Store	98	34	✓				397	93
485	4/24	Clark's Clothing	87	26					310	67

The following checks are outstanding:

Check No.	Date	Payee
480	4/12	Amer. Cancer Society
485	4/24	Clark's Clothing

Current checkbook balance	310.67
− Service charge	5.00
= Adjusted checkbook balance	305.67
+ Outstanding checks	102.26
= Bank balance	407.93

There are no deposits in transit. Peter has not yet recorded a service charge of $5.

PRACTICE ASSIGNMENT

Plan: Complete the following check stubs. Use the shortcut method to reconcile a personal account, and the two-part method to reconcile a business account.

1. Use the given transaction data to complete the following stubs. Check no. 516 was written on August 12, 199X, in the amount of $45.60 to United Parcel Service for sample shipments. (Previous balance was $758.92.)

a.

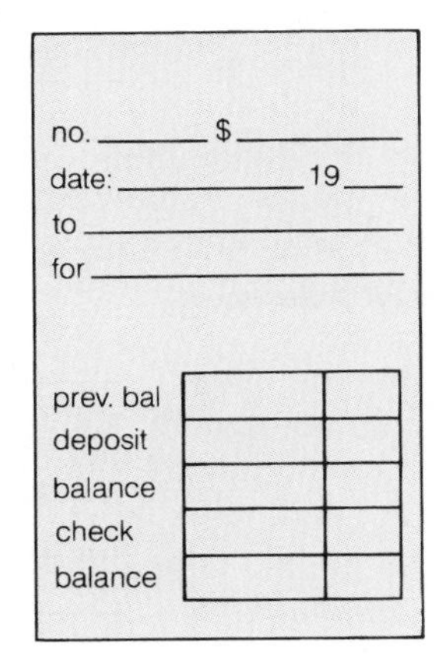

no. ______ $ ________
date: __________ 19 ____
to ____________________
for ____________________

prev. bal
deposit
balance
check
balance

b. Check no. 517 was written on August 19, 199X, in the amount of $73.32 to Public Utilities for electricity.

no. ______ $ ________
date: __________ 19 ____
to ____________________
for ____________________

prev. bal
deposit
balance
check
balance

2. Reconcile the following bank statement with the check register using the shortcut method for personal accounts. Checks not marked with a check mark are outstanding.

Bank Statement

Date	Check No.	Transaction Amount	Balance
			581.44
9/02	596	275.00	306.44
9/05	597	100.00 ATM	206.44
9/15		368.71 DP	575.15
9/18	599	100.00	475.15
9/18	598	38.27	436.88
9/28	601	17.48	419.40
9/30		4.50 SC	414.90

No.	Date	Transaction	Payment Debit		✓	Fee if any	Deposit Credit		Balance	
									581	44
596	9/01	Terrace Apartments	275	00	✓				306	44
597	9/05	Cash	100	00	✓				206	44
598	9/11	Jeans Galore	38	27	✓				168	17
	9/15				✓		368	71	536	88
599	9/18	Cash	100	00	✓				436	88
600	9/21	Dan's Shop-at-Home	79	95					356	93
601	9/22	Music Master	17	48	✓				339	45
	9/29						368	71	708	16
602	9/30	Northern Bell	34	18					673	98

Checkbook balance ________
− Service charge ________
= Adjusted checkbook balance ________
+ Outstanding checks ________
− Deposits in transit ________
= Bank balance ________

3. Minata's Notions received a bank statement listing a service charge of $15.00, an interest payment of $4.05, and a balance of $592.56. Minata's checkbook showed a balance of $671.96. There were four outstanding checks: $38.55, $143.60, $52.07, and $72.33. Deposits of $200 and $175 were in transit. Use the two-part method to complete the reconciliation.

Checkbook Adjustment

Balance ________
+ Interest ________
____
− Service charge ________
= Adj. balance ________

Bank Statement Adjustment

Balance ________
+ Dep. in trans. ________
____
− Outstand. checks ________
= Adj. balance ________

Challenger

The adjusted cost of a minimum balance checking account is the service charge plus any interest you could have earned but lost because you did not deposit the minimum balance in an interest-bearing account.

a. Complete the following schedule.

Minimum Daily Balance	Monthly Charge	Monthly Interest of 5 % Not Received on Smallest Allowed Balance	Adjusted Cost
300.00 and higher	None	________	________
200.00–299.99	3.00	________	________
100.00–199.99	4.00	________	________
0.00–99.99	5.00	________	________

b. Reconsider the Cornerstone Club in table 17.2 of section 17.1. The plan pays five percent interest on the average daily balance. If the balance falls below the minimum of $2,500, there is a service charge of $6. Interest rates on savings accounts may vary to eight percent or higher.
Assume an average daily balance of $500 if the customer does not maintain the minimum balance. What are the adjusted costs?

	Minimum Balance not Maintained			Minimum Balance Maintained	
Charge	Interest Received 5% on 500	Adjusted Cost	Interest Lost 8% on 2,500	Interest Received 5% on 2,500	Adjusted Cost
6.00	________	________	________	________	________

CHAPTER 17 SUMMARY

Concept	Example	Procedure	Formula
SECTION 17.1 Checking; Automated Teller Machines			
Check	Pay $235.50 to the order of Reliable Plumbing.	A check is an order to a bank to pay the amount specified to the payee.	
Transaction charge	$0.10 for each check $0.05 for each deposit 20 checks 2.00 2 deposits 0.10 charge 2.10	A bank may charge a small fee for each transaction.	
Flat monthly fee	$5.00 per month	A bank may charge a flat monthly fee.	
Minimum balance	No service charge if daily balance of $300 is maintained. If the daily balance drops below $300 at anytime during a month, the service charge is $4.00.	A bank may charge a monthly fee if the daily balance falls below a certain amount.	
Interest on checking accounts	Minimum balance is $1,000. Service charge is $6.00 if minimum balance is not maintained. 4.5% interest is paid on average daily balance. Average daily balance = 516.23 Interest = 1.94 Charge = 6.00 Cost for month = 6.00 − 1.94 = 4.06	Checking accounts that are interest-bearing may require a minimum balance.	
Automated teller machine (ATM)		Withdrawals, deposits, or transfers between accounts may be transacted through an automated teller machine.	
SECTION 17.2 Endorsements			
Endorsements	Signature Pay to the order of J. Smith Signature For deposit only in the account of J. Smith Signature	Checks must be endorsed on the back when cashed, deposited, or transferred to another party. An endorsement may be blank, full, or restrictive.	
SECTION 17.3 Records and Reconciliation			
Check stub Check register	Apr 12 balance 451.63 Apr 15 deposit 234.25 Apr 15 balance 685.88 Apr 17 check 65.78 Apr 17 balance 620.10	Record all transactions and keep running balance either on stubs or in register.	Current balance = prev. bal. + deposit − check
Bank reconciliation: Two-part reconciliation	Checkbook balance 313.26 − Charges 5.00 = Adj. checkbook bal. 308.26 Bank record balance 291.59 + Deposits in transit 100.00 391.59 − Outstanding checks 83.33 = Adj. bank balance 308.26	Each month compare checkbook records with the bank statement. Due to time lags, the two balances are not equal at first. After adjustments, they will be equal unless there are errors.	Adj. checkbook balance = checkbook balance + interest received − charges
Three-step shortcut	Checkbook balance 313.26 − Charges 5.00 = Adj. checkbook bal. 308.26 + Outstanding checks 83.33 391.59 − Deposits in transit 100.00 = Bank balance 291.59		Adj. bank balance = bank balance + deposits in transit − outstanding checks Bank balance = checkbook balance − unrecorded charges + unrecorded interest + outstanding checks − deposits in transit

Chapter 17 **Spreadsheet Exercise**

Check Register with Reconciliation Balance

Checkbook transactions for April and May have been recorded on this spreadsheet. Complete the program so that it will update the balances. You will be able to mark cancelled checks and deposits posted by the bank so that the projected bank balance will be computed. The chapter 17 template appears as follows:

```
A1: PR [W6]                                                          READY

        A      B             C                D    E   F        G        H
1
2                              CHECK REGISTER
3     -----------------------------------------------------------------------
4     number date       transaction       payment  x  fee   deposit   balance
5     -----------------------------------------------------------------------
6                  carried forward                                     411.30
7        477  4/03  Terrace Apartments     275.00
8        478  4/05  City Utilities          34.58
9        479  4/07  Acme Home Supplies      67.39
10            4/10                                          405.47
11       480  4/12  American Cancer Soc.    15.00
12       481  4/15  Family Dollar           12.38
13       482  4/16  Charles' Shoe Store     71.15
14       483  4/19  TV Cable Service        20.00
15       484  4/23  College Book Store      98.34
16       485  4/24  Clark's Clothing        87.26
17            4/27                                  5.00
18            4/30                                          405.47
19       486  5/02  City Utilities          32.17
20       487  5/05  Al's Service Station    47.53
```

Cell H6 contains the opening balance.

Rows 7–28 contain the transactions.

Cell H31 will display the projected bank balance.

Spreadsheet Exercise Steps

1. Column H: Balance
   Next balance = previous balance − payment − fee + deposit.
   The previous balance is in H6, a possible payment in D7, a possible fee in F7, and a possible deposit in G7.
   If a spreadsheet formula begins with a cell address, it must be preceded by +.
   The spreadsheet formula is
   + ______ − ______ − ______ + ______ .
   Complete the formula, store it in H7, and copy it from H7 to H8..H28.

Outstanding checks are accumulated in column I.
Deposits in transit are accumulated in column J.
Be aware that when you move columns I and J onto the screen, columns A, B, and C will temporarily move off the screen.

2. Record the May fee (service charge) of $5.00 in cell F29 and a date of 5/30 in B29. Precede the 5 in the date with one space.
   (This fee has been recorded by the bank but will not affect your checkbook balance at this time.)
3. Cell H31: Projected Bank Balance
   Projected bank balance =
   checkbook balance − fee + outstanding checks − deposits in transit.
   The checkbook balance is in H28; the last fee is in F29; the sum of the outstanding checks is in I28; the sum of the deposits in transit is in J28.
   If a spreadsheet formula begins with a cell address, it must be preceded by +.
   The spreadsheet formula is
   + ______ − ______ + ______ − ______ .
   Complete the formula and store it in H31.
4. Mark all checks as cancelled and all payments as recorded with an *x* in column E. The projected bank balance should now be five dollars less than the checkbook closing balance, taking into account the May fee.

5. Suppose the following transactions have not yet cleared:
checks 487, 491, 492, 494, and the last deposit. Erase the *x* in column E for these transactions. This is easy to do by entering one space with the spacebar. Verify the projected bank balance.

Outstanding checks:	#487:	_________
	#491:	_________
	#492:	_________
	#494:	_________
Total:		_________
Checkbook closing balance		_________
− Fee		
+ Outstanding checks		_________
− Deposits in transit		_________
Bank balance		_________

NAME ______________________________

CLASS SECTION ______________ DATE ______________

SCORE __________ MAXIMUM 90 PERCENT __________

Chapter 17 Assignment

1. Barlow Real Estate has a checking account that charges $0.15 per check, $0.10 per deposit, and a monthly service charge of $2.50. During a month, the firm wrote 23 checks and made 2 deposits. What was the total monthly charge? ***(5 points)***

2. Write a check dated September 10, 1991, in the amount of $147.59 to Darryl's Sound Shop. ***(5 points)***

A. Student
1234 Main Street
College Town, USA

208
29-184
1035

__________ 19 ____

PAY TO THE
ORDER OF ______________________________ $ [ ]

______________________________ DOLLARS

NATIONAL BANK AND TRUST

MEMO ______________ ______________

1234 5678 9012 3456 1234567

3. Use the given transaction data to complete the following check stub. Balance of $972.36. Check no. 753 written on August 30, 199X, in the amount of $289.62 to Worldwide Travel Agency for airline tickets. ***(5 points)***

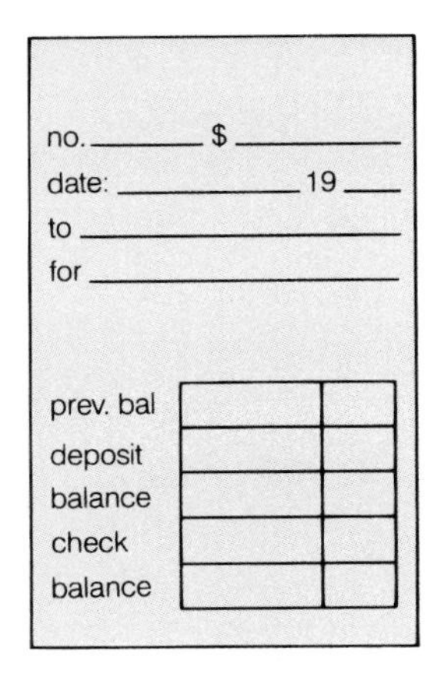
no. ______ $ ______
date: ______ 19 ____
to ______
for ______

prev. bal
deposit
balance
check
balance

4. Linda's Beauty Salon uses a checking account that charges $0.18 per check and $0.10 per deposit. There is no monthly service charge if a $300 minimum balance is maintained. Otherwise, the charge is $4. Linda wrote 31 checks in March and made 3 deposits. Her balance never dropped below $350. What was her monthly charge? ***(5 points)***

5. A small business opened a checking account that charges $0.20 per check and $0.10 per deposit. The bank pays a credit of $0.25 for each $100 of average daily balance. If this firm wrote 15 checks, made one deposit, and maintained an average daily balance of $375, what was the total monthly charge? ***(5 points)***

6. Enter the following transactions into the check register: (The balance on October 27 was $197.61.)

10/28/91 Check no. 630 to Magic Costume Rental, amount: $28.75
10/31/91 Check no. 631 to Al's Service Station, amount: $67.53
11/1/91 Deposit of $262.83
11/8/91 Check no. 632 to Cut 'n Trim, amount: $12.50
11/15/91 Deposit of $262.83
11/16/91 Check no. 633 to S & D Styles, amount: $57.24
11/20/91 Check no. 634 to Universal Shoes, amount: $30.88 *(35 points)*

No.	Date	Transaction	Payment Debit		✓	Fee if any	Deposit Credit		Balance	

7. Use the following data to perform a two-part reconciliation.

Checkbook balance:	$742.38
Total outstanding checks:	318.59
Total deposits in transit:	350.00
Bank statement balance:	704.86
Interest credited:	3.89
Service charge:	10.00

Checkbook Adjustment

Balance ________
+ Interest ________
____
− Service charge ________
= Adj. balance ________

Bank Statement Adjustment

Balance ________
+ Deposits in transit ________
____
− Outstanding checks ________
= Adj. balance ________

(30 points)

Chapter 18 Stocks and Bonds

OBJECTIVES

- Determine the cost of buying securities
- Compute par values and earnings of securities
- Determine the proceeds of selling securities
- Calculate the price/earnings ratio and current yield of stocks
- Calculate current yield and approximate yield to maturity of bonds

SECTIONS

INTRODUCTION

Businesses need money. Corporations finance their operations, including expansion, restructuring, and maintenance through the sale of securities—stocks and bonds.

Section 18.1 Common and Preferred Stock

KEY TERMS **Securities** **Common stock** **Shares** **Par value** **Dividends** **Preferred stock**

What Are Securities?

Securities are stocks and bonds that corporations offer for sale to the public. Stocks represent part ownership by investors, while bonds represent loans from investors.

Common stock issued by a corporation can be divided into any number of equal **shares,** each share representing a portion of ownership. Investors who own stock are called stockholders or shareholders; they are part owners of the companies whose stock they own, and they risk their investment on the successes or failures of those companies. The **par value** of a stock indicates the initial dollar value assigned to a share. Par value should not be confused with market value, which is covered in section 18.2.

Outstanding common stock of Resco, Inc. = \$2,500,000

25,000 shares of common stock are authorized.

$$\text{Par value} = \frac{2500000}{25000} = 100$$

Stockholders hold certificates as evidence of stock ownership. When there is a high volume of trading activity, stock certificates are kept in one brokerage house and computers keep track of ownership.

Stock Certificate

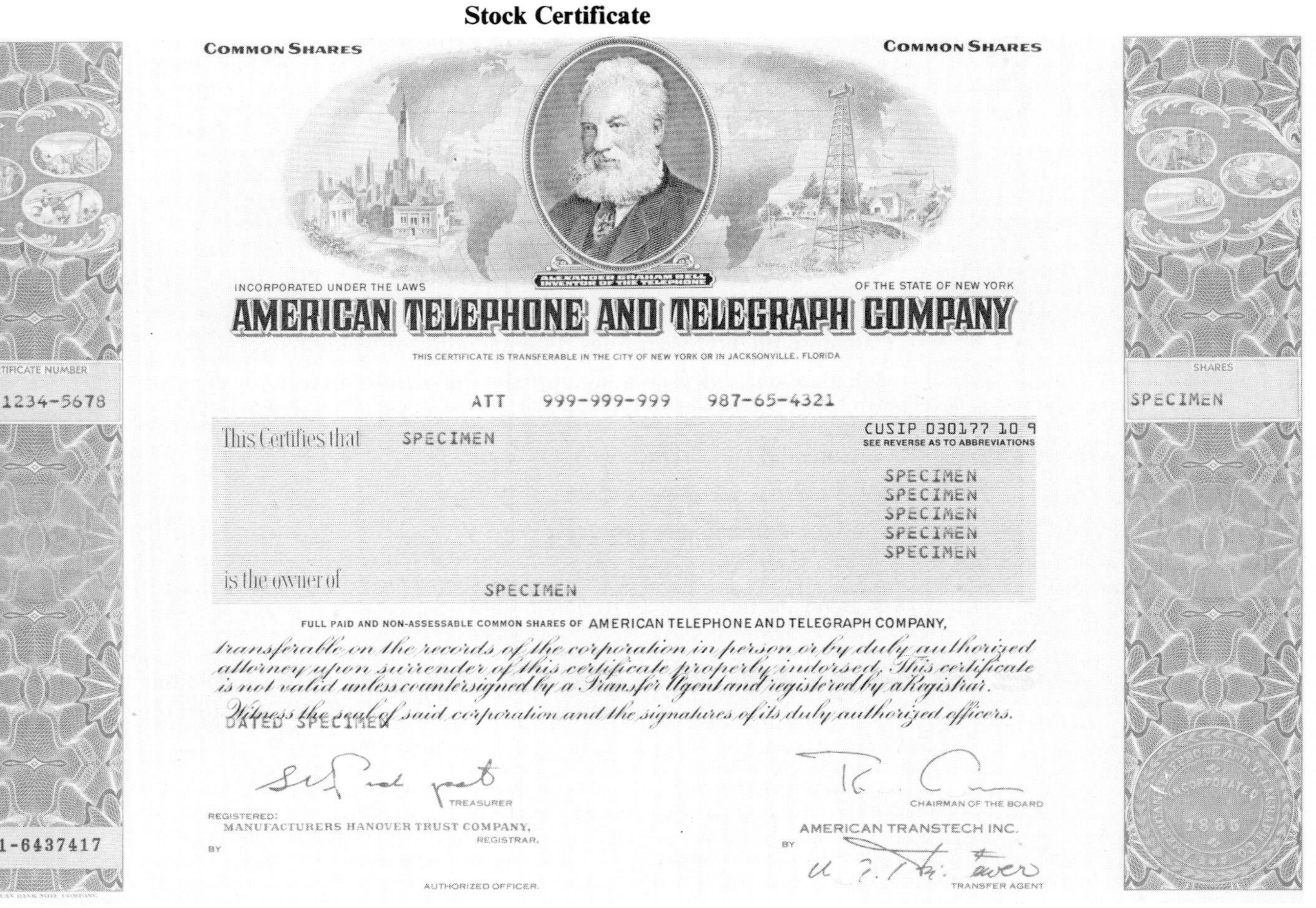

COMMON SHARES COMMON SHARES

ALEXANDER GRAHAM BELL INVENTOR OF THE TELEPHONE

INCORPORATED UNDER THE LAWS OF THE STATE OF NEW YORK

AMERICAN TELEPHONE AND TELEGRAPH COMPANY

THIS CERTIFICATE IS TRANSFERABLE IN THE CITY OF NEW YORK OR IN JACKSONVILLE, FLORIDA

CERTIFICATE NUMBER ZQ 1234-5678

SHARES SPECIMEN

ATT 999-999-999 987-65-4321

This Certifies that SPECIMEN

CUSIP 030177 10 9
SEE REVERSE AS TO ABBREVIATIONS

SPECIMEN
SPECIMEN
SPECIMEN
SPECIMEN
SPECIMEN

is the owner of SPECIMEN

FULL PAID AND NON-ASSESSABLE COMMON SHARES OF AMERICAN TELEPHONE AND TELEGRAPH COMPANY,
transferable on the records of the corporation in person or by duly authorized attorney upon surrender of this certificate properly indorsed. This certificate is not valid unless countersigned by a Transfer Agent and registered by a Registrar.
Witness the seal of said corporation and the signatures of its duly authorized officers.

DATED SPECIMEN

TREASURER CHAIRMAN OF THE BOARD

REGISTERED:
MANUFACTURERS HANOVER TRUST COMPANY,
REGISTRAR,
BY AUTHORIZED OFFICER.

AMERICAN TRANSTECH INC.
BY TRANSFER AGENT

001-6437417

Used with permission of American Telephone and Telegraph Company.

As part owners, stockholders vote annually to elect a board of directors who oversee the operations of a corporation. Some corporate profits are distributed to stockholders in the form of **dividends.** It is up to the individual corporation to declare a dividend, and some may choose not to do so, even after a profitable year. Dividends may be paid quarterly, semiannually, or annually, and may be a percent of par value or a dollar value per share.

With the approval of the board of directors and the common stockholders, corporations may offer shares of **preferred stock.** Owners of preferred stock receive dividends before any are paid to owners of common stock. If a corporation is liquidated, preferred stockholders have priority over common stockholders. Preferred stockholders, however, usually do not have voting rights.

The Martin family owns 2,000 shares of Dynamic Hi-Tec and 2,000 shares of General Stores. Last year, Dynamic Hi-Tec paid out a five percent dividend on shares with par value of $100; General Stores paid out $2.90 per share. How large was each annual dividend?

Dynamic Hi-Tec:
5% of $100 \times 2000 = 10000$

General Stores:
$2.90 \times 2000 = 5800$

Although preferred stock may earn less than common stock, earnings from preferred stock are somewhat more dependable. Convertible preferred stock, which is gaining popularity, can be converted to a specified number of common stock shares at the holder's option.

Example 1 Wymart Corp. declared a quarterly dividend of $108,000. There are 30,000 shares of common stock outstanding.

What is the quarterly dividend check for a stockholder who owns 5,000 shares?

$$\text{Dividend per share} = \frac{108000}{30000} = 3.60$$

$$\text{Dividend for 5000 shares} = 3.6 \times 5000 = 18000$$

Example 2 Compu-Tape, Inc. earned $500,000 last quarter after taxes. The board of directors authorized $300,000 in quarterly dividends. Outstanding seven percent preferred stock represents capital of $1,500,000, and outstanding common stock represents capital of $4,500,000. The par value of all stock (preferred and common) is $100. What are the dividends on preferred and common stock?

There are 15,000 shares of preferred stock.
(1500000 ÷ 100 = 15000)

Since the preferred stock dividend is 7% on the par value, it amounts to $7 per share annually or $1.75 quarterly.
The total preferred dividend is 15000 × 1.75 = 26250.
The remainder of the declared dividend is
300000 − 26250 = 273750.
$273,750 are paid to holders of common stock.
45,000 shares of common stock are outstanding.
(4500000 ÷ 100 = 45000)
273750 ÷ 45000 = 6.08333 . . .
The quarterly dividend for common stock is $6.08333 . . . per share.

PRACTICE ASSIGNMENT

Plan: $\text{Par value} = \dfrac{\text{value of outstanding stock}}{\text{number of authorized shares}}$

The preferred stock dividend is specified in percent (per year). It is the percent of total investment or the percent of a share's par value. The quarterly dividend is the annual dividend divided by four. The common stock dividend is the remainder of the declared dividend after the preferred dividend (if any) is paid. To obtain the dividend per share, divide the total dividend allocated to common stockholders by the number of outstanding shares.

1. Compute par value.

	Capital Stock	Number of Shares Authorized	Par Value
a.	730,000	73,000	________
b.	750,000	25,000	________
c.	840,000	60,000	________
d.	2,500,000	40,000	________
e.	846,000	18,000	________
f.	3,250,000	200,000	________

2. Assuming preferred stock, compute the quarterly dividend per share and the total quarterly dividend to holders of the number of shares indicated. (Do not round dividend per share.)

	Par Value	Annual Rate of Return	Annual Dividend per Share	Quarterly Dividend per Share	Number of Shares Held	Size of Quarterly Dividend Check
a.	10	8%	________	________	200	________
b.	8	5%	________	________	350	________
c.	40	7%	________	________	500	________
d.	25	8%	________	________	80	________
e.	15	5.5%	________	________	450	________
f.	75	6.7%	________	________	725	________

3. Assuming common stock, compute the dividend per share. (Do not round.)

	Total Dividend for Common Stockholders	Number of Outstanding Shares	Dividend per Share		Total Dividend for Common Stockholders	Number of Outstanding Shares	Dividend per Share
a.	450,000	60,000	________	**d.**	51,000	60,000	________
b.	1,225,000	175,000	________	**e.**	1,461,600	203,000	________
c.	247,000	95,000	________	**f.**	65,325	67,000	________

4. An investor owns 380 shares of International Metals preferred stock. If the annual dividend rate is $7\frac{1}{2}\%$ on a par value of \$15, how much is the quarterly dividend for these shares?

5. A corporation acquired \$850,000 in capital by issuing common stock with par value of \$5 per share. At the end of the year the company authorized dividends totaling \$73,950. What was the annual dividend per share?

Section **18.2 The Stock Market**

KEY TERMS **Stockbrokers** **Market value** **Round lot** **Odd lot** **Broker's fee**

How are Stocks Bought and Sold?

Most stocks are traded on securities exchanges, such as the New York Stock Exchange (NYSE) and the American Stock Exchange (AMEX). **Stockbrokers** buy and sell stocks, bonds, and other securities on behalf of investors. Investors who wish to buy or sell securities place an order through a broker.

Diversification refers to investments in different companies in order to offset poor investments with good investments. One way this may be accomplished is through buying shares of investment companies called mutual funds. These shares themselves are traded.

The **market value** of a stock deviates from its par value in response to supply and demand. Traders bargain for the best price, considering earnings, dividends, management, and national trends.

Because the prices of stocks fluctuate daily, a stockholder can gain in two ways: by collecting dividends or by selling shares at a price higher than the buying price. (There is no guarantee of either occurrence, however.)

Most metropolitan newspapers list stock market quotations on the financial page. Over-the-counter stock prices are quoted by *bid* and *asked* prices. The bid price is the selling price; the asked price is the purchase price.

In addition to listed securities, many unlisted stocks are traded "over the counter," meaning they are not traded through an exchange but by dealers who are willing to trade at any time. Over-the-counter market activity is reported by the National Association of Security Dealers' Automated Quote (NASDAQ).

Key to Stock Market Quotations	
52–weeks Hi	Highest price of stock in past year
52–weeks Lo	Lowest price of stock in past year
Stock, Sym	Abbreviated stock name and symbol
Div	Dividend paid per share
Yld %	Current yield (rate of return)
PE	Price/earnings ratio
Vol 100s	Sales volume for the day in hundreds of shares
Hi	High price of stock for the day
Lo	Low price of stock for the day
Close	Closing price of stock
Net chg	Change from closing price of prior trading day

Prices on the stock exchange and over the counter are quoted in dollars per share, stated in mixed numbers; the fractional part is usually 1/8 or one of its multiples:

1/8 1/4 3/8 1/2 5/8 3/4 7/8

Shares may be sold in a **round lot** (ordinarily a multiple of 100), an **odd lot** (less than 100), or a combination of round and odd.

The smallest fraction is $\frac{1}{8}$. A stock market quotation of $42\frac{3}{8}$ means that the price of one share is \$42.375.

100 shares make up a round lot. 28 shares make up an odd lot.

300 shares are traded as a round lot.

330 shares form a combination: 300 are traded as a round lot and 30 are traded as an odd lot.

Portion of *The Wall Street Journal* Stock Quotation

52 Weeks Hi	Lo	Stock	Sym	Div	Yld %	PE	Vol 100s	Hi	Lo	Close	Net Chg
48	36⅜	Hillenbrnd	HB	.55	1.3	20	57	42⅞	42⅝	42⅞	+ ⅛
9⅜	3	HillDeptStr	HDS		...	27	90	3¼	3⅛	3¼	...
115½	45½	Hilton	HLT	1.20	2.5	20	4582	48¾	47½	48¼	−1¼
115⅛	91⅝	Hitachi	HIT	.70r	.7	24	60	96⅛	95⅜	96⅛	− ¼
n 7⅝	4	Holnam	HLN		...	125	106	5	4⅞	5	+ ⅛
s 43½	18¾	HomeDepot	HD		...	36	5718	38⅛	36⅝	37⅞	− ¼
21½	13⅝	HomeIns pf		2.95	16.1	...	27	18⅜	18⅛	18⅜	+ ¼
9¼	3	HomeShop	HSN		...	156	3006	6⅜	5⅞	6¼	− ⅛
47½	13⅝	HomeFed	HFD	.20	1.5	...	1780	14⅛	13⅝	13¾	− ¼
4⅞	2	HomeplxMtgInv	HPX	.50e	14.3	5	129	3⅝	3⅜	3½	...
23⅝	12⅝	Homestake	HM	.20	1.1	31	3011	18⅞	18⅝	18⅞	+ ¼
4⅜	**11/16**	**HomestdFnl**	**HFL**		...	...	**288**	1¼	1⅛	1⅛	− ¼
30⅜	20⅜	HondaMotor	HMC	.17e	.7	21	96	23¼	23	23⅛	+ ¼
112⅜	73¼	Honeywell	HON	2.75	2.7	14	3010	103¾	101⅛	102⅞	+ ⅞
26⅝	17	HongKongTelcm	HKT	1.08e	4.2	...	469	25½	25⅛	25½	+ ⅝
11⅜	2¾	HopperSol	HS		...	...	1	3	3	3	+ ⅛
s 18⅞	13	HormelGeo	HRL	.26	1.5	18	129	17⅜	17	17⅛	+ ⅛
n 10⅝	7¼	Horsham	HSM		...	...	2132	9¼	9⅛	9⅛	− ⅛
10¼	3¼	HotelInvTr	HOT	.52	14.3	...	61	3⅝	3½	3⅝	+ ⅛
43⅜	26	HougtnMif	HTN	.70	2.6	17	107	27⅜	26½	27⅛	+ ⅝
27⅝	15½	HouseFab	HF	.48	1.9	14	465	25¾	25¼	25¼	− ¼
61¾	41⅝	HouseInt	HI	2.14	4.7	7	1230	46⅜	45¾	46	...
25¼	21⅝	HouseInt pfA			...	...	54	23⅞	23¾	23⅞	+ ⅛
35⅞	31¼	HoustnInd	HOU	2.96	9.1	13	1481	32½	31¾	32½	+ ⅛
1⅝	⅞	HoustnOilR	RTH	.10e	8.9	...	34	1⅛	1	1⅛	...
15⅞	8⅜	HowellCp	HWL	.49e	3.4	15	87	14¾	14½	14½	− ¼
16½	8¾	HudsonFood	HFI	.12	1.4	6	403	9	8¾	8¾	− ¼
25⅝	16⅜	Huffy	HUF	.40	1.8	11	398	23½	22¾	22¾	− ⅝
s 20⅛	14⅜	HughsSply	HUG	.36	2.5	11	20	14½	14⅜	14½	− ⅛
50⅞	33	Humana	HUM	1.20	2.5	16	2114	48½	46⅞	48½	+1½
24⅞	11¾	HuntMfg	HUN	.31	2.1	12	218	12¾	12½	12⅝	− ⅛
34¼	21	HuntgInt	HTD	.30e	.9	32	36	33½	33⅛	33½	...
13¼	5¾	Huntway	HWY	1.38	14.3	...	372	9¾	9⅝	9⅝	...
28¼	22	Hydraulic	THC	1.60	7.2	11	58	22¼	22¼	22¼	...
12⅛	9⅞	HyperionFd	HTR	1.26	12.4	...	574	10⅛	10	10⅛	+ ⅛

-I-I-I-

52 Weeks Hi	Lo	Stock	Sym	Div	Yld %	PE	Vol 100s	Hi	Lo	Close	Net Chg
19¼	13½	IBP Inc	IBP	.60	3.5	22	601	17¾	17⅜	17⅜	− ⅜
8½	6	ICM Prop	ICM	.48e	6.3	...	69	7⅝	7½	7⅝	...
6¾	2½	ICN Pharm	ICN		...	...	114	3⅜	3⅛	3¼	...
28⅛	24½	IE Ind	IEL	2.06	8.3	12	130	25	24⅞	24⅞	...
39¾	31⅛	IMC Fertlzr	IFL	1.08	3.0	11	1570	37¼	36½	36½	− ⅞
18¼	15¼	INA Invest	IIS	1.52	9.7	...	16	15⅝	15⅝	15⅝	...
23¾	19	IP Timber	IPT	2.88	13.9	7	91	20¾	20¼	20¾	+ ½
15⅝	10½	IRT Prop	IRT	1.16	10.1	18	136	11½	11⅜	11½	...
64½	51⅜	ITT Cp	ITT	1.60	2.8	8	2809	57⅜	55⅝	57⅜	+ ⅝
104	85	ITT Cp pfK		4.00	4.3	...	10	93⅞	92	93⅞	+ ⅜
95⅞	81⅛	ITT Cp pfO		5.00	5.8	...	8	86¾	86⅛	86⅛	− ⅝
79⅞	65	ITT Cp pfN		2.25	3.1	...	23	72	71¾	71¾	− ¼
30	23⅜	IdahoPwr	IDA	1.86	7.7	11	68	24¼	24	24⅛	− ⅛
17½	10⅝	IdexCp	IEX		...	12	22	15¾	15⅝	15¾	...
19⅜	14½	IllPwr	IPC		...	121	2090	15⅞	15½	15¾	+ ¼
22¾	17¼	IllPwr pf		2.13	10.7	...	z200	19⅞	19⅞	19⅞	+ ⅞
24	17	IllPwr pf		2.21	10.9	...	z600	20¼	20¼	20¼	...
24½	17½	IllPwr pf		2.35	11.1	...	z1760	21¼	21⅛	21¼	...
42½	31	IllPwr pf		4.12	11.6	...	z100	35½	35½	35½	− ⅜
39¾	28½	IllPwr pf		3.78	11.5	...	z[illegible]	33	32¾	33	+ ¼

52 Weeks Hi	Lo	Stock	Sym	Div	Yld %	PE	Vol 100s	Hi	Lo	Close	Net Chg
26¾	23¾	LincNatlnco	LND	2.28a	9.1	...	5	25	25	25	+ ¼
98	68⅞	Litton	LIT		...	10	532	72½	70¾	72¼	+ ¾
25½	22	Litton pf		2.00	8.9	...	2	22½	22½	22½	...
25	12¾	LIVE Entn	LVE		...	13	48	22¼	21⅞	22⅛	+ ⅛
▼ 54¾	29¾	Lockheed	LK	1.80	6.2	243	1749	30⅜	29⅛	29⅛	−1¼
61⅜	40⅞	Loctite	LOC	1.20	2.1	18	110	58⅜	58	58¼	+ ⅛
135	100	Loews Cp	LTR	1.00	.9	9	1506	108½	105¼	107	−1⅛
24⅜	15	Logicon	LGN	.36	2.1	9	121	17½	17⅜	17⅜	− ⅛
8⅝	**¼**	**vjLomasFnl**	**LFC**		...	...	**405**	½	7/16	7/16	− 1/32
17¼	2⅜	LomasNetMtg	LOM	.84	20.4	...	564	4½	4	4⅛	− ⅛
11¾	10	LomasMtgFd	LSF	1.26	11.6	...	394	10⅞	10⅝	10⅞	+ ⅛
37	**9⅜**	**LoneStar**	**LCE**		...	...	**586**	**10⅜**	**9⅝**	**10¼**	+ ½
21⅜	16⅝	LILCo	LIL	1.00	4.9	...	1369	20¾	20½	20½	− ¼
64	40½	LILCo pfE		4.35	10.1	...	z120	43¼	43¼	43¼	− ¾
26⅝	24⅝	LILCo pfY		2.65	10.2	...	921	26⅛	25⅞	26	...
42¼	25⅜	LILCo pfT		3.31	12.6	...	626	26⅜	26¼	26¼	− ⅛
35	22½	LILCo pfP		2.43	9.9	...	15	24¾	24⅝	24⅝	− ⅛
36⅛	23⅜	LILCo pfO		2.47	10.0	...	3	24¾	24¾	24¾	...
48½	36¾	LongsDrg	LDG	1.04	2.6	13	72	40⅛	39¾	39¾	− ½
s 15¼	9¾	LongvwFibr	LFB	.48	4.2	10	336	11⅝	11⅜	11½	...
36½	24⅛	Loral	LOR	.80	3.1	8	2429	26⅜	25	26	+ ⅝
21	14¼	LaGenlSvcs	LGS	.76	3.8	...	45	19⅞	19⅝	19¾	− ⅛
46½	37	LaLandExpl	LLX	1.00	2.3	31	312	43½	42¼	43¼	+1
▼ 45¼	35½	LaPacific	LPX	1.00	2.8	8	399	35⅞	35⅜	35⅞	+ ¼
28	26⅜	LaP&L pf		3.16	11.8	...	12	26⅞	26¾	26¾	...
41¾	36¼	LouvlG&E	LOU	2.78	7.5	12	239	37	36¾	36⅞	...
49⅝	25¼	Lowes Cos	LOW	.52	1.5	16	2960	36	34½	34⅞	− ¼
45⅜	32½	Lubrizol	LZ	1.44	3.3	16	2780	43⅛	42⅜	43	+ ½
31⅞	23⅛	LubysCafe	LUB	.64	2.1	18	82	30⅝	30⅛	30⅝	+ ⅜
45¼	26	Lukens Inc	LUC	1.40	3.6	7	125	40	39	39	− ¾
n 28½	18¾	LuxottGp	LUX	.68e	2.6	...	126	27⅛	26½	26½	− ⅜
36⅝	25½	Lydall	LDL		...	10	16	29¼	29¼	29¼	+ ⅛
22¼	14⅞	Lyondell	LYO	1.60a	7.9	5	2586	20½	20¼	20¼	+ ⅛

-M-M-M-

52 Weeks Hi	Lo	Stock	Sym	Div	Yld %	PE	Vol 100s	Hi	Lo	Close	Net Chg
8	3¼	M/A Com	MAI		...	...	279	4¾	4⅝	4⅝	...
5⅞	1¾	MAIBasic	MBF		...	...	28	2	1⅞	2	...
45⅜	25	MBIA	MBI	.40	1.0	14	417	39⅞	39⅝	39⅝	− ⅜
71⅜	48⅝	MCA	MCA	.68	1.3	19	1495	51⅜	50⅝	50¾	− ⅛
23⅞	19¾	MCN	MCN	1.57	7.7	9	173	20⅝	20⅜	20⅜	...
▼ 2¾	⅝	**MDC Cp**	**MDC**		...	...	151	11/16	7/16	7/16	− 5/16
23	19⅞	MDU Res	MDU	1.42a	7.1	11	53	20	19⅞	19⅞	− ⅛
7¼	4	MEI Divrs	MEI		...	...	90	5¾	5⅝	5¾	+ ⅛
12⅛	9⅛	MFS Charter	MCR	1.51	14.7	...	1467	10¼	10⅛	10¼	...
9¼	7¼	MFS Intermd	MIN	1.05	13.3	...	3414	7⅞	7¾	7⅞	...
10⅛	7	MFS MultInco	MMT	1.23	15.6	...	1486	7⅞	7¾	7⅞	+ ⅛
10½	9¼	MFS MuniTr	MFM	.77	8.1	...	490	9½	9⅜	9½	+ ⅛
9¼	7⅝	MFS MultTot	MFT	.84a	10.0	...	226	8⅜	8⅛	8⅜	+ ⅛
10⅜	7¾	MFS GvMkTr	MGF	1.18	14.5	...	1867	8⅛	8	8⅛	...
10⅜	7½	MFS Inco	MFO	1.20	15.5	...	227	8	7¾	7¾	− ⅛
n 15⅛	11¼	MFS SpcVal	MFV	1.65	13.9	...	93	12	11⅞	11⅞	...
16¼	10⅝	MGI Prop	MGI	1.12	10.0	15	89	11¼	11⅛	11¼	...
23⅛	10¾	MGM/UA	MGM	4.00e	28.8	...	75	13⅞	13¾	13⅞	...
n 16⅜	12⅝	MGMGrand	MGG		...	16	118	15½	15⅜	15⅜	...
1⅜	½	MHI Gp	MH		...	25	29	¾	11/16	¾	...
s 29¼	10¾	MNC Fnl	MNC	1.16	10.1	4	1890	11⅝	11⅜	11½	− ⅛
14¼	7¾	MagneTek	MAG		...	8	179	10⅞	10½	10⅝	...

Source: The Wall Street Journal.

Stockbrokers charge a **broker's fee** (commission) on all transactions. Since the federal government deregulated brokers' fees in 1975, these fees have not been limited. Competition, however, has held brokers' fees in check, and they generally range between one and two percent of the transaction amount. (The rate may be as high as 100 percent on very small transactions or below one percent on very large transactions.) Some states impose an additional fee or tax on all transactions.

Odd-lot orders are usually passed on to an odd-lot dealer, who buys round lots and parcels them out into odd lots. Therefore, odd-lot transactions cost more than round-lot transactions. Again, fees may be higher on small transactions and lower on large transactions.

Suppose you buy 200 shares of Sperry stock at $62\frac{1}{8}$. The broker's fee is 1.5% of the transaction price.

$$\begin{aligned} 200 \times 62.125 &= 12425.00 \quad \text{price} \\ 12425 \times 0.015 &= \underline{186.38} + \text{fee} \\ & 12611.38 = \text{cost} \end{aligned}$$

Suppose you sell 150 shares of Rockwell at $73\frac{3}{4}$. The broker's fee for 100 shares is \$80.55 and \$48.50 for the odd lot.

$$\begin{aligned} 150 \times 73.75 &= 11062.50 \quad \text{price} \\ 80.55 + 48.50 &= \underline{129.05} - \text{fee} \\ & 10933.45 = \text{proceeds} \end{aligned}$$

Price = quote × number of shares
Cost = price + fees
Proceeds = price − fees

Table 18.1 Brokers' Fees (example)

Round-lot Fees

Transaction Amount		Rate of Fee
up to	2,500.00	4%
2,500.01–	10,000.00	3%
10,000.01–	24,000.00	2%
24,000.01–	50,000.00	1.5%
50,000.01–	100,000.00	1%
more than	100,000.00	negotiable

Odd-lot Fees

Transaction Amount		Rate of Fee
up to	25.00	10%
25.01–	100.00	8%
100.01–	500.00	6%
500.01–	1,000.00	5%
1,000.01–	1,500.00	4.5%
more than	1,500.00	4%

Example 1 What is the fee on 500 shares at $43\frac{3}{8}$? (See table 18.1.)

$43\frac{3}{8} = 43.375$

Price = 500 × 43.375 = 21687.50

Round-lot fee = 2% = 21687.5 × 0.02 = 433.75

key: 500 [×] 43.375 [=]
display: 21687.5 **price** **fee = 2%**
key: [×] .02 [=]
display: 433.75
edited: $433.75 **fee**

Example 2 What is the fee on 550 shares at $43\frac{3}{8}$? (See table 18.1.)

The fee for 500 shares at $43\frac{3}{8}$ = 433.75 (example 1)

Odd-lot price = 50 × 43.375 = 2168.75

Odd-lot fee = 4% = 2168.75 × 0.04 = 86.75

Total fee = 433.75 + 86.75 = 520.50

key: 50 [×] 43.375 [=]
display: 2168.75 **odd-lot price** **fee = 4%**
key: [×] .04 [=]
display: 86.75 **odd-lot fee**
key alg: [+] 433.75 [=]
key arith: [+] 433.75 [+]
display: 520.5 **total fee**
edited: $520.50

Example 3 The Fallons purchased 300 shares of Armico at $30\frac{1}{8}$. Six months later they sold these shares at $33\frac{1}{4}$. What was their gain (or loss) on this trade? (Fees are listed in table 18.1.)

Price of purchase	= 300 × 30.125	= 9037.50
Fee	= 3% of 9037.50	= 271.13
Cost		= 9308.63
Price of sale	= 300 × 33.25	= 9975.00
Fee	= 3% of 9975.00	= 299.25
Proceeds		= 9675.75
Proceeds of sale		= 9675.75
Cost of purchase		= 9308.63
Gain		= 367.12

key: 300 [×] 33.25 [=]
display: 9975. **selling price**
key: [M+] [×] .03 [=]
display: 299.25 **selling fee**
key: [M−] 300 × 30.125 [=]
display: 9037.5 **purchase price**
key: [M−] [×] .03 [=]
display: 271.125
rounded: 271.13 **purchase fee**
key: [M−] [MR]
display: 367.12
edited: $367.12 **gain**

s. price − s. fee − p. price − p. fee = gain

PRACTICE ASSIGNMENT

Plan: Find the cost of a purchase or the proceeds of a sale.

Price = quote × number of shares

Round-lot/odd-lot combinations must be split.

Cost = price + fees
Proceeds = price − fees

To obtain the gain or loss on a trade, subtract the cost of the purchase from the proceeds of the sale.

1. Use table 18.1 to compute the costs of purchases or proceeds of sales.

	Quote	Number of Shares	Price	Round-lot Fee	Odd-lot Fee	Cost of Purchase	Proceeds of Sale
a.	$17\frac{5}{8}$	500	____	____	____	____	
b.	62	75	____	____	____	____	
c.	$63\frac{1}{8}$	280	____	____	____	____	
d.	$24\frac{7}{8}$	1500	____	____	____		____
e.	$56\frac{1}{2}$	60	____	____	____		____
f.	$37\frac{5}{8}$	160	____	____	____		____

2. Two hundred fifty shares of Kroger were sold at $19\frac{3}{4}$. If the broker charged $150 for the round lot and $52.50 for the odd lot, what were the proceeds?

3. A broker charges 3.5% on all round-lot transactions below $10,000 and 8% on all odd lots. What is the total brokerage fee for a purchase of 225 shares of stock at $41\frac{1}{4}$?

4. Mitchell Corsi purchased 230 shares of American Mills at $42\frac{3}{4}$. He later sold the shares at $48\frac{1}{2}$ through the same broker. (Use table 18.1 for fees.)

a. What was the cost of the purchase?

b. What were the proceeds of the sale?

c. What was the gain (or loss)?

Section 18.3 Stock Performance Criteria

KEY TERMS **Price/earnings ratio** **Current yield**

How Is a Stock's Performance Judged?

Investors interested in earning profits by trading (buying and selling) read the stock exchange quotations for the **price/earnings (P/E) ratio** of stock. Generally, the lower the price/earnings ratio, the better the prospects for future price increases. Investors compare the P/E ratios of several stocks, however, before making a decision.

Investors interested in earning dividend income consult stock exchange quotations for the **current yield** (Yld. %), comparing dividend with price. Generally, the higher the dividend, the better the investment.

There are other criteria that enter into the judgment of stocks, but P/E ratio and current yield are fundamental.

$$\text{price/earnings ratio} = \frac{\text{price per share}}{\text{annual earnings per share}}$$

$$\text{current yield} = \frac{\text{annual dividend per share}}{\text{price per share}}$$

stock: ComAir
quote: $53\frac{1}{2}$
dividend: 2.34 per share
earnings: 9.73 per share

$$\text{P/E ratio} = \frac{53.5}{9.73} = 5$$

$$\text{Current yield} = \frac{2.34}{53.5} = 4.4\%$$

Example 1 Melcorp reported annual earnings of $650,000 and has 80,000 shares of outstanding stock. What is the P/E ratio if the current quote is $57\frac{7}{8}$?

Earnings per share = annual earnings ÷ number of shares
= 650000 ÷ 80000
= 8.125

P/E ratio = price per share ÷ earnings per share
= 57.875 ÷ 8.125
= 7.1230769

key: 650000 [÷] 80000 [=]
display: 8.125 **earnings per share**
key: [M+] 57.875 [÷] [MR] [=]
display: 7.1230769
edited: 7 **P/E ratio**

Example 2 Telexpress, Inc. pays an annual dividend of $3.15 per share. What is the current yield if a share sells for $38\frac{1}{2}$?

Current yield = annual dividend per share ÷ price per share
= 3.15 ÷ 38.5
= 0.0818181

key: 3.15 [÷] 38.5 [=]
display: 0.0818181
edited: 8.2%

PRACTICE ASSIGNMENT

Plan: Look up the stock exchange quotations in a current major newspaper (*The Wall Street Journal,* if available). The Dow Jones Industrial Average is an index of thirty blue chips (most prestigious stocks) and is an indicator of how the major stocks—and perhaps the entire United States economy—are doing.
Find price/earnings ratios and annual yields.
Given total annual earnings and the number of outstanding shares:

Earnings per share = total annual earnings ÷ outstanding shares.

Round the P/E ratio to the nearest whole number.
Round the current yield to the nearest tenth of a percent.
Given price and P/E ratio, solve for approximate annual earnings:

$$\text{P/E ratio} = \frac{\text{price}}{\text{earnings}}$$

$$\frac{\text{earnings}}{\text{price}} = \frac{1}{\text{P/E}}$$

$$\text{Earnings} = \frac{\text{price}}{\text{P/E}}$$

1. Check the New York Stock Exchange (NYSE) report in a major newspaper.

   a. What was the Dow Jones Industrial Average?

   ____

   b. List the ten most active stocks.

   1. ______________
   2. ______________
   3. ______________
   4. ______________
   5. ______________
   6. ______________
   7. ______________
   8. ______________
   9. ______________
   10. ______________

2. Find the P/E ratio and current yield of the following companies listed on the New York Stock Exchange.

	P/E	Yield %
a. American Telephone and Telegraph (ATT)	____	____.____
b. International Business Machines (IBM)	____	____.____
c. Loews	____	____.____
d. Phillips Petroleum	____	____.____
e. Sears	____	____.____
f. Chrysler	____	____.____
g. Texas Instruments	____	____.____
h. Xerox	____	____.____
i. General Dynamics	____	____.____
j. Upjohn	____	____.____

3. Compute the following P/E ratios and current yields.

	Price	Total Earnings	Number of Shares	Earnings per Share	Dividend per Share	P/E Ratio	Current Yield %
a.	27	278,000	50,000	________	3.75	____	____.____
b.	$56\frac{3}{4}$	850,000	95,000	________	5.87	____	____.____
c.	$13\frac{1}{8}$	385,000	140,000	________	2.17	____	____.____
d.	$39\frac{7}{8}$	302,000	88,000	________	2.35	____	____.____
e.	$9\frac{5}{8}$	189,000	75,000	________	1.60	____	____.____
f.	$84\frac{1}{4}$	3,450,000	275,000	________	10.30	____	____.____

4. A corporation earned $1,300,000 last year. If its stock currently sells for $52\frac{1}{2}$ and there are 110,000 shares outstanding, what is the price/earnings ratio?

5. Steiger Chemicals distributed quarterly dividends of $265,000, $281,000, $237,000, and $305,000. There are 123,000 shares of stock outstanding, and the current price is $59\frac{3}{4}$. What is the current yield?

6. A corporation with 80,000 shares of stock has a price/earnings ratio of 6. If the stock is listed at $28\frac{1}{2}$, what are the approximate annual earnings?

Section 18.4 The Bond Market

KEY TERMS **Corporate bonds** **Government bonds** **Municipal bonds** **Coupon bonds** **Accrued interest**

How Do Bonds Differ from Stocks?

In addition to issuing stocks, corporations may borrow money from investors by issuing **corporate bonds.** Stockholders own part of a business, while bondholders lend money on a long-term basis to an organization. (Sometimes convertible bonds are issued, which can be converted to common stock shares.)

A bond represents a long-term agreement by a corporation or government to repay a loan on the maturity date and to pay interest at a specified rate at specified intervals. (Callable bonds can be called in by a corporation and paid up prior to maturity.)

Government bonds may be issued by all levels of government. Bonds issued by the federal government (treasury bonds) are considered the safest possible investments because they are backed by the United States. Bonds issued by states, counties, cities, and other entities such as school districts, are called **municipal bonds.**

Because municipal bonds are considered safer investments than corporate bonds and because their interest payments are usually exempt from federal tax, they generally pay a lower rate of interest than do corporate bonds. On the other side of the bond spectrum is the "junk bond," a high-risk corporate bond that pays a high interest rate.

The bond ratings published by Moody's and Standard and Poor's are widely recognized. These ratings measure the quality of bonds and rate them AAA to C or D. Triple A bonds are safest; C or D bonds are speculative (risky).

Comparison of Common Stocks and Bonds

Common Stocks	Bonds
Represent ownership	Represent loans
May pay dividend per share No guarantee No fixed rate	Pay interest on face value Guaranteed Fixed rate*
Shareholders elect corporate board of directors	Bondholders exercise no control over management
Not usually repaid by corporation (Occasionally a corporation may buy back stocks.)	Repaid on maturity date (or before, if bonds are callable)
Issued by all corporations Not issued by governments Prices and yields may fluctuate sharply Riskier than bonds	Not issued by all corporations May be issued by governments Prices and yields fluctuate more slowly, influenced by interest rate trends Safer than stocks, especially government bonds or bonds issued by prominent corporations in times of stability (not "junk bonds")

*Most bonds feature fixed interest rates, although some feature variable rates. This chapter considers only fixed-rate bonds.

The par value, or face value, of a bond is called the denomination and is the amount lent by the bondholder. Most bonds are issued in $1,000 denominations, although municipal bonds are now issued in $5,000 denominations. Bonds pay stated annual interest rates and may specify accumulated interest payable upon maturity. Bonds that pay interest semiannually on specified dates are **coupon bonds,** so named for interest coupons attached to the bond certificate. Today actual coupons may not be attached to the bonds, but the term "coupon bond" persists.

The Martinez family owns five $7\frac{1}{2}\%$ bonds that pay interest on June 1 and December 1.

How much is each interest coupon worth?
How much interest does the family collect every six months?

$$1000 \times 0.075 \times \tfrac{1}{2} = 37.50$$
$$5 \times 37.5 = 187.50$$

U.S. Savings Bonds are a special case. People with relatively little money to invest can buy them at banks and savings and loan associations or through payroll savings plans in denominations as small as $25. The interest accumulates to maturity and is exempt from state and local taxes. If U.S. Savings Bonds are cashed in prior to maturity, yields are reduced.

How Are Bonds Traded?

Most bonds, like stocks, are traded through brokers on exchanges or over the counter. Market values, which may differ from par values, fluctuate each day. Bond investors may gain or lose through trading (buying and selling), but many investors hold bonds for interest income. The market value of a bond is stated as a percent of par value. Bonds that sell below par value, are said to sell at a discount; bonds that sell above par value are said to sell at a premium.

Recall that a share of stock quoted at $84\frac{1}{2}$ can be bought or sold for \$84.50.

A bond quoted at $84\frac{1}{2}$ can be bought or sold for 84.5% of par value. Since par value of most bonds is \$1,000, the price is \$845 (at a discount).

A bond quoted at $104\frac{1}{8}$ (at a premium) can be bought or sold for \$1,041.25.

A broker's fee for trading bonds must be added to cost or subtracted from proceeds.

Example 1 Bond quote is $72\frac{3}{4}$ (at a discount)

Bond price is 72.75% of 1000 = 1000 × 0.7275 = 727.5
Or 10 × 72.75 = 727.5

Example 2 Bond quote is $106\frac{3}{8}$ (at a premium)

Bond price is 106.375% of 1000 = 1000 × 1.06375 = 1063.75
Or 10 × 106.375 = 1063.75

An investor earns interest on a bond for the period of time during which it is held. A bond sold between interest coupon dates pays to the buyer the next scheduled interest payment in full even though the buyer did not own the bond for the entire interest period. The seller of the bond, however, receives none of the interest. For this reason, it is customary for the buyer to pay **accrued interest** to the seller; that is, the buyer pays to the seller the interest accumulated from the last payment date to the date preceding the sale.

When the stability of future interest is uncertain, coupon bonds may be traded without accrued interest (flat). Bonds that pay interest at maturity, called zero coupon bonds, are also traded flat.

A bond pays $8\frac{1}{4}$% interest on March 1 and September 1. What accrued interest does a buyer pay the seller on June 5?

Accrued interest = $I = PRT$
P = 1,000 (not market value)
R = 8.25%
$T = d/360$ (customarily ordinary interest)
d = days from March 1 to June 4
= 30 + 30 + 31 + 4 = 95

$I = PRT$
= 1000 × 0.0825 × 95 ÷ 360
= 21.770833

If five bonds are bought, accrued interest equals 5 × 21.770833 = 108.85

Cost = price + accrued interest + fee
Proceeds = price + accrued interest − fee

Example 3 On May 10, the Antelope Lodge purchased five $8\frac{1}{4}$% bonds at $95\frac{3}{4}$ with interest payable on March 15 and September 15. The broker's fee was \$7.50 for the first bond and \$5 for each additional bond. What was the cost of this purchase?

Price per bond = 10 × 95.75 = 957.50
Price for five bonds = 5 × 957.50 = 4787.50
Days for accrued interest from March 15 to May 9
= 16 + 30 + 9 = 55 (or day 129 − day 74 = 55)
Accrued interest per bond
= 1000 × 0.0825 × 55 ÷ 360 = 12.604167
Accrued interest for five bonds = 5 × 12.604167 = 63.02

Fee = 7.50 + 4 × 5 = 27.50

Cost = price + accrued interest + fee
= 4787.50 + 63.02 + 27.50 = 4878.02

key: 5 [×] 957.5 [=] [M+] **price**
5 [×] 1000 [×] .0825 [×] 55
[÷] 360 [=] [M+] **accrued interest**
27.5 [M+] **fee**
[MR]
display: 4878.0208
edited: \$4,878.02

The Antelopes sold their five bonds the following September 1. The bond was quoted at 102, and the broker's fee remained the same. What was their gain on the trade?

Price per bond = 10 × 102 = 1020
Price for five bonds = 5 × 1020 = 5100
Days for accrued interest from March 15 to August 31
= 16 + 30 + 31 + 30 + 31 + 31 = 169
(or 243 − 74 = 169)
Accrued interest per bond
= 1000 × 0.0825 × 169 ÷ 360 = 38.729167
Accrued interest for five bonds = 5 × 38.729167
= 193.65

Fee = 27.50

Proceeds = price + accrued interest − fee
= 5100.00 + 193.65 − 27.50 = 5266.15
Gain = 5266.15 − 4878.02 = 388.13

key: 5 [×] 1020 [=] [M+] **price**
5 [×] 1000 [×] .0825 [×] 169 [÷]
360 [=] [M+] **accrued interest**
27.5 [M−] **fee**
[MR]
display: 5266.1458
edited: $5,266.15

PRACTICE ASSIGNMENT

Plan: Assume bonds with interest at fixed annual rates payable semiannually. Assume par values of $1,000.

$$\text{Semiannual interest per bond} = 1000 \times \text{rate} \times \tfrac{1}{2}$$

Quotes are stated as a percent of par value (face value). If par value is $1,000, use the shortcut of multiplying the decimal form of the quote by 10. A bond priced below $1,000 is at a discount. A bond priced above $1,000 is at a premium. The accrued interest on a bond traded between interest coupon dates is the ordinary interest from the last date interest was paid until the day prior to the trading date.

The cost of a purchase equals:

Number of bonds × price per bond
+ Number of bonds × accrued interest per bond
+ Transaction fee

The proceeds of a sale equal:

Number of bonds × price per bond
+ Number of bonds × accrued interest per bond
− Transaction fee

Obtain the gain or loss on a bond trade by subtracting the cost of the purchase from the proceeds of the sale.

1. Compute the semiannual interest on the following bond investments.

	Interest Rate	Annual Interest per Bond	Semiannual Interest per Bond	Number of Bonds Held	Semiannual Interest
a.	8%	________	________	2	________
b.	$7\frac{3}{4}$%	________	________	2	________
c.	9.1%	________	________	5	________
d.	7.85%	________	________	10	________
e.	8.9%	________	________	7	________
f.	10.05%	________	________	8	________

2. Compute the market values of the following bond investments (disregarding fees and accrued interest).

	Quote	Price per Bond	Number of Bonds Held	Value of Investment
a.	97	________	2	________
b.	104	________	5	________
c.	$102\frac{1}{2}$	________	2	________
d.	$98\frac{3}{4}$	________	7	________
e.	$95\frac{7}{8}$	________	10	________
f.	$105\frac{1}{8}$	________	5	________

3. Given the following rates and interest dates, compute the accrued interest.

	Interest Dates	Date of Sale	Number of Days	Interest Rate	Accrued Interest
a.	Mar 1 and Sept 1	Apr 2	_______	9%	_______
b.	Apr 15 and Oct 15	Aug 2	_______	$8\frac{1}{2}$%	_______
c.	June 15 and Dec 15	Aug 1	_______	9.1%	_______
d.	May 1 and Nov 1	Sept 21	_______	$10\frac{1}{4}$%	_______
e.	June 1 and Dec 1	July 29	_______	8.4%	_______
f.	June 1 and Dec 1	Oct 22	_______	7.95%	_______

4. On November 7, 1990, Brenda and Ray ordered their broker to sell six $8\frac{1}{2}$% bonds. Interest dates were March 1 and September 1. How much accrued interest did they receive?

5. Michelle Rose sold five bonds quoted at $101\frac{7}{8}$. Accrued interest was $19.67 per bond, and her broker's fee was $8 per bond. What were the proceeds?

6. An investment club purchased seven $9\frac{1}{4}$% bonds at $97\frac{3}{4}$ on April 25, paying accrued interest and a transaction fee of $6 per bond. The bonds were sold on August 2 at $103\frac{1}{8}$. Accrued interest was received. The transaction fee remained the same. If the interest dates were March 1 and September 1, what was the gain from the sale?

Section 18.5 Bond Performance Criteria

KEY TERMS **Current yield** **Yield to maturity**

The **current yield** of a bond is the ratio of annual interest to current price, and is listed in newspapers' securities exchange reports.

$$\text{Current yield} = \frac{\text{annual interest}}{\text{current price}}$$

A 7.8% bond is quoted at 84.5

$$\text{Current yield} = \frac{78}{845} = 0.0923076$$
$$= 9.2\%$$

A bond has a maturity date, and an investor needs to know the yield rate over time if the bond is kept until the maturity date. The yield rate of a bond purchased at par value is the stated rate. The yield rate of a bond bought at a discount is greater than the stated rate. The yield rate of a bond bought at a premium is smaller than the stated rate.

Example 1 The quoted price for Medco Corporation's $8\frac{1}{2}\%$ bonds is $101\frac{1}{4}$. What is the current yield of these bonds?

Annual interest = 1000 × 0.085 = 85.00

Current price = 101.25 × 10 = 1012.50

$$\frac{\text{annual interest}}{\text{current price}} = \frac{85}{1012.5} = 0.0839506 = 8.4\%$$

The **yield to maturity** is another common measure of bond yield. A precise determination requires an advanced formula. The following approximation formula, called the "Bond Salesman's Method," is fairly simple to do on a calculator and leads to reasonably accurate results.

$$\text{Approximate yield to maturity} = \frac{\text{annual interest} + \dfrac{\text{par} - \text{price}}{\text{years}}}{\frac{1}{2} \times (\text{par} + \text{price})}$$

Years stands for the number of years remaining to maturity. The denominator is called the average investment.

Example 2 An 8.4% bond is bought at $103\frac{3}{4}$ with five years remaining to maturity. What is the approximate yield to maturity?

Annual interest = 1000 × 0.084 = 84.00

Price = 103.75 × 10 = 1037.50

$$\frac{\text{annual interest} + \dfrac{\text{par} - \text{price}}{\text{years}}}{\frac{1}{2} \times (\text{par} + \text{price})} = \frac{84 + \dfrac{1000 - 1037.5}{5}}{(1000 + 1037.5) \div 2} = 0.075092$$

The algebraic calculator sequence must be entered carefully with the order of operations in mind. The parenthesis keys may be used if available, but in this case the [=] key and the M-memory are more advantageous. Compute the denominator first.

The usual arithmetic sequence will not yield the correct result on some desk calculators. The most simple procedure is to compute with partial results.

key algebraic:
1000 [+] 1037.5 [=] [÷] 2 [=] [M+]
1000 [−] 1037.5 [=] [÷] 5 [+] 84 [=] [÷]
[MR] [=]

key arithmetic:
1000 [+] 1037.5 [+] [÷] 2 [=]
display: 1018.75 record, clear all
1000 [+] 1037.5 [−] [÷] 5 [=]
display: −7.5 record, clear all
84 [+] 7.5 [−] [÷] 1018.75 [=]
display: 0.075092
edited: 7.5%

PRACTICE ASSIGNMENT

Plan: Compute current yields and approximate yields to maturity. Obtain current yield with a single division. Obtain yield to maturity as described above.

If a buyer purchases a bond at par, the yield is the stated rate. If a buyer pays less than par (at a discount), the yield is greater than the stated rate. If a buyer pays more than par (at a premium), the yield is less than the stated rate. Round yields to the nearest tenth of a percent.

1. Compute the current yields and the approximate yields to maturity for the following five bonds.

Quote	Price	Interest Rate	Annual Interest	Current Yield	Years to Maturity	Yield to Maturity
a. 100	______	9%	______	______	6	______
b. $98\frac{1}{2}$	______	7%	______	______	11	______
c. $102\frac{1}{4}$	______	$8\frac{1}{2}$%	______	______	8	______
d. $104\frac{1}{8}$	______	8.6%	______	______	$7\frac{1}{2}$	______
e. $97\frac{5}{8}$	______	7.85%	______	______	$9\frac{1}{2}$	______

2. Eight and one-half percent bonds of Nexfax, Inc. are quoted at $103\frac{3}{4}$. What is the current yield of these bonds?

3. Tina Palermo bought a 7.9% bond at $95\frac{3}{4}$ with six years left until maturity. What is the approximate yield to maturity?

4. Twenty-year bonds of the Kalcom Corporation are quoted at $98\frac{1}{4}$ and pay $8\frac{3}{4}$%. What is the approximate yield to maturity of an eleven-year-old bond?

5. Robert Hamer purchased one 8.3% bond at $98\frac{1}{2}$ and a second 7.8% bond at $95\frac{3}{8}$. Which bond offers the higher current yield?
Current yield of bond 1: ______%
Current yield of bond 2: ______%
Bond with higher current yield: ________

Challenger

Investors may wish to determine the maximum price they should pay for a bond in order to secure a particular yield. The equation for the approximate yield to maturity can be manipulated to produce the maximum bond price given the yield. The derived equation is:

$$\text{Price} = \frac{\text{interest} \times \text{years} + \text{par} \times (1 - 0.5 \times \text{years} \times \text{yield}}{1 + 0.5 \times \text{years} \times \text{yield}}$$

The Richland Corporation issued 8.9% twenty-year coupon bonds in October 1981 (par = $1,000). In April 1990, Gary and Francine Nesmith requested their broker to purchase a Richland bond if the price would guarantee them an approximate 11% yield to maturity. What was the highest *quote* the Nesmiths would accept? (Round to eighths.)

CHAPTER 18 SUMMARY

Concept	*Example*	*Procedure*	*Formula*
SECTION 18.1 Common and Preferred Stock			
Par value of stock	Common stock issued = \$2,000,000 capital 20,000 shares have been authorized. Par value = $\frac{2000000}{20000}$ = 100	Divide the capitalization by the total number of authorized shares.	Par value = $\frac{\text{capital}}{\text{authorized shares}}$ capital[÷]shares[=] capital/shares
Dividends for preferred stock	Par value of 5% preferred stock = \$100 Number of shares held = 1,000 Annual dividend per share = 0.05 × 100 = 5.00 Quarterly dividend per share = 5 ÷ 4 = 1.25 Quarterly dividend check = 1.25 × 1000 = 1250.00	Dividend rate is specified. Preferred shareholders receive first claim to any declared dividends. To obtain annual dividend per share, multiply par value by rate. Divide by four to obtain quarterly dividend per share. Multiply by the number of shares owned to determine the quarterly dividend check.	Annual dividend per share = rate × par rate[×]par[=] rate*par Quarterly dividend per share = $\frac{\text{rate} \times \text{par}}{4}$ rate[×]par[÷]4[=] rate*par/4
Dividends for common stock	50,000 shares of common stock outstanding Total dividend = 200,000 Dividend allocated to preferred stockholders = 150,000 Number of shares held = 1,000 Total common stock dividends = 200000 − 150000 = 50000 Dividend per share = 50000 ÷ 50000 = 1.00 Dividend check = 1.00 × 1000 = 1000.00	After dividends have been allocated to preferred stockholders, the remainder is distributed to common stockholders. To obtain dividend per share, divide the available dividend by the total number of common shares outstanding. Multiply by the number of shares owned to determine the dividend check.	Common dividend = total dividend − preferred dividend alg: totdiv[−]prefdiv[=] arith: totdiv[+]prefdiv[−] totdiv−prefdiv Dividend per share = $\frac{\text{common dividend}}{\text{common shares}}$ div[÷]shares[=] div/shares
SECTION 18.2 The Stock Market			
Stock quotation	Stock quoted at $42\frac{3}{8}$ One share of stock costs \$42.375.	The price of one share of stock is stated as a mixed number; the fractional part is $\frac{1}{8}$ or a multiple of $\frac{1}{8}$. Stock market quotations are listed in most major newspapers.	
Round lot Odd lot	300 shares are traded as a round lot. 30 shares are traded as an odd lot. 325 shares are traded as one round lot and one odd lot.	Shares are traded in a round lot (multiple of 100) or an odd lot (less than 100). Combinations are split into a round lot and an odd lot.	
Broker's fee (commission)	350 shares are purchased at $64\frac{7}{8}$. Fees set by broker are: round lot: 2% odd lot: 4% Price = 350 × 64.875 = 22706.25 Round lot fee = 0.02 × 300 × 64.875 = 389.25 Odd lot fee = 0.04 × 50 × 64.875 = 129.75 Cost = 23225.25	Brokers can set any fee or odd-lot differential. Fees may be commission rates or fixed amounts. The cost of a purchase is price + fees. The proceeds of a sale are price − fees, where price = number of shares × quote.	Price = no. of shares × quote shares[×]quote[=] shares*quote Cost of purchase = price + fees alg: shares[×]quote[+]fees[=] arith: shares[×]quote[=][+]fees[+] shares*quote+fees Proceeds of sale = price − fees alg: shares[×]quote[−]fees[=] arith: shares[×]quote[=][+]fees[−] shares*quote−fees
Gain or loss from trading	Proceeds of lot = 26811.17 Cost of lot = 23225.25 Gain = 3585.92	To determine gain or loss, subtract cost from proceeds.	Gain (or loss) = proceeds − cost alg: proceeds[−]cost[=] arith: proceeds[+]cost[−] proceeds−cost

CHAPTER 18 SUMMARY—*(Continued)*

Concept	*Example*	*Procedure*	*Formula*
SECTION 18.3 Stock Performance Criteria			
Price/earnings ratio	Price per share = $68\frac{1}{8}$ Total annual earnings = \$1,500,000 Number of outstanding shares = 100,000 Earnings per share = 1500000 ÷ 100000 = 15 P/E ratio = $\frac{68.125}{15}$ = 5	The P/E ratio gives an indication of how favorable a stock is to a potential investor. The lower the ratio, the greater a bargain the stock seems to be. The P/E ratio is rounded to a whole number in the stock market reports.	Earnings per share = $\frac{\text{total earnings}}{\text{number of shares}}$ totearn[÷]shares[=] totearn/shares P/E ratio = $\frac{\text{price per share}}{\text{earnings per share}}$ quote[÷]earn[=] quote/earn
Current yield	Price per share = $68\frac{1}{8}$ Annual dividend per share = 5.50 Current yield = $\frac{5.5}{68.125}$ = 0.0807339 = 8.1%	The current yield gives an indication of how much income is received from a stock. The ratio of annual dividend to price is listed as percent (rounded to one-tenth of a percent) in the stock market reports.	Current yield = $\frac{\text{annual dividend per share}}{\text{price per share}}$ andiv[÷]quote[=] andiv/quote
SECTION 18.4 The Bond Market			
Par value of a bond	Par value = \$1,000	The par value (face value) of a bond represents the amount lent by the bondholder to the organization.	
Bond interest	Interest rate = 7% Annual interest = \$70 Semiannual interest (coupon interest) = \$35	Most bonds pay interest semiannually. To determine each interest payment, multiply the rate by the par value (usually \$1,000) and divide by 2.	Semiannual interest = $\frac{\text{par} \times \text{rate}}{2}$ par[×]rate[÷]2[=] par*rate/2
Current price	Quote = $84\frac{1}{4}$ = 84.25 84.25% of 1000 = 1000 × 0.8425 = 842.50 Or: 10 × 84.25 = 842.50	A quote represents a percent of par. If par = 1,000, a shortcut method for calculating the current price is to multiply the decimal form of the quote by 10.	Price = $\frac{\text{quote}}{100} \times$ par quote[÷]100[×]par[=] quote/100*par
Accrued interest	8% bond paying interest on Jan. 15 and July 15 is bought on May 24 at $93\frac{3}{4}$. Broker's fee = \$6.00 Time for accrued interest = Jan. 15–May 23 = 128 days Accrued interest = 1000 × 0.08 × $\frac{128}{360}$ = 28.44 Cost of purchase = 937.50 + 28.44 + 6.00 = 971.94	The buyer pays the seller ordinary interest for the period from the last interest date to the day before the transfer date. There is a broker's fee for trading bonds. To determine the cost of a purchase, add the accrued interest plus broker's fee to the current price. To determine the proceeds of a sale, add the accrued interest to the current price and subtract the broker's fee.	Accrued interest = par × rate × days/360 par[×]rate[×]days[÷]360[=] par*rate*days/360 Cost = price + accrued interest + fee alg: price[+]accrint[+]fee[=] arith: price[+]accrint[+]fee[+] price+accrint+fee Proceeds = price + accrued interest − fee alg: price[+]accrint[−]fee[=] arith: price[+]accrint[+]fee[−] price+accrint−fee

CHAPTER 18 SUMMARY—(Continued)

Concept	*Example*	*Procedure*	*Formula*
SECTION 18.5 Bond Performance Criteria			
Current yield	7.8% bond is quoted at $84\frac{1}{2}$ Current yield $= \frac{78}{845}$ $= 0.0923076 = 9.2\%$	The current yield indicates how the interest compares to the current price.	Current yield $= \frac{\text{annual interest}}{\text{current price}}$ quote[÷]100[×]par[=][M+] int[÷][MR][=] or int[÷] [(]quote[÷]100[×]par[)][=] int/(quote/100*par)
Approximate yield to maturity	8.4% bond bought at $103\frac{3}{4}$ has 5 years remaining to maturity Annual interest = $84 Approximate yield to maturity $= \frac{84 + \frac{1000 - 1037.5}{5}}{\frac{1}{2} \times (1000 + 1037.5)}$ $= 0.075092 = 7.5\%$	The approximate yield to maturity indicates how interest compares to price if the bond is kept to maturity. If the bond is bought at par, yield to maturity is the stated rate; if bought at a premium, the yield to maturity is lower; if bought at a discount, the yield to maturity is higher.	Approximate yield to maturity $= \frac{\text{annual interest} + \frac{\text{par} - \text{price}}{\text{years}}}{\frac{1}{2} \times (\text{par} + \text{price})}$ alg: par[+]price[=][÷]2[=][M+] par[−]price[=][÷]years[+] int[=][÷][MR][=] (int+(par−price)/years)/ (.5*(par+price))

Chapter 18 **Spreadsheet Exercise**

Yields to Maturity

An investor who keeps bonds for high yields needs to know how close the current yield is to the long-term yield (yield to maturity), and how yields to maturity vary as the number of remaining years varies.

Use the power of the computer and the approximate yield to maturity formula to answer these questions.

Your chapter 18 template will appear as follows:

```
A1: PR [W8]                                                          READY

        A           B     C     D         E      F      G      H      I      J
1                                                    YIELDS TO MATURITY
2
3                                                    years to maturity
4   interest                 current       ---------------------------------------
5     rate        quote       yield             5     10     15     20     25
6   -------------------------------------------------------------------------------
7    10.00%        120
8    10.00%        110
9    10.00%        100
10   10.00%         90
11   10.00%         80
12
13    9.00%        120
14    9.00%        110
15    9.00%        100
16    9.00%         90
17    9.00%         80
18
19    8.00%        120
20    8.00%        110
```

Column A contains bond interest rates: 10% in rows 7–11, 9% in rows 13–17, 8% in rows 19–23.

Column B contains bond quotes ranging from 80 to 120.

Spreadsheet Exercise Steps

1. Column D: Current Yield

$$\text{Current yield} = \frac{\text{interest}}{\text{price}} = \frac{1000 \times \text{rate}}{10 \times \text{quote}} = \frac{100 \times \text{rate}}{\text{quote}}$$

The rate is in A7; the quote is in B7.
The spreadsheet formula is
_______ * ______ / ______.
Complete the formula and store it in D7.
Copy the formula from D7 to D8..D11, from D7 to D13..D17 and from D7 to D19..D23.

2. Columns F, G, H, I, J: yields to maturity

Approximate yield to maturity

$$= \frac{\text{annual interest} + \dfrac{\text{par} - \text{price}}{\text{years}}}{\frac{1}{2} \times (\text{par} + \text{price})}$$

The interest rate is in A7, par = 1000, the quote is in B7, and the years begin in F5.
The column addresses of A7 and B7 and the row address of F5 must be absolute.
The spreadsheet formula is:

(1000*$A7+(1000−10*$B7)/F$5)/
(.5*(1000+10*$B7))

Enter this formula very carefully (including all parentheses and symbols). Store the formula in F7; copy it from F7 to F7..J11, from F7 to F13..J17 and from F7 to F19..J23.
When the price is at par, all yields are the same as the bond rate.
When the price is at a premium, all yields are lower than the bond rate.
When the price is at a discount, all yields are higher than the bond rate.
However, when the period to maturity is long, the effect of the purchase price is somewhat diminished, because the difference between par and price is spread over the years.

3. Change the interest rates in rows 7–11 to 12% (enter as .12) and complete the following chart:

Quote	Current Yield	Yield to Maturity Bond Kept 5 Years	Bond Kept 25 Years
120	______%	______%	______%
100	______%	______%	______%
80	______%	______%	______%

4. What is the approximate yield to maturity (rounded to the nearest percent) of a five-year bond that pays 15% interest and sells at 80?

NAME ______________________________

CLASS SECTION ______________ DATE ______________

SCORE __________ MAXIMUM 140 PERCENT __________

Chapter 18 Assignment

Round all rates in your answers to the nearest tenth of a percent.

1. Dennis Minter owns 125 shares of Healthcare, Inc. common stock. The company declared an annual dividend of $476,560 for its 148,000 outstanding shares of common stock. What quarterly dividend should Dennis receive? *(5 points)*

2. Astrocorp distributed $224,250 in common stock dividends for the year. There are 130,000 shares of common stock outstanding. Dawn Rojas received a quarterly dividend check of $120.75. How many shares does she own? *(5 points)*

3. Kristin Grosz ordered her broker to purchase 400 shares of Northrup at $23\frac{5}{8}$. The broker charged a $2\frac{3}{4}$% fee. What was the cost of the purchase? *(5 points)*

4. Bruce Daniels purchased four $8\frac{3}{4}$% bonds on July 2. The interest dates are March 15 and September 15. How much accrued interest must be paid? *(5 points)*

5. Consolidated Utilities has 95,000 shares of stock outstanding. What is the current yield if the corporation distributed $598,000 in dividends and its price is $36\frac{5}{8}$? *(5 points)*

6. An investor purchased several $7\frac{3}{4}$% bonds at $98\frac{3}{8}$. What is the current yield of these bonds? *(5 points)*

7. Carl Shimura ordered his broker to sell 160 shares of Paramount at $65\frac{7}{8}$. The broker charged commissions of 1.8% for the round lot and 4.5% for the odd lot. What were the proceeds of the sale? *(10 points)*

8. An investor purchased three 8.3% coupon bonds on May 5, 1988. Interest dates were April 1 and October 1. How much interest had the investor received from the corporation by October 15, 1991? *(5 points)*

9. On September 1 the Mershacks's broker purchased six 9.2% bonds at $98\frac{1}{4}$. They paid the accrued interest to the seller. Interest dates are April 15 and October 15. The broker charged $6.50 per bond. What was the cost of this purchase? *(10 points)*

10. The Murchisons purchased a 9.1% bond at $101\frac{7}{8}$. The bond will mature in $12\frac{1}{2}$ years.
What is the current yield?
What is the approximate yield to maturity? *(10 points)*

11. Precision Technology issued 10,000 shares of common stock (par value $80) and 13,000 shares of 6.5% preferred stock (par value $50). The corporation declared a quarterly dividend of $90,000. How was this dividend apportioned between preferred and common stockholders?

Total preferred dividend: ____________
Preferred dividend per share: ___________
Total common dividend: ____________
Common dividend per share: ___________ *(20 points)*

12. Last May, Dana Hamlin's broker purchased 55 shares of stock at $83\frac{1}{2}$. The broker charges an odd-lot fee of 6% on amounts below $5,000 and 5% on amounts of $5,000 and above. In September, the 55 shares were sold at $96\frac{1}{4}$ through the same broker.

a. What was the cost of the purchase?

b. What were the proceeds of the sale?

c. What was the gain (or loss)? *(15 points)*

13. Pan-World Airlines has 125,000 shares of stock listed at $31\frac{1}{4}$. Microman Software has 110,000 shares of stock listed at $23\frac{1}{8}$. Last year Pan-World earned $475,000 and Microman earned $441,000. Which company has the better price/earnings ratio?
Pan-World P/E ratio: _____
Microman P/E ratio: _____
Stock with better P/E ratio: ___________ *(15 points)*

14. On May 9, 1991, a broker sold four $8\frac{1}{4}$% bonds on behalf of her client. The quote was $102\frac{3}{8}$ and the buyer paid the seller accrued interest. The interest dates were January 15 and July 15. The broker's fee was $8 for the first bond and $5.50 for the others. What were the proceeds from the sale?
(15 points)

15. An investor purchased a 9% bond of PDQ Enterprises at $101\frac{3}{8}$ and an $8\frac{3}{4}$% bond of Lite-Way, Inc. at $99\frac{1}{4}$. The PDQ bond matures in seven and a half years, while the Lite-Way bond matures in nine years. Which bond has the higher approximate yield to maturity?
PDQ approximate yield to maturity: _______%
Lite-Way approximate yield to maturity: _______%
Bond with higher approximate yield to maturity:
___________ *(10 points)*

Chapter 19 Insurance: Life and Auto

OBJECTIVES

- Compare basic life insurance plans
- Analyze nonforfeiture options
- Learn features of interest-sensitive life insurance
- Understand auto insurance coverages

SECTIONS

INTRODUCTION

Financial security is important to families, individuals, and businesses. Purchasing insurance is one way of coping with financial uncertainty. There are insurance policies for everything from homes to health, but this chapter focuses on two of the more common types of coverage—life and auto.

Section 19.1 Life Insurance

KEY TERMS **Policy** **Beneficiary** **Premium** **Term insurance** **Straight life** **Limited payment life** **Endowment** **Interest-sensitive life insurance**

What Is the Purpose of Life Insurance?

A life insurance **policy** is a contract between the insurer and the insured that provides financial protection for the **beneficiary** of the policy in the event of the policyholder's premature death. A policyholder may be a parent who wishes to provide for children or a businessperson whose partner is the beneficiary.

A policyholder buys a policy by paying a **premium** to an insurer. Paying premiums at regular intervals keeps a policy in force. The premium is small compared to policy benefits because the insurance company spreads the risk of loss over a large group of policyholders.

A stock insurance company is owned by shareholders. A mutual insurance company is a cooperative association owned by the policyholders rather than stockholders. Premiums must cover all possibilities. If an excess of funds accumulates, refunds may be paid to policyholders in the form of dividends. Policies that pay dividends are called participating policies. Most mutual companies issue participating policies; most stock companies issue nonparticipating policies.

What Are the Classical Life Insurance Plans?

Face value is the predetermined policy benefit. Younger people pay lower premiums than older people pay because young peoples' premiums have more time to accumulate. Women pay lower premiums than men pay because a woman's life expectancy exceeds a man's. Premiums that remain the same each year are called level premiums.

Four basic types of life insurance have developed over the years—**term, straight life, limited payment life,** and **endowment.**

Policy	How Long Premiums Are Paid	How Long Coverage Lasts	Explanation and Comparative Cost
Term	During the term	During the term	Premiums can be low because the benefit is not necessarily paid. Example: 10-year term.
Straight life (whole life, ordinary)	For a lifetime	For a lifetime	Premiums are higher than for term insurance because the benefit is always paid.
Limited payment life	During a specific period	For a lifetime	Premiums are higher yet. Payments stop at the end of the specified period. The benefit is always paid. Example: 30-year period.
Endowment	During a specific period	During period; benefit is paid to holder if period is survived	Most expensive policy. Combines insurance with savings. Benefit is always paid. Example: 20-year period. Benefit paid at age 65.

What Are the Investment Options of Life Insurance Plans?

With the exception of term insurance, the classical plans not only purchased protection but also built up an investment. These plans entitled policyholders to certain nonforfeiture options if they decided to turn in their policies and discontinue the insurance. The principal nonforfeiture options were cash value, paid-up insurance, and extended term.

Nonforfeiture Options of Classical Life Insurance Policies (except term)

Suppose a policyholder takes out $30,000 of straight life at age 31 and then discontinues the policy at age 51.

Cash Value	If the insured discontinues the policy after the first few years, he or she can receive cash payments, increasing each year but often amounting to less than the paid-in premiums. Assume a cash value of $5,120. The policyholder terminates the insurance and receives $5,120 in cash.
Paid-up Insurance	The policyholder can use the cash value to purchase a reduced level of paid-up insurance. Assume a paid-up insurance policy of $15,840. Without paying any further premiums, the former policyholder will have coverage of $15,840 for the remainder of his or her life.
Extended Term	The policyholder can use the cash value to purchase insurance for the full benefit for a limited period at no additional cost. Assume an extended term of 15 years. The former policyholder will receive the same coverage for 15 more years without paying any additional premiums.

How Do Modern Life Insurance Plans Differ from Classical Plans?

Fluctuating interest rates along with inflation have challenged insurers to maintain attractive plans. For example, when educational expenses were stable, a 20-year endowment policy was a sound method for financing higher education, whether or not the policyholder survived. Today, pure endowment policies, which have always been fairly expensive, are no longer economical because the buying power of a long-term endowment may decrease by 50 percent or more.

In response to a rapidly changing financial climate, new insurance policies stress flexibility. Basic plans are still available, but now insurance agents sell **interest-sensitive life insurance** (flexible whole life policies). The term "interest-sensitive" means that premiums and benefits are subject to change with interest rates. In theory, interest-sensitive policies allow policyholders to purchase custom designed policies.

Interest-sensitive Insurance Policies

Benefits	Range from low to high. Can be increased or decreased periodically.
Cash Value	Can increase or decrease.
Premiums	Flexible. May change over the life of the policy. Nonsmokers often pay lower premiums than smokers. Younger people pay lower premiums than older people. Women pay lower premiums than men.
Yield	May change from time to time. A bottom interest rate is usually guaranteed.
Other Variations	Different nonforfeiture options and settlement options exist that vary among companies.
Disadvantages	Reduced sense of security. Planning is difficult and is shifted from company to client.

The following table illustrates a portion of a universal life insurance plan that begins with a face value of $50,000, an assumed yield of 8.3%, and certain assumed operating costs. Keep in mind that there are many variations on this plan between insurance companies.

Table 19.1 Universal Life Insurance Premiums

Age		*Planned Annual Premium*	*Projected Cash Values At End of Year*				*At Age*
			5	*10*	*15*	*20*	*65*
22	M	216.50	551	1437	3147	4857	40382
	MNS	193.21	492	1283	2810	4336	36050
	F	170.50	460	1114	2276	3437	25906
	FNS	155.65	420	1017	2078	3138	23652
24	M	236.50	645	1681	3676	5671	39660
	MNS	211.06	576	1501	3282	5063	35406
	F	184.00	505	1228	2543	3858	24275
	FNS	167.97	461	1121	2322	3522	22163
26	M	256.00	731	1914	4142	6370	37103
	MNS	228.46	652	1709	3698	5687	33123
	F	197.00	542	1321	2752	4183	23758
	FNS	179.84	495	1206	2512	3819	21691
28	M	275.50	805	2132	4541	6949	33473
	MNS	245.86	718	1904	4054	6206	29882
	F	230.00	665	1664	3543	5421	23157
	FNS	209.97	607	1519	3234	4949	21142
30	M	302.00	900	2416	5092	7767	31240
	MNS	269.51	803	2157	4546	6935	27889
	F	256.00	745	1926	4089	6252	22732
	FNS	233.70	680	1759	3733	5708	20754
32	M	341.00	1056	2823	6114	9104	30952
	MNS	304.32	942	2521	5459	8128	27632
	F	282.00	825	2192	4621	7050	22195
	FNS	257.44	753	2001	4219	6436	20264
34	M	374.00	1179	3110	6549	9987	28183
	MNS	333.77	1052	2777	5847	8917	25160
	F	315.00	946	2536	5331	8125	21708
	FNS	287.56	864	2315	4867	7418	19819
35	M	400.00	1271	3346	7086	10825	27055
	MNS	356.97	1134	2988	6327	9665	24153
	F	341.00	1056	2823	5970	9120	21260
	FNS	311.30	964	2578	5450	8329	19410
40	M	544.50	1691	4547	9815	15083	24738
	MNS	485.93	1509	4060	8763	13467	22084
	F	446.00	1400	3693	7887	12080	19670
	FNS	407.15	1278	3372	7200	11028	17959
45	M	715.00	2180	5955	12825	19694	19694
	MNS	638.08	1945	5317	11451	17583	17583
	F	603.50	1838	5018	10842	16665	16665
	FNS	550.93	1678	4582	9898	15214	15214
50	M	918.50	2750	7425	15705	23984	15705
	MNS	819.69	2454	6630	14022	22414	14022
	F	787.00	2402	6526	13974	21421	13974
	FNS	718.45	2193	5959	12757	19556	12757
55	M	1239.50	3634	9739	20602	31464	9739
	MNS	1106.16	3243	8696	18394	28297	8696
	F	1030.00	3069	8280	17478	26675	8280
	FNS	940.28	2802	7560	15956	24345	7560

Key:
M = male
MNS = male nonsmoker
F = female
FNS = female nonsmoker

Example 1 Barbara Ochoa does not smoke. At age 30 she bought a universal life policy with an annual premium of $233.70, a projected cash value at age 45 (after 15 years) of $3,733, and a projected cash value at age 65 of $20,754.

The projected net cost of a nonparticipating policy over a number of years is

Years × annual premium − projected cash value

Projected net cost over 15 years is

$$15 \times 233.70 - 3733 = -227.50$$

One school of thought would say that since the net cost over 15 years is negative, barring drastic changes in interest rates Barbara will be insured during those years at no cost. The net costs are low because continuing high interest rates are assumed. Another school of thought maintains that there really is a net cost because the paid-in premiums could be drawing interest from other investments. (See the spreadsheet exercise at the end of this chapter.)

PRACTICE ASSIGNMENT

Plan: Use table 19.1 for premiums and projected cash values. Problem 3 pertains to cash value and nonforfeiture options of extended term insurance. Compare the benefits of extended term coverage to invested cash.

1. Find the following values.

	Age at Issue	Sex	Smoker or Nonsmoker	Annual Premium	After 10 Years	Projected Cash Values After 20 Years	At Age 65
a.	35	M	S	________	________	________	________
b.	35	F	NS	________	________	________	________
c.	40	M	NS	________	________	________	________

	Age at Issue	Sex	Smoker or Nonsmoker	Annual Premium	Projected Cash Values After 10 Years	After 20 Years	At Age 65
d.	26	F	S	________	________	________	________
e.	22	M	NS	________	________	________	________
f.	55	F	NS	________	________	________	________

2. Roger and Marilyn Cernas each purchased a $50,000 universal life policy, each naming the other as beneficiary. At the time they purchased the policies, Roger, a smoker, was 40, and Marilyn, a nonsmoker, was 32. At age 55, Roger died, and at that time, Marilyn cashed in her own policy. How much did she receive from the insurance company? $54,219

3. After 25 years, Walter Zimmer terminated his $30,000 straight life policy which featured a cash value of $12,150 and an extended term option of 14 years. He could take advantage of the extended term option or invest the cash value at approximately 9% each year. Complete the following option schedule rounding all amounts to the nearest dollar.

End of Year	Extended Term Insurance	Available Cash	+	9 %	=	Total Cash Reserve
1	30,000	12,150		________		________
2	30,000	________		________		________
3	30,000	________		________		________
4	30,000	________		________		________
5	30,000	________		________		________
6	30,000	________		________		________
7	30,000	________		________		________
8	30,000	________		________		________

End of Year	Extended Term Insurance	Available Cash	+	9 %	=	Total Cash Reserve
9	30,000	________		________		________
10	30,000	________		________		________
11	30,000	________		________		________
12	30,000	________		________		________
13	30,000	________		________		________
14	30,000	________		________		________
15	0	________		________		________

Section 19.2 Auto Insurance

KEY TERMS **Bodily injury liability** **Property damage liability** **Collision** **Medical expense** **Comprehensive**

Auto insurance is essentially a special type of property insurance that overlaps business and personal needs. Without insurance, an auto accident can lead to lawsuits for the automobile owner as well as the driver and may result in financial disaster.

Most states require that motor vehicle owners carry minimum insurance coverage. Requirements may vary from state to state, but the following definitions are fairly standard:

Auto Insurance Coverage Options	
Bodily Injury Liability	Pays claims arising from injuries or deaths caused by the insured vehicle.
Property Damage Liability	Pays claims arising from damage to property caused by the insured vehicle.
Collision	Reimburses the owner of the insured vehicle if it is damaged.
Medical Expense	Pays medical expenses incurred by any injured passenger in the insured vehicle.
Comprehensive	Protects against damage to the insured vehicle by fire, theft, wind, hail, or vandalism. In some cases, the insured is protected when driving a rented vehicle.

Special coverages pay for injuries incurred in accidents caused by uninsured motorists. Under a no-fault auto insurance system, insurers need not determine who was at fault in settling accident claims, which avoids legal action and expense.

If a motor vehicle is damaged beyond repair (total loss), insurance compensation is determined by the vehicle's market value at the time of the accident. Most accidents, however, cause injuries and/or property damage. Policies specify claims settlement limits for maximum loss payments. If a policy pays a maximum of $25,000 for injuries to any one person and $75,000 for injuries to all persons involved in an accident, the limits are listed on the policy as 25000/75000 or 25/75. A policy that also pays a maximum of $30,000 for all property damage caused by an accident may list this coverage as 25000/75000/30000 or 25/75/30. Some policies have deductible clauses that reduce premiums. A deductible of $200, for example, means that the insured pays the first $200 of a claim.

An owner/driver carries 30/75 in bodily injury liability.

An accident causes injuries to a pedestrian in the amount of $29,000.

The insurance company pays $29,000.

Example 1 A pickup truck rammed a car. The car's driver and passenger filed bodily injury claims of $30,000 and $35,000, as well as a property damage claim of $3,050. The truck driver was not injured, but damage to the truck amounted to $1,200. The truck driver carried liability coverage of 40/100/20, but no collision coverage.

a. How much did the truck driver's insurance company pay?
$65,000 for injuries and $3,050 for damages.

b. How much did the accident cost the truck driver?
$1,200 because he carried no collison coverage.

What Factors Affect Auto Insurance Premiums?

The factors that affect premiums vary widely between insurers. Good driving records reduce premiums. Owners who insure more than one vehicle with the same insurance company may receive multiple car discounts.

Auto insurance premiums are usually quoted for semiannual policies. Because table 19.2 is a sample, the rates may seem low. Insurance rates rise steadily with inflation, are affected by population density, and vary widely from state to state.

Auto Insurance Premium Criteria

Type of vehicle: compact car, "muscle car," van, truck

Vehicle usage: private, commercial, commuting

Principal driving territory: metropolitan area, city, small town, rural area

Number of miles driven per year

Age of car

Principal driver's age, sex, marital status, occupation, safety record

Table 19.2 Automobile Insurance Premiums

Semiannual Rates		*Bodily Injury*				*Property Damage*		
Territory Class	*Driver Class*	*15/35*	*25/60*	*50/100*	*75/150*	*20*	*50*	*75*
A Rural	1	81.90	87.73	97.21	107.48	35.93	41.51	43.92
	2	83.67	89.60	98.34	108.78	36.43	41.92	44.36
	3	86.32	92.42	101.55	110.14	37.30	42.61	45.45
	4	87.70	94.52	103.05	112.34	37.69	43.81	46.56
B Small Town	1	82.60	88.48	97.97	108.31	36.10	42.11	44.85
	2	84.45	90.44	99.30	109.57	36.61	42.72	45.40
	3	86.98	93.55	102.25	110.97	37.59	43.21	46.29
	4	88.08	95.35	104.90	113.21	38.00	44.89	47.32
C Suburban	1	83.35	89.27	98.67	109.22	36.60	42.61	45.38
	2	85.62	90.82	100.50	110.48	37.11	43.71	46.17
	3	87.30	94.21	103.07	112.58	37.82	44.83	47.09
	4	88.85	95.91	105.75	114.61	39.38	45.54	48.12
D City	1	83.50	89.75	99.50	110.58	37.51	44.10	46.30
	2	86.37	91.13	101.27	112.52	38.54	44.60	47.92
	3	88.28	94.80	103.95	114.39	39.53	45.35	48.64
	4	89.39	96.26	106.10	117.05	40.49	46.20	49.76

Table 19.2 *Continued*

Collision and Comprehensive		*Deductibles*					
Driver Class	*Vehicle Age*	*0*	*0*	*100*	*100*	*200*	*200*
		Col	*Comp*	*Col*	*Comp*	*Col*	*Comp*
1	0–1	107.82	40.00	88.10	35.50	77.10	32.00
	2	104.17	38.50	86.21	34.50	75.60	31.00
	3–4	99.65	35.50	83.70	31.80	74.15	28.00
	5–	97.19	35.00	80.07	31.00	72.28	27.50
2	0–1	109.20	45.00	89.36	37.50	78.90	33.00
	2	107.52	43.25	87.00	35.25	77.35	32.50
	3–4	105.14	40.75	84.70	33.50	75.50	30.75
	5–	101.12	38.50	82.72	32.80	74.21	29.20
3	0–1	111.17	46.75	90.72	39.00	79.79	36.60
	2	109.20	45.00	88.75	37.40	77.80	34.00
	3–4	107.30	42.50	85.15	35.75	76.51	32.60
	5–	104.18	40.25	83.27	34.35	75.27	31.10
4	0–1	118.31	48.25	91.81	41.00	80.72	38.00
	2	117.17	46.00	90.77	38.50	79.92	35.25
	3–4	114.70	43.50	86.61	36.00	77.55	33.00
	5–	109.19	41.75	84.33	35.00	76.23	32.25

Key:
1 = 25 or over, married
2 = 25 or over, unmarried
3 = under 25 female
4 = under 25 male

Example 2 Determine the semiannual premium for the following coverages using table 19.2.

Liability: $15,000 bodily injuries per person
$35,000 total bodily injuries
$50,000 property damages

The principal driver is 34 years old, unmarried, and lives in a suburban area. Driver class is 2; territory class is C.

Semiannual premium for bodily injury 15/35:	$85.62
Semiannual premium for property damage 50:	43.71
Total semiannual premium for limits 15/35/50:	$129.33

PRACTICE ASSIGNMENT

Plan: With regard to coverages, 100/300/100 means:

$100,000 bodily injury liability, one victim;

$300,000 bodily injury liability, all victims (one accident);

$100,000 all property damage liability (one accident).

Compare the victim's expenses to what he or she can recover. Is there a deductible? If there was damage to the owner's vehicle, check the collision coverage.

With regard to premiums, check status territory and vehicle. Look up the premiums in table 19.2 for each coverage, and derive the total semiannual premium.

1. Don's Pizza House employs college students to deliver pizzas in small pickup trucks. On a delivery, one of the truck drivers ran a stop sign and collided with a car, whose driver sued Don's Pizza for $45,000 in personal injuries. The car, valued at $15,000, was damaged beyond repair. The truck driver escaped injury, but the truck damages amounted to $4,500. Don's carries 50/100/50 insurance coverage on each truck in addition to collision coverage with a $500 deductible.
   a. How much did the insurer pay?
   b. How much did the accident cost Don's?

2. Repairs on an automobile damaged during a storm amounted to $442.70. The owner carried collision coverage with a $300 deductible and comprehensive coverage with a $200 deductible. How much did the insurer pay on the damage claim?

3. Determine the semiannual premium for the following coverages.

a. Liability: $25,000 bodily injuries per person; ________
$60,000 total bodily injuries: ________
$50,000 property damages: ________
The principal driver is 48 years old, married, and lives in a small town.

b. Collision, $200 deductible: ________
Comprehensive, $100 deductible: ________
The vehicle is four years old; the principal driver is female and under 25. ________

4. Pelk's Department Store purchased three new panel trucks that will be driven by store personnel, all of whom are over 25 and married. Pelk's is located in a city and will purchase the following insurance coverage:

$50,000 liability for each personal injury
$100,000 liability for total personal injuries
$75,000 liability for total property damage
collision and comprehensive with no deductibles

What will the semiannual premium be per vehicle if the rates are 18% higher for panel trucks?

5. Refer to problem 4. Suppose Pelk's purchases the following additional coverage for each panel truck:

Coverage	Semiannual Premium
Uninsured motorists	67.15
No-fault addendum	54.34
Towing and emergency road service	29.25

What will the total annual insurance premium cost for all three vehicles?

Challenger

In a world where finances are so important, it is never too early to establish a personal financial plan. Whether you are single with no family responsibilities or a parent with children to support, you need a financial plan to secure your future. Ask yourself, "Where do I expect to be in another ten years?" Fill in the following personal financial statement with your estimates for ten years from now. Base your estimates on your expected earnings, being as realistic as possible.

Personal Financial Statement

Assets	
Home (full market value)	$________
Life Insurance	
Cash value of your life insurance	________
Cash value of family members' life insurance	________
Cash or Equivalent Funds	
Cash on hand	________
Checking account	________
Savings accounts	________
Certificates of deposit	________
Investments	
Bonds	________
Stocks	________
Mutual funds	________
Personal Property	
Home furnishings	________
Car	________
Jewelry, clothes, silver, china, furs, art, etc.	________
Miscellaneous	________
Total Assets	$________
Liabilities	
Mortgages	$________
Installment Loans	________
Education Loans	________
Other Loans	________
Charge Accounts	________
Other Debts	________
Total Liabilities	$________
Total Assets	$________
− Total Liabilities	________
= Net Worth	$________

CHAPTER 19 SUMMARY

Concept	*Example*	*Procedure*
SECTION 19.1 Life Insurance		
Life insurance Basic plans	$40,000 straight life pays $40,000 to beneficiary if policyholder dies.	Straight life (whole life, ordinary) provides protection as long as premiums are paid.
	$40,000 30-year limited payment policy pays $40,000 to beneficiary if policyholder dies. Premiums are paid for 30 years.	Limited payment life provides protection for lifetime; premiums are discontinued after a certain number of years.
	$40,000 10-year term policy pays $40,000 to beneficiary if policyholder dies during the 10-year term.	Term insurance provides protection for the duration of the term; premiums are discontinued after term ends.
	$40,000 20-year endowment pays $40,000 to beneficiary if policyholder dies during the 20-year term. If policyholder survives the 20 years, he or she receives $40,000.	Endowment insurance provides protection for a certain number of years and pays the face value to the policyholder if he or she survives those years.
Nonforfeiture options Cash value	$30,000 straight life policy bought at age 25. Cash value is $8,520 at age 45.	If policy is discontinued after first few years, some cash is returned.
Paid up insurance	Face value = $30,000 Paid up insurance = $14,250 Benefit of $14,250 for lifetime	If policy is discontinued, a smaller benefit can remain in effect for lifetime without additional premiums.
Extended term	Face value = $30,000 $30,000 benefit remains in effect for 15 more years.	If policy is discontinued, the full benefit can remain in effect for limited time without additional premiums.
Interest-sensitive plans	Face value = $50,000 Issue age = 30 Annual premium = $302 Cash values: **5 years**: 900; **10 years**: 2,416; **20 years**: 7,767; **age 65**: 31,240 Lower premiums and cash values for nonsmokers.	Provides flexible protection. Premiums, benefits, and cash values may change periodically. Policies can be individualized. Additional rating factors are used.
SECTION 19.2 Auto Insurance		
Auto insurance coverages	30/100/25 means limits of $30,000 injury liability per person, $100,000 injury liability per accident, $25,000 property damage per accident.	Bodily injury liability and property damage liability coverages are required. Other coverages are collision, comprehensive, uninsured motorist protection, medical expenses, towing and emergency road service.
Auto insurance premiums	Territory class B, driver class 3: 25/60/50 costs 93.55 + 43.21 = 136.76 for half a year.	Premium factors are type of vehicle, usage, driving territory, miles driven, age, sex, marital status, occupation, record, etc. Ratings and premiums vary from state to state and between carriers. Premiums tend to rise steadily.

Chapter 19 **Spreadsheet Exercise**

Interest-Adjusted Net Cost of Insurance

The net cost formula (premiums − dividends − cash value) used to calculate the cost of insurance policies is misleading because the policyholder could have been earning interest on invested premiums.

The interest-adjusted method for estimating cost applies an interest factor to the premiums paid and dividends received. The chapter 19 template appears as follows:

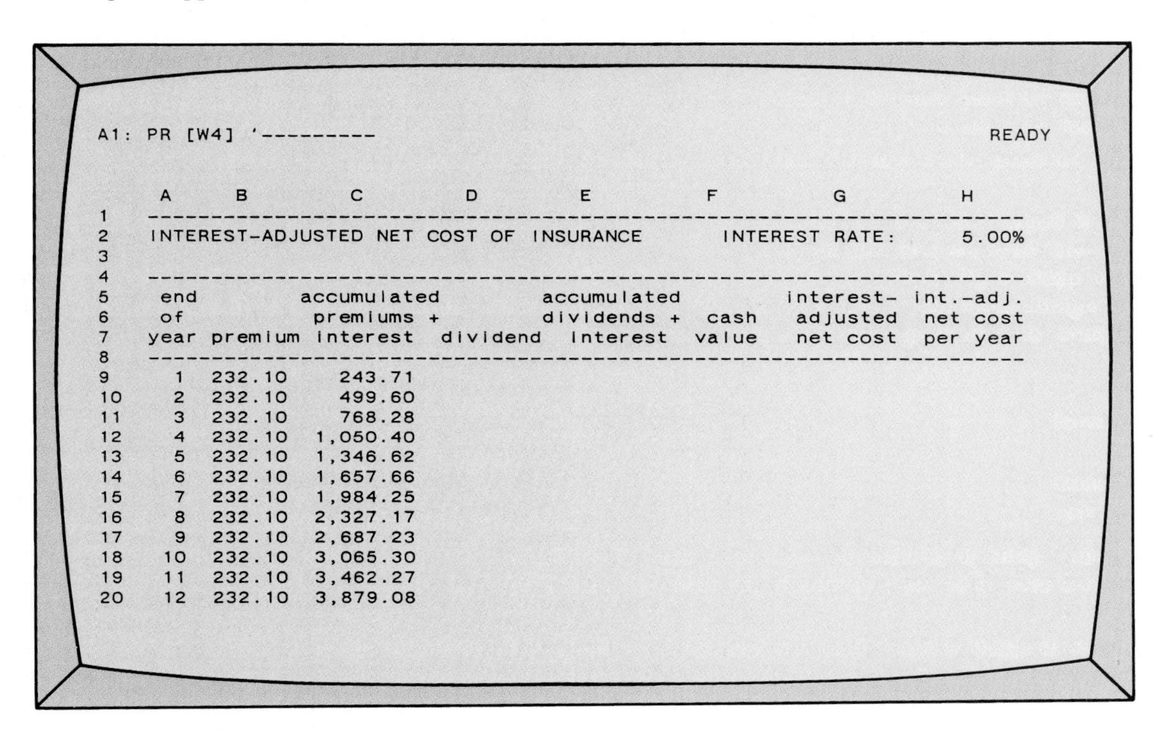

A1: PR [W4] '---------- READY

	A	B	C	D	E	F	G	H
1	------	------	------	------	------	------	------	------
2	INTEREST-ADJUSTED NET COST OF INSURANCE					INTEREST RATE:		5.00%
3								
4	------	------	------	------	------	------	------	------
5	end		accumulated		accumulated		interest-	int.-adj.
6	of		premiums +		dividends +	cash	adjusted	net cost
7	year	premium	interest	dividend	interest	value	net cost	per year
8	------	------	------	------	------	------	------	------
9	1	232.10	243.71					
10	2	232.10	499.60					
11	3	232.10	768.28					
12	4	232.10	1,050.40					
13	5	232.10	1,346.62					
14	6	232.10	1,657.66					
15	7	232.10	1,984.25					
16	8	232.10	2,327.17					
17	9	232.10	2,687.23					
18	10	232.10	3,065.30					
19	11	232.10	3,462.27					
20	12	232.10	3,879.08					

Column B displays level premiums.

Column C displays the interest-adjusted premiums; it is assumed that each premium, if kept, could have received interest at the end of the year. The interest rate is in H2.

Column D will contain the received dividends.

Column E will display the interest-adjusted dividends after you enter the received dividends; it is assumed that each dividend is invested at the end of the year at the same interest rate.

Column F will contain the cash values upon surrender of the policy.

Spreadsheet Exercise Steps

1. Enter the following dividends into column D, rows 9–28:

year	dividend	year	dividend	year	dividend	year	dividend
1	25.13	6	45.31	11	62.81	16	78.87
2	25.13	7	45.31	12	68.30	17	78.87
3	32.40	8	49.82	13	68.30	18	85.50
4	38.40	9	53.82	14	73.76	19	85.50
5	42.28	10	58.91	15	73.76	20	91.23

2. Enter the following cash values into column F, rows 9–28:

year	cash value	year	cash value	year	cash value	year	cash value
1	0.00	6	603.46	11	1872.85	16	2959.28
2	0.00	7	910.34	12	2170.40	17	3156.56
3	0.00	8	1150.06	13	2367.42	18	3353.85
4	285.00	9	1366.87	14	2564.71	19	3551.13
5	461.64	10	1617.34	15	2761.99	20	3748.42

3. Column G: Interest-adjusted Net Cost

accumulated premiums − accumulated dividends − cash value

The accumulated premiums are in C9, accumulated dividends in E9, and the cash value in F9.
The spreadsheet formula must be preceded by +.
The spreadsheet formula is + ______ − ______ − ______ .
Complete the formula, store it in G9, and copy it from G9 to G10..G28.

4. Column H will display the interest-adjusted net cost per year: the amount per year invested at the assumed interest rate which would have covered the net cost.
An insurance agent might point out that after a number of years the net cost will be zero. But interest becomes increasingly significant when rates rise. Vary the interest rate in H2, initially set at 5%. (Enter the rates in decimal form.) Complete the following table for a policy that is kept for 20 years.

	5%	6%	7%	8%	9%	10%
Interest-adjusted net cost	______	______	______	______	______	______
Interest-adjusted net cost per year	______	______	______	______	______	______

Although money spent on premiums could have been invested elsewhere, this should not be interpreted as an argument against life insurance. The early premium payments provide protection, and an instrument for forced savings may be useful.

5. If interest rates dropped steeply, money would not earn much. The insurance agent would be correct because the policy would cost very little. How much would the policy cost each year if kept for 15 years and the interest rate were 2%?

6. On the other hand, if interest rates rose steeply, the policy would cost substantially more because money could earn more. How much would the policy cost each year if kept for 15 years and the interest rate were 12%?

7. What have insurance companies done to protect themselves against extreme interest rate fluctuation?

NAME ____________________

CLASS SECTION ____________________ DATE ____________________

SCORE ____________ MAXIMUM 150 PERCENT ____________

CHAPTER 19 ASSIGNMENT

Refer to table 19.1 for the first four problems.

1. Joyce Evans is a smoker who purchased a $50,000 universal life policy at age 30. Susan Wynder, who does not smoke, purchased the same type of policy at age 35. How do the two premiums compare? How do the projected cash values compare after 20 years?

	Evans	Wynder
Annual premium	______	______
Projected cash value	______	______

(10 points)

2. Stuart Macmillan is a nonsmoker who purchased a $50,000 universal life policy at age 28. After 15 years, he cancelled it. What was his net cost?

(10 points)

3. Janet Gross is buying a $50,000 universal life policy at age 35. She is an occasional smoker.

a. What is her annual premium?

b. What is her projected cash value at age 55?

c. How much would she save in premiums over the next 20 years if she were a nonsmoker?

d. What would her projected cash value be at age 55 if she were a nonsmoker? *(20 points)*

4. What is the projected net cost of a life insurance policy purchased by a nonsmoking female at age 30 who discontinues the policy at age 50?

(10 points)

5. Which coverage applies to the theft of hubcaps?

a. Personal injury liability

b. Property damage liability

c. Collision

d. Comprehensive

e. None of the above *(5 points)*

6. When a life insurance company pays a benefit to a policyholder or beneficiary, it need not be received in one lump sum; there are various settlement options. Therefore, insurance proceeds may be used to enhance retirement income. Suppose that the proceeds of a life insurance policy amount to $50,000 and the beneficiary wants to receive $5,000 at the beginning of each year in order to augment retirement funds. For a face value of $50,000 such an arrangement would last 10 years. But this insurance company guarantees 8% interest on funds left with them, so they pay $5,000 annually and add the 8% interest earned on the remaining balance. How long will the income from this $50,000 policy last? Complete the following schedule, rounding to the nearest dollar.

Beginning of Year	Balance	Benefit Payment	Balance	8% Interest on Balance at End of Year	New Balance
1	50000	5000	45000	3600	48600
2	48600	5000	43600	3488	47088
3	47088	5000	42088	______	______
4	______	5000	______	______	______
5	______	5000	______	______	______
6	______	5000	______	______	______
7	______	5000	______	______	______
8	______	5000	______	______	______

Beginning of Year	Balance	Benefit Payment	Balance	8% Interest on Balance at End of Year	New Balance
9	______	5000	______	______	______
10	______	5000	______	______	______
11	______	______	______	______	______
12	______	______	______	______	______
13	______	______	______	______	______
14	______	______	______	______	______
15	______	______	______	______	______
16	______	______	______	______	______
17	______	______	______	______	______
18	______	______	______	______	______
19	______	______	______	______	______
20	______	______	______	______	______

(50 points)

How many years will this income last?

7. Frank Lee's car sustained damage of $2,400 when he struck another vehicle, causing total loss to that vehicle, valued at $18,000. The two occupants of the other car claimed $10,000 each for medical expenses. Lee's liability limits are 20/40/20, and he carries collision and comprehensive with $200 deductibles.
   a. What will Mr. Lee's insurer pay?
   b. What will Mr. Lee be obligated to pay?

   (10 points)

8. Bill, a salesman for Wesco Cleaning Supplies, was injured in an auto accident caused by an uninsured driver. Bill's medical expenses amounted to $8,500, and his car, valued at $15,000, was damaged beyond repair.

   Wesco carries 40/80/40, collision with $400 deductible, and uninsured motorist coverage containing the same limits as liability coverage. How much did Wesco's insurer pay on the claim?

   (10 points)

Refer to table 19.2 for the remaining exercises.

9. Determine the semiannual premium for the following coverages. The vehicle is 3 years old; the principal driver is 28 years old and unmarried.

   Collision, $100 deductible: ______

   Comprehensive, no deductible: ______

   ______

   (5 points)

10. Jeff Fitzpatrick is a twenty-four-year-old city resident who wants maximum liability protection for his new Mercury to cover injuries and property damage. He wants collision and comprehensive with no deductibles. What is his annual premium? *(10 points)*

11. Vera Escobar is a 38-year-old married woman who lives in a rural area and drives a seven-year-old Chevrolet. She wants minimum protection for liability (injuries and property damage) with the highest deductible for comprehensive and no collision. What is her annual premium? *(10 points)*

P A R T

5 PAYROLL AND BUSINESS EXPENSES

Chapter 20 Payroll

OBJECTIVES

- Determine compensation based on salaries, hourly wages, and commissions
- Compute withholdings from gross earnings: federal income tax, FICA, and others

SECTIONS

INTRODUCTION

All companies, whether they have one employee or thousands of employees, must process payroll on a regular basis. While large companies have converted manual payrolls to computerized systems, many small businesses continue to process their payrolls manually. In this chapter you will learn about gross pay, payroll deductions and tax withholdings, and how take-home pay is determined.

Section 20.1 Salary

KEY TERMS **Gross earnings** **Gross salary**

Many employees—managers, professionals, and others—are paid a fixed periodic payment called a salary. Salaried personnel can count on regular, equal payments but usually do not receive extra compensation for overtime work.

The cash compensation an employee and employer agree to is called **gross earnings** or gross pay and is subject to payroll deductions. Salary is usually expressed as an annual gross. Annual **gross salary** is actually divided into allotments paid periodically throughout a year. Sometimes salary is quoted as a monthly figure. An annual salary of $15,000, for example, is equivalent to a monthly salary of $1,250, 15000 ÷ 12 = 1250.

$$\text{Periodic gross salary} = \frac{\text{annual gross salary}}{\text{number of payments per year}}$$

Advertisement for Salaried Employees

COMPUTER OPERATORS
2-year degree in computer-related field required; IBM main frame; multiple-task on-line environment; OS/VS; training for operating system given. Mostly day work, some night work and weekends. Secure positions for conscientious employees. Starting Salary $19,000–$22,000. Fee paid. Kent Employment Agency, 325 West Main Street.

Monthly payroll: Employee is paid once each month, 12 payments per year

Semimonthly payroll: Employee is paid every half-month (twice each month), 24 payments per year

Weekly payroll: Employee is paid once each week, 52 payments per year

Biweekly payroll: Employee is paid once every two weeks, 26 payments per year

Example 1 Kim Beasley earns a gross salary of $24,960 per year. What is her periodic gross salary?

Pay Period	Number of Pay Periods	Periodic Gross Salary
Monthly	12 times per year	$\frac{24960}{12} = 2080$
Semimonthly (every half-month)	24 times per year	$\frac{24960}{24} = 1040$
Biweekly (every two weeks)	26 times per year	$\frac{24960}{26} = 960$
Weekly	52 times per year	$\frac{24960}{52} = 480$

PRACTICE ASSIGNMENT

Plan: To calculate periodic salary, divide the annual salary by the number of payments per year. To obtain the annual salary, multiply the periodic salary by the number of payments per year.

1. Convert the following salaries from annual to periodic.

	Annual Salary	Pay Period	Periodic Salary
a.	9,000.00	Monthly	________
b.	15,000.00	Semimonthly	________
c.	23,400.00	Weekly	________
d.	29,775.00	Monthly	________
e.	15,613.00	Weekly	________
f.	26,701.22	Biweekly	________

2. Convert the following salaries from periodic to annual.

	Periodic Salary	Pay Period	Annual Salary
a.	1,250.00	Monthly	________
b.	375.00	Weekly	________
c.	742.00	Biweekly	________
d.	2,135.50	Monthly	________
e.	297.16	Weekly	________
f.	1,029.25	Semimonthly	________

3. Ashley Frank earns $648.75 every two weeks. What is her annual salary?

4. The assistant manager of a retail outlet earns an annual gross salary of $31,575. What is the monthly gross salary?

Section **20.2 Hourly Wage**

KEY TERMS **Hourly rate** **Minimum wage** **Overtime** **Time-and-a-half rate method** **Overtime rate** **Overtime premium method**

Many employers pay a certain amount of money for each hour an employee works, called an hourly wage or **hourly rate.**

Employees working in manufacturing and service industries, full- or part-time, are often paid by the hour.

Advertisement for Hourly Employees

> **CASHIERS**
> Full Time. Must be over 21. Vacations and fringe benefits. Steady work. Occasional overtime. Starting rate: $6.75 per hour. Call 820-7737, 9AM to 5PM.

> Gross pay
> = time × rate
> = hours worked × hourly rate
>
> 35 hours worked in a week
> at $7.15 per hour earn
> 35 × 7.15 = 250.25 gross pay

Federal law requires a minimum hourly rate, called the **minimum wage,** which applies to most hourly workers. At the time of this writing, the federal minimum wage was raised to $4.25.

Federal law that applies to most hourly workers also stipulates that work in excess of 40 hours per week is **overtime** and must be compensated at time and a half, one and one-half times the hourly wage. Gross pay can be computed by the **time-and-a-half rate method,** by the **overtime premium method,** or by other procedures. Some companies adopt more generous plans that may include overtime for over 8 hours of work per day, double time for weekend and holiday work, and shift differentials.

Overtime Pay Computations

Time-and-a-half Rate Method:
Gross = regular hours × regular rate
+ overtime hours × overtime rate
where **overtime rate** = 1.5 × regular rate

Overtime Premium Method:
Gross = total hours × regular rate
+ overtime hours × 0.5 × regular rate

Example 1 George Callas is employed by Quick Burger at a rate of $4.25 per hour. If he worked 46 hours during one week, what was his gross pay?

Time-and-a-half Rate Method:

Regular rate = 4.25

Overtime rate = 4.25 × 1.5 = 6.375

Gross = regular hours × regular rate
+ overtime hours × overtime rate
= 40 × 4.25 + 6 × 6.375
= 170.00 + 38.25
= 208.25

key: 40 [×] 4.25 [+] 6 [×] 6.375 [=]

On arithmetic calculators and inexpensive algebraic calculators that do not execute operations in the correct order, use memory:

key: 40 [×] 4.25 [=] [M+] 6 [×] 6.375 [=] [M+] [MR]
display: 208.25
edited: $208.25

Overtime Premium Method:

Gross = total hours × regular rate
+ overtime hours × 0.5 × regular rate
= 46 × 4.25 + 6 × 0.5 × 4.25
= 46 × 4.25 + 6 × 2.125
= 195.50 + 12.75
= 208.25

key: 46 [×] 4.25 [+] 6 [×] 2.125 [=]
or key: 46 [×] 4.25 [=] [M+] 6 [×] 2.125 [=] [M+] [MR]
display: 208.25
edited: $208.25

PRACTICE ASSIGNMENT

Plan: All problems pertain to hourly wage earners. If the number of hours worked in a week is not greater than 40,

gross pay = number of hours worked × hourly rate.

Apply the standard federal overtime rule that any hours over 40 worked in a week must be compensated at time and a half. Both methods of calculating gross pay with overtime, the time-and-a-half rate method and the overtime premium method, should produce the same gross. Calculated pay rates may contain more than two decimal places. Do not round until you have obtained the gross.

1. Complete the following weekly pay schedule. Use the time-and-a-half rate method.

Employee	Regular Rate	Overtime Rate	Hours	Regular Pay	Overtime Pay	Total Pay
Abrams	4.95	______	39	______	______	______
Bono	4.80	______	42	______	______	______
Ernst	4.95	______	46	______	______	______
Farley	5.04	______	42.5	______	______	______
Jones	5.12	______	46.5	______	______	______

2. Complete the following hourly payroll schedule. Use the overtime premium method.

Name	M	T	W	Th	F	Rate	0.5 x Rate	Total Hours	Overtime Hours	Regular Wages	Overtime Wages	Total Gross Wages
Lee	8	7	8	6	8	8.02						
Reeves	9	8	10	5	10	7.85						
Martin	8	8	8	8	8	10.25						
Stein	8	8	8	9	10	9.45						
Sanchez	8	9	10	10	8	8.75						

3. Ben Anderson is a construction worker who earns \$18.05 per hour. If he worked $37\frac{1}{2}$ hours last week, what was his gross pay?

4. Refer to problem 3. Anderson worked 43.5 hours during another week.

a. What was his overtime rate?

b. What was his weekly gross?

Section 20.3 Commission

KEY TERMS **Straight commission** **Graduated commission**

People employed in sales are often paid a percent of the sales as a performance incentive. **Straight commission** is a straight percentage of the sales.

> Commission rate = 6%
> Monthly sales = \$5,635.28
> Commission for month = 6% of 5635.28 = 338.12

Advertisement for Sales Employees

> **SALES.** Large publisher of text and trade books seeks experienced salesperson for Southeastern region. Commission and paid expenses. Extensive travel required. May lead to salary + commission. Excellent opportunity for aggressive, sales-minded person. Call Merit Publishing, Human Resources Department, (123) 456-7890

Some salespersons receive a salary plus commission. In some companies, the commission may apply only to sales that exceed a certain amount.

Salary: $1,500 per month
Commission: 2% of sales over $10,000
Sales for month: $12,430
Gross = 1500 + 2% of 2430
= 1500 + 48.6
= 1548.60

Example 1 Manuel Vargas earns $1,375 per month plus 7 percent commisson on all sales over $15,000. If his monthly sales total $26,800, what are his gross earnings?

Gross = 1375 + 7% of (26800 − 15000)
= 1375 + 11800 × 0.07
= 1375 + 826
= 2201

key alg: 26800 [−] 15000 [=] [×] .07 [+] 1375 [=]
key arith: 1375 [M+] 26800 [+] 15000 [−] [×] .07 [=] [M+] [MR]
display: 2201.
edited: $2,201

A commission rate that increases with increasing sales is called a sliding scale commission or **graduated commission.** A sliding scale is feasible for a company whose sales expenses remain constant while sales are rising.

First $5,000 of sales: 6.5%
Next $5,000 of sales: 7.0%
Sales of over $10,000: 8.0%

Monthly sales: $7,680.50

5000.00 × 0.065 = 325.00
+ 2680.50 × 0.07 = 187.64
Commission = 512.64

Example 2 Calculate the commission for sales of $6,450.

First $1,000 of sales: 5%	1000 × 0.05	= 50.00
Next $1,500 of sales: 5.5%	+ 1500 × 0.055	= 82.50
Next $2,500 of sales: 6%	+ 2500 × 0.06	= 150.00
Sales over $5,000: 7%	+ 1450 × 0.07	= 101.50
	Commission	= 384.00

key: 1000 [×] .05 [=] [M+]
1500 [×] .055 [=] [M+]
2500 [×] .06 [=] [M+]
1450 [×] .07 [=] [M+] [MR]
display: 384.
edited: $384

The sliding scale can also be shown as:

0.01–1000.00	5%
1000.01–2500.00	5.5%
2500.01–5000.00	6%
5000.01–	7%

PRACTICE ASSIGNMENT

Plan: Straight commission is a percent of sales or of net sales. (Net sales are sales less credits or returns.) Travel expenses may be reimbursed. If a salesperson receives a salary and a commission,

gross = salary + commission.

In some cases, the commission may be a percent of sales above a certain amount. If a problem involves graduated commissions or salaries plus graduated commissions, you must break down the sales amount into commission classes. Then compute the commission for each class. The sales amount in the last class is the excess over the previous classes. Sum the commissions.

1. A sales representative for Stratford Cosmetics receives an 8% commission. If the representative sells 160 bottles of bath oil at $3.85 per bottle, what is the commission?

2. After her last sales trip, Karen Spahn submitted travel expense vouchers amounting to $275.50 for transportation and $195.75 for lodging. Her sales totaled $29,458. If she receives a 6% commission and is reimbursed for all travel expenses, how much should she be paid?

3. Richard Suzuki earns an annual salary of $19,500. In addition, he receives a commission of 7.5% on monthly sales exceeding $25,000. During the last month, his sales totaled $37,510.68. What were his total earnings for the month?

4. Use the following commission scale to compute commissions for monthly sales.

First $2,000 of sales	3%
Next $2,000 of sales	4%
Next $1,000 of sales	5%
Sales over $5,000	6%
Or	
0.01–2000.00	3%
2000.01–4000.00	4%
4000.01–5000.00	5%
5000.01–	6%

a. Monthly sales = 975.89
Commission = ________

b. Monthly sales = 3,200.00
Commission = ________

c. Monthly sales = 2,175.99
Commission = ________

d. Monthly sales = 8,070.00
Commission = ________

e. Monthly sales = 4,820.30
Commission = ________

5. Wendy Santana works for Household Telemarketing. She is paid $175 per week plus a commission of 6% on the first $500 of telephone orders and 7.5% on orders exceeding $500. Cancelled orders are deducted. During one week she received orders totaling $889.65. Cancellations amounted to $74.29. What was her gross pay for the week?

6. John Taylor, a senior representative at Putnam Home Improvements, receives a monthly salary of $1,500 plus expenses plus a commission of 4% on the first $5,000 of net sales and 5.5% on net sales in excess of $5,000. During one month, he submitted orders in the amount of $18,290.71 and vouchers for reimbursable travel expenses in the amount of $520.64. Merchandise worth $750.85 was returned. What was his gross pay for the month?

Section **20.4** Federal Income Tax Withholdings

KEY TERMS **Deductions** **Net pay** **Withholding** **Percentage tables** **Withholding allowance** **Withholding tables**

Why Is Take-Home Pay Less Than Gross Pay?

Deductions are those portions of gross earnings that employers withhold for taxes, retirement plans, additional voluntary savings, health insurance, and professional dues. The remainder is **net pay** or take-home pay,

Net pay = gross pay − deductions.

The federal government requires that individuals pay the bulk of their income tax obligations as they earn their income. Each employee must file a Withholding Allowance Certificate, Form W-4, that is used to calculate income tax **withholding** amounts.

Cut here and give the certificate to your employer. Keep the top portion for your records.

Form **W-4** Department of the Treasury Internal Revenue Service

Employee's Withholding Allowance Certificate

► For Privacy Act and Paperwork Reduction Act Notice, see reverse.

OMB No. 1545-0010 19

1 Type or print your first name and middle initial: ERIN L. Last name: SIMPSON

2 Your social security number: 991-01-0000

Home address (number and street or rural route): 571 AVENUE H

City or town, state, and ZIP code: GREENDALE, FL 31007

3 Marital status: ☑ Single ☐ Married ☐ Married, but withhold at higher Single rate.
Note: *If married, but legally separated, or spouse is a nonresident alien, check the Single box.*

4 Total number of allowances you are claiming (from line G above or from the Worksheets on back if they apply) . . . 4 1

5 Additional amount, if any, you want deducted from each pay . . . 5 $

6 I claim exemption from withholding and I certify that I meet ALL of the following conditions for exemption:

- Last year I had a right to a refund of **ALL** Federal income tax withheld because I had **NO** tax liability; **AND**
- This year I expect a refund of **ALL** Federal income tax withheld because I expect to have **NO** tax liability; **AND**
- This year if my income exceeds $500 and includes nonwage income, another person cannot claim me as a dependent.

If you meet all of the above conditions, enter the year effective and "EXEMPT" here . . . ► 6 19

7 Are you a full-time student? (**Note:** *Full-time students are not automatically exempt.*) . . . 7 ☐ Yes ☑ No

Under penalties of perjury, I certify that I am entitled to the number of withholding allowances claimed on this certificate or entitled to claim exempt status.

Employee's signature ► Erin Simpson **Date ►** JANUARY 17 19 9X

8 Employer's name and address (**Employer:** Complete 8 and 10 **only if sending to IRS**) | 9 Office code (optional) | 10 Employer identification number

Source: Department of the Treasury, Internal Revenue Service.

By the end of the year, withholdings should be close to an employee's yearly tax obligation, and by April 15 each employee must file an income tax return which reconciles the amount of tax withheld with the amount owed. If too much was withheld, the government pays the employee a refund. If not enough was withheld, the employee must pay the government the additional amount owed.

Because income tax laws change and some values are indexed to the annual inflation rate, tax obligations and withholdings change from year to year. As you study this chapter you may find that figures are not entirely current. Withholding methods, however, remain fairly stable. An employee's tax obligation is generally based on three criteria:

1. Amount of gross earnings
2. Marital status
3. Number of withholding allowances

Percentage Table Method for Federal Tax Withholdings

Many businesses use the withholding **percentage tables** provided by the federal government. Some computer programs also incorporate these tables. To use these tables, the correct **withholding allowance** must first be subtracted from gross earnings. The term *wages* in the tax tables means earnings subject to withholdings, including salaries, hourly wages, commissions, or other compensation plans.

Table 20.1 Withholding Allowances

Payroll	*One Withholding Allowance*
Weekly	38.46
Biweekly	76.92
Semimonthly	83.33
Monthly	166.67
Quarterly	500.00
Semiannually	1,000.00
Annually	2,000.00
Daily	7.69

Table 20.2

Tables for Percentage Method of Withholding

1—If the Payroll Period With Respect to an Employee Is Weekly

(a) SINGLE person—including head of household:

If the amount of wages (after subtracting withholding allowances) is: / *The amount of income tax to be withheld shall be:*

Not over $21 0

Over—	But not over—		of excess over—
$21	—$378	15%	—$21
$378	—$885	$53.55 plus 28%	—$378
$885	—$2,028	$195.51 plus 33%	—$885
$2,028		$572.70 plus 28%	—$2,028

(b) MARRIED person—

If the amount of wages (after subtracting withholding allowances) is: / *The amount of income tax to be withheld shall be:*

Not over $62 0

Over—	But not over—		of excess over—
$62	—$657	15%	—$62
$657	—$1,501	$89.25 plus 28%	—$657
$1,501	—$3,695	$325.57 plus 33%	—$1,501
$3,695		$1,049.59 plus 28%	—$3,695

2—If the Payroll Period With Respect to an Employee Is Biweekly

(a) SINGLE person—including head of household:

If the amount of wages (after subtracting withholding allowances) is: / *The amount of income tax to be withheld shall be:*

Not over $42 0

Over—	But not over—		of excess over—
$42	—$756	15%	—$42
$756	—$1,769	$107.10 plus 28%	—$756
$1,769	—$4,055	$390.74 plus 33%	—$1,769
$4,055		$1,145.12 plus 28%	—$4,055

(b) MARRIED person—

If the amount of wages (after subtracting withholding allowances) is: / *The amount of income tax to be withheld shall be:*

Not over $123 0

Over—	But not over—		of excess over—
$123	—$1,313	15%	—$123
$1,313	—$3,002	$178.50 plus 28%	—$1,313
$3,002	—$7,389	$651.42 plus 33%	—$3,002
$7,389		$2,099.13 plus 28%	—$7,389

3—If the Payroll Period With Respect to an Employee Is Semimonthly

(a) SINGLE person—including head of household:

If the amount of wages (after subtracting withholding allowances) is: / *The amount of income tax to be withheld shall be:*

Not over $46 0

Over—	But not over—		of excess over—
$46	—$819	15%	—$46
$819	—$1,917	$115.95 plus 28%	—$819
$1,917	—$4,393	$423.39 plus 33%	—$1,917
$4,393		$1,240.47 plus 28%	—$4,393

(b) MARRIED person—

If the amount of wages (after subtracting withholding allowances) is: / *The amount of income tax to be withheld shall be:*

Not over $133 0

Over—	But not over—		of excess over—
$133	—$1,423	15%	—$133
$1,423	—$3,252	$193.50 plus 28%	—$1,423
$3,252	—$8,005	$705.62 plus 33%	—$3,252
$8,005		$2,274.11 plus 28%	—$8,005

4—If the Payroll Period With Respect to an Employee Is Monthly

(a) SINGLE person—including head of household:

If the amount of wages (after subtracting withholding allowances) is: / *The amount of income tax to be withheld shall be:*

Not over $92 0

Over—	But not over—		of excess over—
$92	—$1,638	15%	—$92
$1,638	—$3,833	$231.90 plus 28%	—$1,638
$3,833	—$8,786	$846.50 plus 33%	—$3,833
$8,786		$2,480.99 plus 28%	—$8,786

(b) MARRIED person—

If the amount of wages (after subtracting withholding allowances) is: / *The amount of income tax to be withheld shall be:*

Not over $267 0

Over—	But not over—		of excess over—
$267	—$2,846	15%	—$267
$2,846	—$6,504	$386.85 plus 28%	—$2,846
$6,504	—$16,010	$1,411.09 plus 33%	—$6,504
$16,010		$4,548.07 plus 28%	—$16,010

Example 1 Judy Lykins is a single woman earning $465 weekly. She claims two withholding allowances on her W-4 form. How much is withheld from her weekly pay for federal income taxes?

Weekly gross earnings		= 465.00
One weekly allowance = 38.46		
Two allowances = 38.46 × 2 = 76.92	−	76.92
Earnings subject to withholding		= 388.08

Part 1(a) of table 20.2 applies to a single person on a weekly payroll and specifies that, for adjusted earnings over $378 but less than $885, the weekly withholding is $53.55 plus 28% of the excess over $378.

$$388.08 - 378.00 = 10.08$$
$$53.55 + 28\% \text{ of } 10.08 = 53.55 + 10.08 \times 0.28 = 53.55 + 2.82 = 56.37$$

key alg: 388.08 [−] 378 [=] [×] .28 [+] 53.55 [=]
key arith: 53.55 [M+] 388.08 [+] 378 [−] [×] .28 [=] [M+] [MR]
display: 56.3724
edited: $56.37

The amount withheld each week from Judy Lykins's gross earnings for federal taxes is $56.37 ($2,931.24 per year).

Example 2 John McDermott earns $1,075 every two weeks. He is single and claims three withholding allowances. How much is subject to withholding each pay period? How much is withheld from each paycheck for federal income tax?

Biweekly gross earnings		= 1075.00
One biweekly allowance = 76.92		
Three allowances = 76.92 × 3 = 230.76	−	230.76
Earnings subject to withholding		= 844.24

Part 2(a) of table 20.2 applies to a single person paid biweekly and specifies that, for adjusted earnings over $756 but not over $1,769, the amount of tax to be withheld is $107.10 plus 28% of the excess over $756.

$$844.24 - 756.00 = 88.24$$
$$107.10 + 28\% \text{ of } 88.24 = 107.10 + 88.24 \times 0.28 = 107.10 + 24.71 = 131.81$$

key alg: 844.24 [−] 756 [=] [×] .28 [+] 107.1 [=]
key arith: 107.1 [M+] 844.24 [+] 756 [−] [×] .28 [=] [M+] [MR]
display: 131.8072
edited: $131.81

PRACTICE ASSIGNMENT 1

Plan: Withholding allowances are displayed in table 20.1. Multiply the withholding allowance by the number of claimed allowances, and subtract the product from the gross earnings. The remainder is the amount subject to withholdings. Use table 20.2 to compute the withholdings.

1. Compute the amounts subject to withholding.

	Periodic Gross Earnings	Payroll Period	Marital Status	Number of Withholding Allowances	Amount Subject to Withholding
a.	279.45	Weekly	Single	1	________
b.	1,476.21	Monthly	Single	0	________

	Periodic Gross Earnings	Payroll Period	Marital Status	Number of Withholding Allowances	Amount Subject to Withholding
c.	1,948.09	Semimonthly	Single	2	________
d.	4,125.70	Monthly	Married	3	________
e.	589.65	Biweekly	Married	1	________
f.	3,849.33	Semimonthly	Married	6	________

2. Refer to problem 1 for amounts subject to withholding and compute the periodic withholdings.

	Amount Subject to Withholding	Periodic Withholding Rounded to Cent
a.	________	________ + ______% of ________ = ________
b.	________	________ + ______% of ________ = ________
c.	________	________ + ______% of ________ = ________
d.	________	________ + ______% of ________ = ________
e.	________	________ + ______% of ________ = ________
f.	________	________ + ______% of ________ = ________

3. Joan Loomis is on a monthly payroll, and her gross earnings are $1,726.90. She is married and claims three withholding allowances.

a. What is the amount subject to withholding?

b. How much is withheld each month from her gross earnings for federal income tax?

4. Before her marriage, Tanya Ferguson claimed three withholding allowances. Her semimonthly salary is $907.50. She changed her filing status to *married* and kept the same number of withholding allowances. How much less is withheld from her paycheck each payday?

Withholding Table Method for Federal Tax Withholdings

Small businesses that calculate withholdings manually find **withholding tables** more convenient than percentage tables. Withholding tables are organized by marital status, payroll period, earning brackets, and allowances. Withholding allowances are not subtracted from gross earnings prior to using the tables. Since the withholdings are in whole dollars, the withholding tables do not always produce precisely the same result as the percentage tables. This relatively small difference is reconciled on the annual federal tax return.

> Judy Lykins, single, earns $465 per week. She claims two withholding allowances.
>
> Earnings are at least $460 but less than $470
>
> Table 20.3 indicates $56 to be withheld.
>
> (Compare to $56.37 withheld through the percentage tax table 20.2.)

Example 3 Diane Mellichamp earns a gross monthly salary of $1,791. She is married and claims three withholding allowances. How much is withheld from each paycheck?

See table 20.3 on pages 269 and 270 for married persons paid monthly. Earnings are at least $1,760 but less than $1,800. The column for 3 withholding allowances lists $152 as the amount to be withheld.

Table 20.3

SINGLE Persons–WEEKLY Payroll Period

And the wages are–		And the number of withholding allowances claimed is–										
At least	But less than	0	1	2	3	4	5	6	7	8	9	10
		The amount of income tax to be withheld shall be–										
240	250	34	28	22	16	11	5	0	0	0	0	0
250	260	35	29	24	18	12	6	0	0	0	0	0
260	270	37	31	25	19	14	8	2	0	0	0	0
270	280	38	32	27	21	15	9	3	0	0	0	0
280	290	40	34	28	22	17	11	5	0	0	0	0
290	300	41	35	30	24	18	12	6	1	0	0	0
300	310	43	37	31	25	20	14	8	2	0	0	0
310	320	44	38	33	27	21	15	9	4	0	0	0
320	330	46	40	34	28	23	17	11	5	0	0	0
330	340	47	41	36	30	24	18	12	7	1	0	0
340	350	49	43	37	31	26	20	14	8	2	0	0
350	360	50	44	39	33	27	21	15	10	4	0	0
360	370	52	46	40	34	29	23	17	11	5	0	0
370	380	53	47	42	36	30	24	18	13	7	1	0
380	390	56	49	43	37	32	26	20	14	8	3	0
390	400	58	50	45	39	33	27	21	16	10	4	0
400	410	61	52	46	40	35	29	23	17	11	6	0
410	420	64	53	48	42	36	30	24	19	13	7	1
420	430	67	56	49	43	38	32	26	20	14	9	3
430	440	70	59	51	45	39	33	27	22	16	10	4
440	450	72	62	52	46	41	35	29	23	17	12	6
450	460	75	64	54	48	42	36	30	25	19	13	7
460	470	78	67	56	49	44	38	32	26	20	15	9
470	480	81	70	59	51	45	39	33	28	22	16	10
480	490	84	73	62	52	47	41	35	29	23	18	12
490	500	86	76	65	54	48	42	36	31	25	19	13
500	510	89	78	68	57	50	44	38	32	26	21	15
510	520	92	81	70	60	51	45	39	34	28	22	16
520	530	95	84	73	62	53	47	41	35	29	24	18
530	540	98	87	76	65	54	48	42	37	31	25	19

MARRIED Persons–WEEKLY Payroll Period

And the wages are–		And the number of withholding allowances claimed is–										
At least	But less than	0	1	2	3	4	5	6	7	8	9	10
		The amount of income tax to be withheld shall be–										
320	330	40	34	28	22	16	11	5	0	0	0	0
330	340	41	35	29	24	18	12	6	1	0	0	0
340	350	43	37	31	25	19	14	8	2	0	0	0
350	360	44	38	32	27	21	15	9	4	0	0	0
360	370	46	40	34	28	22	17	11	5	0	0	0
370	380	47	41	35	30	24	18	12	7	1	0	0
380	390	49	43	37	31	25	20	14	8	2	0	0
390	400	50	44	38	33	27	21	15	10	4	0	0
400	410	52	46	40	34	28	23	17	11	5	0	0
410	420	53	47	41	36	30	24	18	13	7	1	0
420	430	55	49	43	37	31	26	20	14	8	3	0
430	440	56	50	44	39	33	27	21	16	10	4	0
440	450	58	52	46	40	34	29	23	17	11	6	0
450	460	59	53	47	42	36	30	24	19	13	7	1
460	470	61	55	49	43	37	32	26	20	14	9	3
470	480	62	56	50	45	39	33	27	22	16	10	4
480	490	64	58	52	46	40	35	29	23	17	12	6
490	500	65	59	53	48	42	36	30	25	19	13	7
500	510	67	61	55	49	43	38	32	26	20	15	9
510	520	68	62	56	51	45	39	33	28	22	16	10
520	530	70	64	58	52	46	41	35	29	23	18	12
530	540	71	65	59	54	48	42	36	31	25	19	13
540	550	73	67	61	55	49	44	38	32	26	21	15
550	560	74	68	62	57	51	45	39	34	28	22	16
560	570	76	70	64	58	52	47	41	35	29	24	18
570	580	77	71	65	60	54	48	42	37	31	25	19
580	590	79	73	67	61	55	50	44	38	32	27	21
590	600	80	74	68	63	57	51	45	40	34	28	22
600	610	82	76	70	64	58	53	47	41	35	30	24
610	620	83	77	71	66	60	54	48	43	37	31	25

Table 20.3 (*Continued*)

SINGLE Persons–MONTHLY Payroll Period

And the wages are–		And the number of withholding allowances claimed is–										
At least	But less than	0	1	2	3	4	5	6	7	8	9	10
		The amount of income tax to be withheld shall be–										
560	580	72	47	22	0	0	0	0	0	0	0	0
580	600	75	50	25	0	0	0	0	0	0	0	0
600	640	79	54	29	4	0	0	0	0	0	0	0
640	680	85	60	35	10	0	0	0	0	0	0	0
680	720	91	66	41	16	0	0	0	0	0	0	0
720	760	97	72	47	22	0	0	0	0	0	0	0
760	800	103	78	53	28	3	0	0	0	0	0	0
800	840	109	84	59	34	9	0	0	0	0	0	0
840	880	115	90	65	40	15	0	0	0	0	0	0
880	920	121	96	71	46	21	0	0	0	0	0	0
920	960	127	102	77	52	27	2	0	0	0	0	0
960	1,000	133	108	83	58	33	8	0	0	0	0	0
1,000	1,040	139	114	89	64	39	14	0	0	0	0	0
1,040	1,080	145	120	95	70	45	20	0	0	0	0	0
1,080	1,120	151	126	101	76	51	26	1	0	0	0	0
1,120	1,160	157	132	107	82	57	32	7	0	0	0	0
1,160	1,200	163	138	113	88	63	38	13	0	0	0	0
1,200	1,240	169	144	119	94	69	44	19	0	0	0	0
1,240	1,280	175	150	125	100	75	50	25	0	0	0	0
1,280	1,320	181	156	131	106	81	56	31	6	0	0	0
1,320	1,360	187	162	137	112	87	62	37	12	0	0	0
1,360	1,400	193	168	143	118	93	68	43	18	0	0	0
1,400	1,440	199	174	149	124	99	74	49	24	0	0	0
1,440	1,480	205	180	155	130	105	80	55	30	5	0	0
1,480	1,520	211	186	161	136	111	86	61	36	11	0	0
1,520	1,560	217	192	167	142	117	92	67	42	17	0	0
1,560	1,600	223	198	173	148	123	98	73	48	23	0	0
1,600	1,640	229	204	179	154	129	104	79	54	29	4	0
1,640	1,680	238	210	185	160	135	110	85	60	35	10	0

MARRIED Persons–MONTHLY Payroll Period

And the wages are–		And the number of withholding allowances claimed is–										
At least	But less than	0	1	2	3	4	5	6	7	8	9	10
		The amount of income tax to be withheld shall be–										
1,440	1,480	179	154	129	104	79	54	29	4	0	0	0
1,480	1,520	185	160	135	110	85	60	35	10	0	0	0
1,520	1,560	191	166	141	116	91	66	41	16	0	0	0
1,560	1,600	197	172	147	122	97	72	47	22	0	0	0
1,600	1,640	203	178	153	128	103	78	53	28	3	0	0
1,640	1,680	209	184	159	134	109	84	59	34	9	0	0
1,680	1,720	215	190	165	140	115	90	65	40	15	0	0
1,720	1,760	221	196	171	146	121	96	71	46	21	0	0
1,760	1,800	227	202	177	152	127	102	77	52	27	2	0
1,800	1,840	233	208	183	158	133	108	83	58	33	8	0
1,840	1,880	239	214	189	164	139	114	89	64	39	14	0
1,880	1,920	245	220	195	170	145	120	95	70	45	20	0
1,920	1,960	251	226	201	176	151	126	101	76	51	26	1
1,960	2,000	257	232	207	182	157	132	107	82	57	32	7
2,000	2,040	263	238	213	188	163	138	113	88	63	38	13
2,040	2,080	269	244	219	194	169	144	119	94	69	44	19
2,080	2,120	275	250	225	200	175	150	125	100	75	50	25
2,120	2,160	281	256	231	206	181	156	131	106	81	56	31
2,160	2,200	287	262	237	212	187	162	137	112	87	62	37
2,200	2,240	293	268	243	218	193	168	143	118	93	68	43
2,240	2,280	299	274	249	224	199	174	149	124	99	74	49
2,280	2,320	305	280	255	230	205	180	155	130	105	80	55
2,320	2,360	311	286	261	236	211	186	161	136	111	86	61
2,360	2,400	317	292	267	242	217	192	167	142	117	92	67
2,400	2,440	323	298	273	248	223	198	173	148	123	98	73
2,440	2,480	329	304	279	254	229	204	179	154	129	104	79
2,480	2,520	335	310	285	260	235	210	185	160	135	110	85
2,520	2,560	341	316	291	266	241	216	191	166	141	116	91
2,560	2,600	347	322	297	272	247	222	197	172	147	122	97
2,600	2,640	353	328	303	278	253	228	203	178	153	128	103

PRACTICE ASSIGNMENT 2

Plan: When you use table 20.3, no arithmetic is necessary. Table 20.3 is organized by marital status, payroll period, earnings, and number of withholding allowances.

1. Determine the withholdings using table 20.3, weekly payroll.

	Weekly Earnings	Marital Status	Number of Withholding Allowances	Weekly Withholding
a.	246.50	Single	1	______
b.	605.42	Married	2	______
c.	477.25	Married	0	______
d.	398.75	Single	3	______
e.	517.00	Single	1	______
f.	495.73	Married	6	______

2. Determine the withholdings using table 20.3, monthly payroll.

	Monthly Earnings	Marital Status	Number of Withholding Allowances	Monthly Withholding
a.	1,035.50	Single	2	______
b.	1,719.00	Married	1	______
c.	1,844.70	Married	0	______
d.	1,520.00	Single	3	______
e.	1,619.00	Single	2	______
f.	2,119.00	Married	7	______

3. Refer to Joan Loomis in practice assignment 1, problem 3. She is on a monthly payroll, has gross earnings of $1,726.90, is married, and claims three withholding allowances. Use table 20.3 to determine how much to withhold each month from her gross earnings. How does the amount from table 20.3 compare to the percentage table method of table 20.2? What is the difference in her annual take-home pay?

Amount withheld using withholding table (table 20.3): __________

Amount withheld using percentage table (table 20.2): __________

Difference in annual take-home pay: __________

Section 20.5 Social Security Withholdings (FICA Contributions)

KEY TERMS **FICA** **FICA Rate** **FICA maximum base** **Year-to-date gross (YTD)**

FICA, the Federal Insurance Contributions Act, was passed during the 1930s to create a fund that would provide a minimum income (social security) for retired or disabled workers, or for widows and dependent children of deceased workers. Later extensions to the act include Medicare and automatic cost-of-living increases. The Social Security Administration system is funded with payroll withholdings and matching employers' contributions.

Congress establishes a fixed rate, called the **FICA rate** that is applied to gross earnings up to a specified ceiling, called the **FICA maximum base.** Initially the FICA rate was low, and the maximum base to which it could be applied was $3,000, but over the years, both the rate and the maximum base have risen considerably.

Payroll systems compute the FICA contribution as a percent of gross earnings. Contributions continue until an employee has paid the maximum annual FICA contribution (that is, until the employee's gross earnings equal FICA maximum base).

> 1987 maximum base: $43,800
> FICA rate: 7.15%
>
> 1990 maximum base: $50,400
> FICA rate: 7.65%
>
> Annual FICA contribution = rate × annual gross (but not greater than rate × maximum base)
>
> Annual gross = $22,500
> Annual FICA contribution = 7.65% of 22500 = 1721.25
>
> Maximum annual FICA contribution = 7.65% of 50400 = 3855.60

A payroll system must recall the accumulated gross up to (but not including) the current payroll, which is called the **year-to-date gross (YTD).** If YTD is greater than maximum base, FICA withholding = 0. If (YTD + new gross) is less than or equal to maximum base,

FICA = rate × new gross.

If (YTD + new gross) is greater than maximum base,

FICA = rate × (maximum base − YTD).

This brings the FICA to its maximum for the year.*

YTD	17796.08
+ new gross	+ 378.64
	18174.72

(YTD + new gross) is less than maximum base.
FICA = 7.65% of 378.64 = 28.97.

YTD	50034.72
+ new gross	+ 905.86
	50940.58

(YTD + new gross) is greater than maximum.
FICA = 7.65% of (50400.00 − 50034.72)
= 7.65% of 365.28 = 27.94.
In this case, no additional FICA contributions will be deducted this year.

Example 1 A data entry clerk earns $269.50 per week. Year-to-date gross is $10,241.00. What is the FICA contribution for the current week?

10241.00 + 269.50 = 10510.50
(10510.50 is less than 50400, maximum base)
FICA contribution = 169.5 × 0.0765 = 20.62

key: .0765 [×] 269.5 [=]
display: 20.61675
edited: $20.62

Example 2 A government official earns $4,637.50 per month. By the end of November, yearly gross is $48,312.50. What is the FICA contribution for the last month of the year?

48312.50 + 4637.50 = 52950.00
52950 is greater than 50400.
FICA contribution = 7.65% of (50400.00 − 48312.50)
= 2087.5 × 0.0765 = 159.69

key alg: 50400 [−] 48312.5 [=] [×] .0765 [=]
key arith: 50400 [+] 48312.5 [−] [×] .0765 [=]
display: 159.69375
edited: $159.69

Example 3 A salaried attorney earned $59,750 by the end of October. What is the FICA contribution for November? YTD of $59,750 is greater than maximum base of $50,400. The attorney has fulfilled FICA obligation and no more FICA contributions are required for the year.

---

*The above rules (except for changing rate and maximum), in force for several decades, were amended by Congress after this writing. The Old Age, Survivors, and Disability (OASDI) portion and the Medicare (HI) portion have been split into separate deductions with different maximum bases. It remains useful to study the above principles, which are still valid, but future editions will reflect the change.

Table 20.4

7.65% Social Security Employee Tax Table

Wages at least	But less than	Tax to be withheld	Wages at least	But less than	Tax to be withheld	Wages at least	But less than	Tax to be withheld	Wages at least	But less than	Tax to be withheld
$0.00	$0.07	$0.00	12.75	12.88	.98	25.56	25.69	1.96	38.37	38.50	2.94
.07	.20	.01	12.88	13.01	.99	25.69	25.82	1.97	38.50	38.63	2.95
.20	.33	.02	13.01	13.14	1.00	25.82	25.95	1.98	38.63	38.76	2.96
.33	.46	.03	13.14	13.27	1.01	25.95	26.08	1.99	38.76	38.89	2.97
.46	.59	.04	13.27	13.40	1.02	26.08	26.21	2.00	38.89	39.02	2.98
.59	.72	.05	13.40	13.53	1.03	26.21	26.34	2.01	39.02	39.16	2.99
.72	.85	.06	13.53	13.67	1.04	26.34	26.48	2.02	39.16	39.29	3.00
.85	.99	.07	13.67	13.80	1.05	26.48	26.61	2.03	39.29	39.42	3.01
.99	1.12	.08	13.80	13.93	1.06	26.61	26.74	2.04	39.42	39.55	3.02
1.12	1.25	.09	13.93	14.06	1.07	26.74	26.87	2.05	39.55	39.68	3.03
1.25	1.38	.10	14.06	14.19	1.08	26.87	27.00	2.06	39.68	39.81	3.04
1.38	1.51	.11	14.19	14.32	1.09	27.00	27.13	2.07	39.81	39.94	3.05
1.51	1.64	.12	14.32	14.45	1.10	27.13	27.26	2.08	39.94	40.07	3.06
1.64	1.77	.13	14.45	14.58	1.11	27.26	27.39	2.09	40.07	40.20	3.07
1.77	1.90	.14	14.58	14.71	1.12	27.39	27.52	2.10	40.20	40.33	3.08
1.90	2.03	.15	14.71	14.84	1.13	27.52	27.65	2.11	40.33	40.46	3.09
2.03	2.16	.16	14.84	14.97	1.14	27.65	27.78	2.12	40.46	40.59	3.10
2.16	2.29	.17	14.97	15.10	1.15	27.78	27.91	2.13	40.59	40.72	3.11
2.29	2.42	.18	15.10	15.23	1.16	27.91	28.04	2.14	40.72	40.85	3.12
2.42	2.55	.19	15.23	15.36	1.17	28.04	28.17	2.15	40.85	40.99	3.13
2.55	2.68	.20	15.36	15.50	1.18	28.17	28.31	2.16	40.99	41.12	3.14
2.68	2.82	.21	15.50	15.63	1.19	28.31	28.44	2.17	41.12	41.25	3.15
2.82	2.95	.22	15.63	15.76	1.20	28.44	28.57	2.18	41.25	41.38	3.16
2.95	3.08	.23	15.76	15.89	1.21	28.57	28.70	2.19	41.38	41.51	3.17
3.08	3.21	.24	15.89	16.02	1.22	28.70	28.83	2.20	41.51	41.64	3.18
3.21	3.34	.25	16.02	16.15	1.23	28.83	28.96	2.21	41.64	41.77	3.19
3.34	3.47	.26	16.15	16.28	1.24	28.96	29.09	2.22	41.77	41.90	3.20
3.47	3.60	.27	16.28	16.41	1.25	29.09	29.22	2.23	41.90	42.03	3.21
3.60	3.73	.28	16.41	16.54	1.26	29.22	29.35	2.24	42.03	42.16	3.22
3.73	3.86	.29	16.54	16.67	1.27	29.35	29.48	2.25	42.16	42.29	3.23
3.86	3.99	.30	16.67	16.80	1.28	29.48	29.61	2.26	42.29	42.42	3.24
3.99	4.12	.31	16.80	16.93	1.29	29.61	29.74	2.27	42.42	42.55	3.25
4.12	4.25	.32	16.93	17.06	1.30	29.74	29.87	2.28	42.55	42.68	3.26
4.25	4.38	.33	17.06	17.19	1.31	29.87	30.00	2.29	42.68	42.82	3.27
4.38	4.51	.34	17.19	17.33	1.32	30.00	30.14	2.30	42.82	42.95	3.28
4.51	4.65	.35	17.33	17.46	1.33	30.14	30.27	2.31	42.95	43.08	3.29
4.65	4.78	.36	17.46	17.59	1.34	30.27	30.40	2.32	43.08	43.21	3.30
4.78	4.91	.37	17.59	17.72	1.35	30.40	30.53	2.33	43.21	43.34	3.31
4.91	5.04	.38	17.72	17.85	1.36	30.53	30.66	2.34	43.34	43.47	3.32
5.04	5.17	.39	17.85	17.98	1.37	30.66	30.79	2.35	43.47	43.60	3.33
5.17	5.30	.40	17.98	18.11	1.38	30.79	30.92	2.36	43.60	43.73	3.34
5.30	5.43	.41	18.11	18.24	1.39	30.92	31.05	2.37	43.73	43.86	3.35
5.43	5.56	.42	18.24	18.37	1.40	31.05	31.18	2.38	43.86	43.99	3.36
5.56	5.69	.43	18.37	18.50	1.41	31.18	31.31	2.39	43.99	44.12	3.37
5.69	5.82	.44	18.50	18.63	1.42	31.31	31.44	2.40	44.12	44.25	3.38
5.82	5.95	.45	18.63	18.76	1.43	31.44	31.57	2.41	44.25	44.38	3.39
5.95	6.08	.46	18.76	18.89	1.44	31.57	31.70	2.42	44.38	44.51	3.40
6.08	6.21	.47	18.89	19.02	1.45	31.70	31.84	2.43	44.51	44.65	3.41
6.21	6.34	.48	19.02	19.16	1.46	31.84	31.97	2.44	44.65	44.78	3.42
6.34	6.48	.49	19.16	19.29	1.47	31.97	32.10	2.45	44.78	44.91	3.43
6.48	6.61	.50	19.29	19.42	1.48	32.10	32.23	2.46	44.91	45.04	3.44
6.61	6.74	.51	19.42	19.55	1.49	32.23	32.36	2.47	45.04	45.17	3.45
6.74	6.87	.52	19.55	19.68	1.50	32.36	32.49	2.48	45.17	45.30	3.46
6.87	7.00	.53	19.68	19.81	1.51	32.49	32.62	2.49	45.30	45.43	3.47
7.00	7.13	.54	19.81	19.94	1.52	32.62	32.75	2.50	45.43	45.56	3.48
7.13	7.26	.55	19.94	20.07	1.53	32.75	32.88	2.51	45.56	45.69	3.49
7.26	7.39	.56	20.07	20.20	1.54	32.88	33.01	2.52	45.69	45.82	3.50
7.39	7.52	.57	20.20	20.33	1.55	33.01	33.14	2.53	45.82	45.95	3.51
7.52	7.65	.58	20.33	20.46	1.56	33.14	33.27	2.54	45.95	46.08	3.52
7.65	7.78	.59	20.46	20.59	1.57	33.27	33.40	2.55	46.08	46.21	3.53
7.78	7.91	.60	20.59	20.72	1.58	33.40	33.53	2.56	46.21	46.34	3.54
7.91	8.04	.61	20.72	20.85	1.59	33.53	33.67	2.57	46.34	46.48	3.55
8.04	8.17	.62	20.85	20.99	1.60	33.67	33.80	2.58	46.48	46.61	3.56
8.17	8.31	.63	20.99	21.12	1.61	33.80	33.93	2.59	46.61	46.74	3.57
8.31	8.44	.64	21.12	21.25	1.62	33.93	34.06	2.60	46.74	46.87	3.58
8.44	8.57	.65	21.25	21.38	1.63	34.06	34.19	2.61	46.87	47.00	3.59
8.57	8.70	.66	21.38	21.51	1.64	34.19	34.32	2.62	47.00	47.13	3.60
8.70	8.83	.67	21.51	21.64	1.65	34.32	34.45	2.63	47.13	47.26	3.61
8.83	8.96	.68	21.64	21.77	1.66	34.45	34.58	2.64	47.26	47.39	3.62
8.96	9.09	.69	21.77	21.90	1.67	34.58	34.71	2.65	47.39	47.52	3.63
9.09	9.22	.70	21.90	22.03	1.68	34.71	34.84	2.66	47.52	47.65	3.64
9.22	9.35	.71	22.03	22.16	1.69	34.84	34.97	2.67	47.65	47.78	3.65
9.35	9.48	.72	22.16	22.29	1.70	34.97	35.10	2.68	47.78	47.91	3.66
9.48	9.61	.73	22.29	22.42	1.71	35.10	35.23	2.69	47.91	48.04	3.67
9.61	9.74	.74	22.42	22.55	1.72	35.23	35.36	2.70	48.04	48.17	3.68
9.74	9.87	.75	22.55	22.68	1.73	35.36	35.50	2.71	48.17	48.31	3.69
9.87	10.00	.76	22.68	22.82	1.74	35.50	35.63	2.72	48.31	48.44	3.70
10.00	10.14	.77	22.82	22.95	1.75	35.63	35.76	2.73	48.44	48.57	3.71
10.14	10.27	.78	22.95	23.08	1.76	35.76	35.89	2.74	48.57	48.70	3.72
10.27	10.40	.79	23.08	23.21	1.77	35.89	36.02	2.75	48.70	48.83	3.73
10.40	10.53	.80	23.21	23.34	1.78	36.02	36.15	2.76	48.83	48.96	3.74
10.53	10.66	.81	23.34	23.47	1.79	36.15	36.28	2.77	48.96	49.09	3.75
10.66	10.79	.82	23.47	23.60	1.80	36.28	36.41	2.78	49.09	49.22	3.76
10.79	10.92	.83	23.60	23.73	1.81	36.41	36.54	2.79	49.22	49.35	3.77
10.92	11.05	.84	23.73	23.86	1.82	36.54	36.67	2.80	49.35	49.48	3.78
11.05	11.18	.85	23.86	23.99	1.83	36.67	36.80	2.81	49.48	49.61	3.79
11.18	11.31	.86	23.99	24.12	1.84	36.80	36.93	2.82	49.61	49.74	3.80
11.31	11.44	.87	24.12	24.25	1.85	36.93	37.06	2.83	49.74	49.87	3.81
11.44	11.57	.88	24.25	24.38	1.86	37.06	37.19	2.84	49.87	50.00	3.82
11.57	11.70	.89	24.38	24.51	1.87	37.19	37.33	2.85	50.00	50.14	3.83
11.70	11.84	.90	24.51	24.65	1.88	37.33	37.46	2.86	50.14	50.27	3.84
11.84	11.97	.91	24.65	24.78	1.89	37.46	37.59	2.87	50.27	50.40	3.85
11.97	12.10	.92	24.78	24.91	1.90	37.59	37.72	2.88	50.40	50.53	3.86
12.10	12.23	.93	24.91	25.04	1.91	37.72	37.85	2.89	50.53	50.66	3.87
12.23	12.36	.94	25.04	25.17	1.92	37.85	37.98	2.90	50.66	50.79	3.88
12.36	12.49	.95	25.17	25.30	1.93	37.98	38.11	2.91	50.79	50.92	3.89
12.49	12.62	.96	25.30	25.43	1.94	38.11	38.24	2.92	50.92	51.05	3.90
12.62	12.75	.97	25.43	25.56	1.95	38.24	38.37	2.93	51.05	51.18	3.91

Table 20.4 (*Continued*)

7.65% Social Security Employee Tax Table

Wages at least	But less than	Tax to be withheld	Wages at least	But less than	Tax to be withheld	Wages at least	But less than	Tax to be withheld	Wages at least	But less than	Tax to be withheld
51.18	51.31	3.92	63.86	63.99	4.89	76.54	76.67	5.86	89.22	89.35	6.83
51.31	51.44	3.93	63.99	64.12	4.90	76.67	76.80	5.87	89.35	89.48	6.84
51.44	51.57	3.94	64.12	64.25	4.91	76.80	76.93	5.88	89.48	89.61	6.85
51.57	51.70	3.95	64.25	64.38	4.92	76.93	77.06	5.89	89.61	89.74	6.86
51.70	51.84	3.96	64.38	64.51	4.93	77.06	77.19	5.90	89.74	89.87	6.87
51.84	51.97	3.97	64.51	64.65	4.94	77.19	77.33	5.91	89.87	90.00	6.88
51.97	52.10	3.98	64.65	64.78	4.95	77.33	77.46	5.92	90.00	90.14	6.89
52.10	52.23	3.99	64.78	64.91	4.96	77.46	77.59	5.93	90.14	90.27	6.90
52.23	52.36	4.00	64.91	65.04	4.97	77.59	77.72	5.94	90.27	90.40	6.91
52.36	52.49	4.01	65.04	65.17	4.98	77.72	77.85	5.95	90.40	90.53	6.92
52.49	52.62	4.02	65.17	65.30	4.99	77.85	77.98	5.96	90.53	90.66	6.93
52.62	52.75	4.03	65.30	65.43	5.00	77.98	78.11	5.97	90.66	90.79	6.94
52.75	52.88	4.04	65.43	65.56	5.01	78.11	78.24	5.98	90.79	90.92	6.95
52.88	53.01	4.05	65.56	65.69	5.02	78.24	78.37	5.99	90.92	91.05	6.96
53.01	53.14	4.06	65.69	65.82	5.03	78.37	78.50	6.00	91.05	91.18	6.97
53.14	53.27	4.07	65.82	65.95	5.04	78.50	78.63	6.01	91.18	91.31	6.98
53.27	53.40	4.08	65.95	66.08	5.05	78.63	78.76	6.02	91.31	91.44	6.99
53.40	53.53	4.09	66.08	66.21	5.06	78.76	78.89	6.03	91.44	91.57	7.00
53.53	53.67	4.10	66.21	66.34	5.07	78.89	79.02	6.04	91.57	91.70	7.01
53.67	53.80	4.11	66.34	66.48	5.08	79.02	79.16	6.05	91.70	91.84	7.02
53.80	53.93	4.12	66.48	66.61	5.09	79.16	79.29	6.06	91.84	91.97	7.03
53.93	54.06	4.13	66.61	66.74	5.10	79.29	79.42	6.07	91.97	92.10	7.04
54.06	54.19	4.14	66.74	66.87	5.11	79.42	79.55	6.08	92.10	92.23	7.05
54.19	54.32	4.15	66.87	67.00	5.12	79.55	79.68	6.09	92.23	92.36	7.06
54.32	54.45	4.16	67.00	67.13	5.13	79.68	79.81	6.10	92.36	92.49	7.07
54.45	54.58	4.17	67.13	67.26	5.14	79.81	79.94	6.11	92.49	92.62	7.08
54.58	54.71	4.18	67.26	67.39	5.15	79.94	80.07	6.12	92.62	92.75	7.09
54.71	54.84	4.19	67.39	67.52	5.16	80.07	80.20	6.13	92.75	92.88	7.10
54.84	54.97	4.20	67.52	67.65	5.17	80.20	80.33	6.14	92.88	93.01	7.11
54.97	55.10	4.21	67.65	67.78	5.18	80.33	80.46	6.15	93.01	93.14	7.12
55.10	55.23	4.22	67.78	67.91	5.19	80.46	80.59	6.16	93.14	93.27	7.13
55.23	55.36	4.23	67.91	68.04	5.20	80.59	80.72	6.17	93.27	93.40	7.14
55.36	55.50	4.24	68.04	68.17	5.21	80.72	80.85	6.18	93.40	93.53	7.15
55.50	55.63	4.25	68.17	68.31	5.22	80.85	80.99	6.19	93.53	93.67	7.16
55.63	55.76	4.26	68.31	68.44	5.23	80.99	81.12	6.20	93.67	93.80	7.17
55.76	55.89	4.27	68.44	68.57	5.24	81.12	81.25	6.21	93.80	93.93	7.18
55.89	56.02	4.28	68.57	68.70	5.25	81.25	81.38	6.22	93.93	94.06	7.19
56.02	56.15	4.29	68.70	68.83	5.26	81.38	81.51	6.23	94.06	94.19	7.20
56.15	56.28	4.30	68.83	68.96	5.27	81.51	81.64	6.24	94.19	94.32	7.21
56.28	56.41	4.31	68.96	69.09	5.28	81.64	81.77	6.25	94.32	94.45	7.22
56.41	56.54	4.32	69.09	69.22	5.29	81.77	81.90	6.26	94.45	94.58	7.23
56.54	56.67	4.33	69.22	69.35	5.30	81.90	82.03	6.27	94.58	94.71	7.24
56.67	56.80	4.34	69.35	69.48	5.31	82.03	82.16	6.28	94.71	94.84	7.25
56.80	56.93	4.35	69.48	69.61	5.32	82.16	82.29	6.29	94.84	94.97	7.26
56.93	57.06	4.36	69.61	69.74	5.33	82.29	82.42	6.30	94.97	95.10	7.27
57.06	57.19	4.37	69.74	69.87	5.34	82.42	82.55	6.31	95.10	95.23	7.28
57.19	57.33	4.38	69.87	70.00	5.35	82.55	82.68	6.32	95.23	95.36	7.29
57.33	57.46	4.39	70.00	70.14	5.36	82.68	82.82	6.33	95.36	95.50	7.30
57.46	57.59	4.40	70.14	70.27	5.37	82.82	82.95	6.34	95.50	95.63	7.31
57.59	57.72	4.41	70.27	70.40	5.38	82.95	83.08	6.35	95.63	95.76	7.32
57.72	57.85	4.42	70.40	70.53	5.39	83.08	83.21	6.36	95.76	95.89	7.33
57.85	57.98	4.43	70.53	70.66	5.40	83.21	83.34	6.37	95.89	96.02	7.34
57.98	58.11	4.44	70.66	70.79	5.41	83.34	83.47	6.38	96.02	96.15	7.35
58.11	58.24	4.45	70.79	70.92	5.42	83.47	83.60	6.39	96.15	96.28	7.36
58.24	58.37	4.46	70.92	71.05	5.43	83.60	83.73	6.40	96.28	96.41	7.37
58.37	58.50	4.47	71.05	71.18	5.44	83.73	83.86	6.41	96.41	96.54	7.38
58.50	58.63	4.48	71.18	71.31	5.45	83.86	83.99	6.42	96.54	96.67	7.39
58.63	58.76	4.49	71.31	71.44	5.46	83.99	84.12	6.43	96.67	96.80	7.40
58.76	58.89	4.50	71.44	71.57	5.47	84.12	84.25	6.44	96.80	96.93	7.41
58.89	59.02	4.51	71.57	71.70	5.48	84.25	34.38	6.45	96.93	97.06	7.42
59.02	59.16	4.52	71.70	71.84	5.49	84.38	84.51	6.46	97.06	97.19	7.43
59.16	59.29	4.53	71.84	71.97	5.50	84.51	84.65	6.47	97.19	97.33	7.44
59.29	59.42	4.54	71.97	72.10	5.51	84.65	84.78	6.48	97.33	97.46	7.45
59.42	59.55	4.55	72.10	72.23	5.52	84.78	84.91	6.49	97.46	97.59	7.46
59.55	59.68	4.56	72.23	72.36	5.53	84.91	85.04	6.50	97.59	97.72	7.47
59.68	59.81	4.57	72.36	72.49	5.54	85.04	85.17	6.51	97.72	97.85	7.48
59.81	59.94	4.58	72.49	72.62	5.55	85.17	85.30	6.52	97.85	97.98	7.49
59.94	60.07	4.59	72.62	72.75	5.56	85.30	85.43	6.53	97.98	98.11	7.50
60.07	60.20	4.60	72.75	72.88	5.57	85.43	85.56	6.54	98.11	98.24	7.51
60.20	60.33	4.61	72.88	73.01	5.58	85.56	85.69	6.55	98.24	98.37	7.52
60.33	60.46	4.62	73.01	73.14	5.59	85.69	85.82	6.56	98.37	98.50	7.53
60.46	60.59	4.63	73.14	73.27	5.60	85.82	85.95	6.57	98.50	98.63	7.54
60.59	60.72	4.64	73.27	73.40	5.61	85.95	86.08	6.58	98.63	98.76	7.55
60.72	60.85	4.65	73.40	73.53	5.62	86.08	86.21	6.59	98.76	98.89	7.56
60.85	60.99	4.66	73.53	73.67	5.63	86.21	86.34	6.60	98.89	99.02	7.57
60.99	61.12	4.67	73.67	73.80	5.64	86.34	86.48	6.61	99.02	99.16	7.58
61.12	61.25	4.68	73.80	73.93	5.65	86.48	86.61	6.62	99.16	99.29	7.59
61.25	61.38	4.69	73.93	74.06	5.66	86.61	86.74	6.63	99.29	99.42	7.60
61.38	61.51	4.70	74.06	74.19	5.67	86.74	86.87	6.64	99.42	99.55	7.61
61.51	61.64	4.71	74.19	74.32	5.68	86.87	87.00	6.65	99.55	99.68	7.62
61.64	61.77	4.72	74.32	74.45	5.69	87.00	87.13	6.66	99.68	99.81	7.63
61.77	61.90	4.73	74.45	74.58	5.70	87.13	87.26	6.67	99.81	99.94	7.64
61.90	62.03	4.74	74.58	74.71	5.71	87.26	87.39	6.68	99.94	100.00	7.65
62.03	62.16	4.75	74.71	74.84	5.72	87.39	87.52	6.69			
62.16	62.29	4.76	74.84	74.97	5.73	87.52	87.65	6.70			
62.29	62.42	4.77	74.97	75.10	5.74	87.65	87.78	6.71			
62.42	62.55	4.78	75.10	75.23	5.75	87.78	87.91	6.72			
62.55	62.68	4.79	75.23	75.36	5.76	87.91	88.04	6.73			
62.68	62.82	4.80	75.36	75.50	5.77	88.04	88.17	6.74			
62.82	62.95	4.81	75.50	75.63	5.78	88.17	88.31	6.75			
62.95	63.08	4.82	75.63	75.76	5.79	88.31	88.44	6.76			
63.08	63.21	4.83	75.76	75.89	5.80	88.44	88.57	6.77			
63.21	63.34	4.84	75.89	76.02	5.81	88.57	88.70	6.78			
63.34	63.47	4.85	76.02	76.15	5.82	88.70	88.83	6.79			
63.47	63.60	4.86	76.15	76.28	5.83	88.83	88.96	6.80			
63.60	63.73	4.87	76.28	76.41	5.84	88.96	89.09	6.81			
63.73	63.86	4.88	76.41	76.54	5.85	89.09	89.22	6.82			

Wages	Taxes
100	$7.65
200	15.30
300	22.95
400	30.60
500	38.25
600	45.90
700	53.55
800	61.20
900	68.85
1,000	76.50

FICA computation is fairly simple since only one rate applies to any periodic gross up to maximum base. The Social Security Administration provides a Social Security Employee Tax table for determining FICA contributions. (See table 20.4)

The income brackets on the table are small enough so that the contribution listed on the table equals the computed contribution rounded to the nearest cent.

Using Table 20.4

Weekly gross = $97.88
Wages are at least 97.85 but less than 97.98.
"Tax to be withheld" column displays: 7.49.

gross earnings = $782.76
See lower right corner of second page.

FICA tax for 700 =	53.55
FICA tax for 82.76 =	6.33
(at least 82.68 but less than 82.82)	59.88

FICA contribution for $782.76 = $59.88.

Note that 7.65% of 782.76
= 782.76 × 0.0765
= 59.88114
= 59.88

PRACTICE ASSIGNMENT

Plan: Except where otherwise indicated, the FICA rate is 7.65% and the maximum base is $50,400. The maximum annual FICA contribution is 7.65% of $50,400. Unless the accumulated gross (YTD) reaches the maximum base, the FICA contribution is 7.65% of periodic gross, which you can compute or look up in table 20.4. To compute the current contribution, given the YTD:

If YTD is greater than maximum base, FICA contribution = 0.
If (YTD + current gross) is less than maximum base, FICA = rate × current gross.
If (YTD + current gross) is greater than maximum base, FICA = rate × (maximum base − YTD).

1. For the following record of an employee's FICA contributions during the first four months of 1987 when the rate was 7.15%, find the corresponding 1990 FICA contributions at the rate of 7.65%. Assume that YTD is not close to maximum base.

Month	Monthly Earnings	Monthly FICA tax in 1987	Monthly FICA Tax in 1990
Jan	4,145	296.37	______
Feb	3,800	271.70	______
Mar	4,250	303.88	______
Apr	3,850	275.28	______

2. Compute the total FICA withholdings for the year.

	Annual Gross Earnings	Annual FICA Withholdings
a.	9,715.32	______
b.	29,450.00	______
c.	14,005.80	______
d.	46,224.00	______
e.	57,620.00	______

3. Compute the current FICA withholdings.

	YTD	New Gross Earnings	YTD + New Gross	FICA Rate Applied on	FICA Withheld
a.	27,614.03	1,404.25	______	______	______
b.	2,476.50	825.50	______	______	______
c.	11,511.99	426.37	______	______	______
d.	17,050.50	947.25	______	______	______
e.	47,725.00	3,975.00	______	______	______

4. By November 30, a machinist's year-to-date earnings amount to $48,314.17. What are the FICA withholdings on the next semimonthly gross of $1,573.01?

5. Sally Wu earns a gross annual salary of $19,400. She also receives a 4% commission on all sales. During the month of March her sales totaled $5,372.68. How much was her FICA contribution for March?

Section 20.6 Other Withholdings, Paycheck, and Paycheck Stub

KEY TERMS **Payroll taxes** **Paycheck** **Paycheck stub** **Electronic fund transfer (EFT)**

Many states impose a state income tax that must be withheld. Additional withholdings may include city and/or county **payroll taxes,** medical insurance, disability insurance, stock purchase plans, and other savings or retirement plans. Miscellaneous payroll deductions are often straight percentages or flat monthly payments.

Examples of Miscellaneous Withholdings

State disability premium = 0.9% of first $21,900

City tax = 0.25% of gross

Retirement plan premium = 5.105% of gross, matched by employer's payment

Employee's share of medical insurance premium = $95 per month

Example 1 Gross pay is $1,876.62 and county tax is 0.5% of gross. What is the payroll deduction for county tax?

0.5% = 0.005
1876.62 × 0.005 = 9.38

key: 1876.62 [×] .005 [=]
display: 9.3831
edited: $9.38

Each **paycheck** has a **paycheck stub** that lists payroll deductions, gross pay, and net pay.

While many employees receive paychecks that they cash or deposit into bank accounts, direct deposit is another option that is gaining popularity. The pay of an employee who authorizes direct deposit is transferred from the employer's bank account directly into the employee's bank account. The process by which banks transfer funds between one another is called **electronic fund transfer (EFT).** Instead of a paycheck, the employee receives a paycheck stub that records all payroll deductions and the net pay that was deposited.

Figure 20.1 Paycheck and Stub

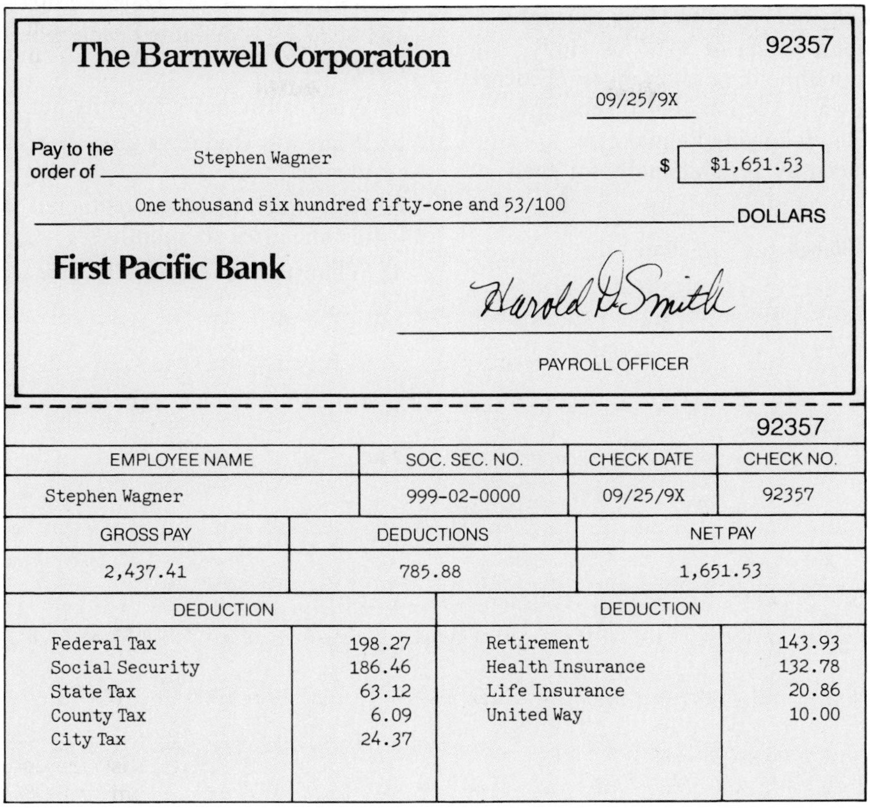

The Barnwell Corporation 92357

09/25/9X

Pay to the order of Stephen Wagner $ $1,651.53

One thousand six hundred fifty-one and 53/100 DOLLARS

First Pacific Bank

Harold D. Smith

PAYROLL OFFICER

92357

EMPLOYEE NAME	SOC. SEC. NO.	CHECK DATE	CHECK NO.
Stephen Wagner	999-02-0000	09/25/9X	92357

GROSS PAY	DEDUCTIONS	NET PAY
2,437.41	785.88	1,651.53

DEDUCTION		DEDUCTION	
Federal Tax	198.27	Retirement	143.93
Social Security	186.46	Health Insurance	132.78
State Tax	63.12	Life Insurance	20.86
County Tax	6.09	United Way	10.00
City Tax	24.37		

PRACTICE ASSIGNMENT

Plan: The computations for miscellaneous deductions are simple because they are often straight percentages of gross or flat amounts, but the rates are sometimes fractions of percent. (Recall, for example, that 0.25% = 0.0025.) Compute the rates or percentages by the usual rate or percentage formulas.

1. Refer to the paycheck stub in figure 20.1 to answer the following questions. Show the rates to the nearest thousandth of a percent where necessary.

a. What percent of gross is the city payroll tax?

b. What percent of gross is the county payroll tax?

c. What percent of gross is the employee's share of the retirement plan premium?

d. If this is a monthly paycheck, how much per year does medical insurance cost the employee?

e. What payroll deductions listed on the stub have not been mentioned so far in this chapter?

__________ __________

2. What is the city tax deduction corresponding to a periodic wage of $853.45 at 0.2% of gross?

3. An employee earns $25,000 annually. What is the annual payroll deduction for a retirement plan at 5.6%?

4. What is an employee's annual share of the state disability insurance premium at 0.9% of the first $21,900 if annual gross earnings are $18,750?

5. One state withholds income tax (after withholding allowances have been subtracted) at the rate of 2% for the first $15,000 of annual income and 4% for any income above $15,000. If an employee's annual taxable earnings total $26,530, how much is deducted from each monthly paycheck for state income tax?

Challenger

Rodney Callaghan is an engineer employed by Starlight Semiconductors at an annual salary of $41,775. He is married and claims three withholding allowances. (Federal withholding rates are shown in section 20.4.) The state withholds income tax at the following annual rate: Gross income is first reduced by $750 annually for each withholding allowance.

Earnings Subject to Withholdings	Rate
first $10,000	3%
second $10,000	4%
above $20,000	5%

In addition to state and federal income taxes and social security tax (7.65%), 6% is deducted for his pension plan, and $135.40 is deducted each month for medical insurance.

a. What is Rodney's monthly net income?

b. What was Rodney's gross year-to-date after six months?

c. How much had been deducted from Rodney's gross income after six months?

d. What percent of gross income is his take-home pay?

CHAPTER 20 SUMMARY

Concept	*Example*	*Procedure*	*Formula*
SECTION 20.1 Salary			
Salary Periodic salary	Salary = 22,500 Monthly salary = 22500 ÷ 12 = 1875.00	Periodic gross is annual salary divided by number of pay periods.	Periodic salary $= \frac{\text{salary}}{\text{no. of pay periods per year}}$ salary[÷]n[=] salary/n
SECTION 20.2 Hourly Wage			
Hourly rate	hours worked = 37 hourly rate = 6.25 gross = 37 × 6.25 = 231.25	Hourly wage is payment per hour worked. Multiply hourly wage by hours worked.	Gross = hours × rate hours[×]rate[=] hours*rate
Weekly overtime	Hours worked = 46 6 hours are overtime	Hours over 40 per week are paid at time and a half.	
Time-and-a-half rate method	Rate = 6.25 Time-and-a-half rate = 6.25 × 1.5 = 9.38 Gross = 40 × 6.25 + 6 × 9.38 = 306.28	Regular hours are paid at regular rate, and overtime hours are paid at overtime rate, where overtime rate = 1.5 × rate.	Gross = reg. hrs. × rate + ov. hrs. × ov. rate alg: reg.hrs[×]rate[+]ov.hrs [×]ov.rate [=] arith. or inexpensive alg: reg.hrs[×]rate[=][M+]ov.hrs[×] ov.rate[=][M+][MR] reghrs*rate+ovhrs*ovrate
Overtime premium method	Rate = 6.25 Half rate = 6.25 ÷ 2 = 3.13 Gross = 46 × 6.25 + 6 × 3.13 = 306.28		Gross = tot. hrs. × rate + ov. hrs. × half rate alg: tot.hrs[×]rate[+]ov.hrs [×]halfrate[=] arith. or inexpensive alg: tot.hrs[×]rate[=][M+]ov.hrs[×] halfrate[=][M+][MR] tothrs*rate+ovhrs*halfrate
SECTION 20.3 Commission			
Commission	Monthly sales = 5635.28 Commission rate = 6% Commission = 6% of 5635.28 = 338.12	Commission equals percent of sales, or	Commission = sales × rate sales[×]rate[=] sales*rate
	Monthly sales = 11,508.50 Commission = 5% of sales over 10,000 = 5% of 1508.50 = 75.43	Commission equals percent of sales above a certain amount.	
Graduated or sliding scale commission	1st 5,000 of sales: 5% 2d 5,000 of sales: 6% Sales above 10,000: 7%	The commission rate increases with increasing sales.	
	Sales = 12,370.00 Commission: 5000 × 0.05 = 250.00 + 5000 × 0.06 = 300.00 + 2370 × 0.07 = 165.90 715.90	Multiply each sales level by its corresponding commission rate. Add the commissions.	

CHAPTER 20 SUMMARY—*(Continued)*

Concept	*Example*	*Procedure*	*Formula*
SECTION 20.4 Federal Income Tax Withholdings			
Federal Income Tax Withholdings Withholding allowances	One withholding allowance for a biweekly payroll is 76.92. For a single employee's biweekly gross of 942.31 with two allowances: Gross = 942.31 2 allowances = 153.84 earnings subject to withholding = 788.47 Withholding = 107.10 + 28% of (788.47 − 756) = 116.19	Federal income tax withholdings can be calculated using the withholding percentage tables after allowances are deducted. These tables are arranged according to pay period and marital status. Withholdings can also be found using the withholding tables that are organized according to pay period, marital status, and number of allowances.	Taxable income = gross − no. of allow. × allow. alg: gross[−]no.allow.[×]allow.[=] arith. or inexpensive alg: gross[M+]no.allow.[×]allow.[=][M−][MR] gross−noallow*allow Tax = result of progressive rates in federal tables
SECTION 20.5 Social Security Withholdings			
FICA contributions	The Social Security contribution is 7.65% of gross. Maximum base = $50,400.	One rate applies, but the maximum annual contribution is rate × the maximum base.	
Year-to-date (YTD)	YTD = 17,432.50 Current gross = 987.30 FICA = 7.65% of 987.30 = 75.53 YTD = 50,140.21 Current gross = 987.30 FICA = 7.65% of (50400 − 50140.21) = 19.87 YTD = 51,720.15 FICA = 0.00	The FICA contribution is a straight percentage. It can also be found in the Social Security Employee Tax Table. Check YTD against maximum base. If YTD is close to maximum base, do not apply rate to all of current gross. If YTD has reached maximum base, there is no further contribution.	FICA = rate × gross rate[×]gross[=] FICA = rate × (max.base − YTD) alg: max.[−]YTD[=][×]rate[=] arith: max.[+]YTD[−][×]rate[=] IF(YTD>max,0,IF((YTD+gross) <=max,rate*gross, rate*(max−YTD)))
SECTION 20.6 Other Withholdings			
Miscellaneous deductions	City tax = 0.25% of gross Medical insurance: $0.00 for individuals, $98.75 for families. Retirement contribution = 5.75% of gross stock purchase plan contribution = $25.00	States, cities, and/or counties may impose payroll taxes. Other payroll deductions may include medical insurance, disability insurance, stock purchase plans, or retirement plans. These withholdings are often a small percent of gross or a flat payment.	

Chapter 20 **Spreadsheet Exercise**

Payroll Register

Payroll registers used to be painstakingly maintained by hand. Many businesses now use computer programs to maintain payroll records. In this exercise, you will complete part of a payroll register.

Your chapter 20 template will appear as follows:

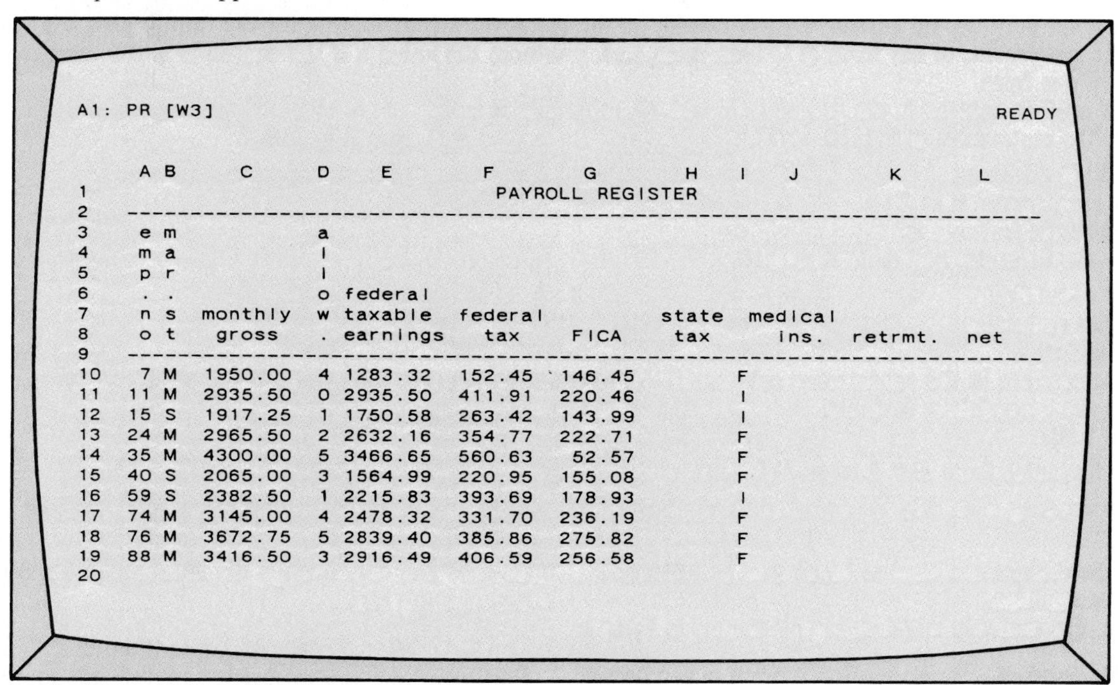

```
A1: PR [W3]                                                                    READY

     A  B      C      D   E          F          G         H     I   J         K       L
1                                        PAYROLL REGISTER
2    ----------------------------------------------------------------------------------
3    e  m             a
4    m  a             l
5    p  r             l
6    .  .             o  federal
7    n  s   monthly   w  taxable    federal              state  medical
8    o  t    gross    .  earnings     tax       FICA      tax      ins.     retrmt.   net
9    ----------------------------------------------------------------------------------
10    7 M   1950.00   4  1283.32    152.45    146.45             F
11   11 M   2935.50   0  2935.50    411.91    220.46             I
12   15 S   1917.25   1  1750.58    263.42    143.99             I
13   24 M   2965.50   2  2632.16    354.77    222.71             F
14   35 M   4300.00   5  3466.65    560.63     52.57             F
15   40 S   2065.00   3  1564.99    220.95    155.08             F
16   59 S   2382.50   1  2215.83    393.69    178.93             I
17   74 M   3145.00   4  2478.32    331.70    236.19             F
18   76 M   3672.75   5  2839.40    385.86    275.82             F
19   88 M   3416.50   3  2916.49    406.59    256.58             F
20
```

Payroll data pertaining to ten employees are displayed in rows 10–19 as follows:

Employee number is in column A.

Marital status is in column B.

Monthly gross is in column C.

Number of withholding allowances is in column D.

Individual or family medical coverage (I or F) is in column I.

The year-to-date gross is in column M and is not visible unless column A is moved off the screen.

Column E: Taxable Earnings
Gross − number of allowances × 166.67
The spreadsheet formula has already been entered; it is:
+C10−D10*166.67.
The values in column E are displayed.

Column F: Federal Tax

Taxable Earnings Over	But not over	Amount to be Withheld
		Single
92	1638	15% of excess over 92
1638	3833	231.90 plus 28% of excess over 1638
		Married
267	2846	15% of excess over 267
2846	6504	386.85 plus 28% of excess over 2846

The spreadsheet formula has already been entered; it is:

@IF(B10="S",

@IF(E10<=1638,.15*(E10−92),
231.9+.28*(E10−1638)),

@IF(E10<=2846,.15*(E10−267),
386.85+.28*(E10−2846))).

The values in column F are displayed.

Column G: FICA
If YTD > 48000, FICA = 0; otherwise:
if (YTD + gross) <= 48000, FICA = 0.0751 × gross,
otherwise: FICA = 0.0751 × (48000 − YTD)

The spreadsheet formula has already been entered; it is

@IF(M10>48000,0,

@IF((M10+C10)<=48000,
.0751*C10,.0751*(48000−M10))).

The values in column G are displayed.

Spreadsheet Exercise Steps

1. Column H: State Tax
   2% of (gross less $100 per year for each tax allowance) (100/12 per month).
   Gross is in C10, the number of allowances in D10.
   The spreadsheet formula is: ______ * (______ − ______ * ______ / ______).
   Complete the formula, store it in H10, and copy it from H10 to H11..H19.
2. Column J: Medical Insurance
   I denotes individual plan, *F* family plan (in column I).
   The individual plan is paid entirely by the employer; the employee's contribution for the family plan is $104.75. Recall that the IF statement, of the form @ IF(condition,x,y), produces the value *x* if the condition is true and the value *y* if the condition is false.
   If there is an *F* in column I, the premium is 104.75, otherwise it is 0.
   Complete the statement: @ IF(I10="F", ______ , ______) and store it in J10.
3. Column K: Retirement
   5.31% of gross; gross is in C10.
   The spreadsheet formula is: ______ * ______ .
   Complete the formula, and store it in K10.
4. Column L: Net Pay
   Gross − all deductions
   Gross is in C10.
   The deductions are in the consecutive columns F–H, in column J, and in column K.
   Complete the spreadsheet formula: + ______ − @ SUM (______ .. ______) − ______ − ______ and store it in L10.
5. Copy the formulas from J10..L10 to J11..L19.
6. Before proceeding with step 7, record each employee's take-home pay under "previous take-home pay" on the chart in step 7.
7. Suppose the deductions need to be altered as follows:
   Federal Income Tax
   The value of a monthly withholding allowance is 170.83.

Taxable Earnings		Amount to be Withheld
Over	But not over	Single
100	1721	15% of excess over 100
1721	4021	243.15 plus 28% of excess over 1721
		Married
283	2988	15% of excess over 283
2988	6817	405.75 plus 28% of excess over 2988

FICA
If YTD > 50400, FICA = 0; otherwise
If (YTD + gross) ≤ 50400, FICA = 0.0765 × gross,
otherwise: FICA = 0.0765 × (50400 − YTD)

State tax
2.5% of (gross less $150 per year for each tax allowance)

Medical Insurance
The employee's contribution for the family plan has increased from $104.75 to $182.35.

Revise all appropriate formulas in row 10 by substituting the new values for the corresponding old values.

Be sure you only change the numbers in the formulas and nothing else.

Use the editing feature of Lotus (F2).
Editing allows you to scan a long formula and to alter constants to corresponding values. For example, 1638 should be changed to 1721. Use the insert key to engage either typeover or insert mode (similar to WordPerfect).

After you have revised the formulas in row 10, you must copy them once more (from E10..H10 to E11..H19 and from J10 to J11..J19).

Record the new nets in the chart.

Emp. No.	Previous Take-home Pay	New Take-home Pay
7	______	______
11	______	______
15	______	______
24	______	______
35	______	______
40	______	______
59	______	______
74	______	______
76	______	______
88	______	______

NAME ______________________________

CLASS SECTION ______________ DATE ______________

SCORE __________ MAXIMUM 200 PERCENT __________

CHAPTER 20 ASSIGNMENT

1. Compute the periodic gross payments that correspond to an annual gross salary of $20,000.

a. Monthly gross: __________

b. Semimonthly gross: __________

c. Biweekly gross: __________

d. Weekly gross: __________ *(4 points)*

2. Larry Magruder earns $29,700 per year. If he is paid semimonthly, what are his gross earnings each payday? *(1 point)*

3. Del Guerra is a student who works part-time in a restaurant and earns $4.30 per hour. What is his weekly gross if he works 25 hours per week? *(5 points)*

4. A salesperson receiving a 7.5% commission has sold 8 sets of encyclopedias at $975 per set. What is the commission? *(5 points)*

5. A clothing store has three employees whose gross salaries total $879 per week. On the average, how much is each employee's annual gross? *(5 points)*

6. Frank Armbruster is paid minimum wage and works $37\frac{1}{2}$ hours each week. If the minimum wage is raised from $3.85 per hour to $4.25 per hour, how much more will he earn each week? *(5 points)*

7. Linda earns a gross biweekly salary of $697.75. Her friend Marsha earns a gross semimonthly salary of $742.50. Who earns more? *(10 points)*

8. Peter Santucci receives a 5% commission plus reimbursement for travel expenses. On his most recent trip, his sales totaled $23,850, and his reimbursable travel expenses amounted to $358.27. How much should he be paid? *(5 points)*

9. A door-to-door magazine salesman receives a commission of 6.5% on weekly sales of $200 or less. He earns a 7% commission on sales of $200.01 to $400.00 and 8% on sales greater than $400. If his weekly sales total $763.75, how much commission will he receive? *(10 points)*

10. During the first six months she worked for Rocky's Pizza, Brenda Meyers received a training wage of $4 per hour for a 38-hour week. After six months, her wage was increased to $4.65, and her hours remained the same. After working a year for Rocky's, how much did she earn in gross wages? *(10 points)*

11. Robin Lundquist works for an office machine supplier and receives an annual salary of $19,380 plus a 4% commission on all sales. On her last monthly sales trip, Lundquist sold 26 copiers at $1,275.50 each and 148 typewriters at $589 each. Her reimbursable travel expenses amounted to $397.38. How much should the company pay her for the month? *(10 points)*

12. Before the new labor contract went into effect, the average worker at Farber Enterprises earned $7.29 per hour and worked an average of 42 hours per week, receiving time and a half for overtime. Under a new agreement reached between the company and the union, each employee is entitled to an $0.85 per hour raise. However, the average work week has been reduced to 40 hours. How much more does the average worker earn per week under the new contract? ***(10 points)***

13. A saleswoman who works in an appliance store receives a monthly salary of $1,380 plus a 3% commission on the first $8,000 of sales plus a 5% commission on sales exceeding $8,000. During one month she sold 11 refrigerators at $627.50 each, 5 refrigerators at $739 each, and 8 freezers at $598.75 each. What was her gross pay for that month? ***(10 points)***

14. Blake Malinowsky is married and claims four withholding allowances. If he is paid a weekly salary of $497.50, how much is withheld from each paycheck for federal income tax using the withholding table (table 20.3, page 269)? ***(5 points)***

15. A clerk-typist who is single and claims one withholding allowance earns $289.35 per week. If the company is on a biweekly payroll schedule, what amount is subject to withholding each pay period? ***(5 points)***

16. David Keller earns $25,389 per year and is paid weekly. He is married and claims five withholding allowances. Using the withholding table (table 20.3, page 269), how much is withheld each pay period? How does this compare to the percentage table method (table 20.2, page 266)?

Amount withheld using withholding table (table 20.3):

____

Amount withheld using percentage table (table 20.2):

________ ***(10 points)***

17. Chad Falcone earns $423.65 per week. He is single and claims two withholding allowances.

a. How much is subject to withholding?

b. How much is withheld weekly from his gross wages for federal income tax? (Use the percentage table method in table 20.2). ***(10 points)***

18. A saleswoman earns $23,082 per year plus sales commissions. Last month's commissions amounted to $289.75. If her filing status is "single" and she claims one withholding allowance, how much was withheld that month? (Use the percentage table method in table 20.2). ***(10 points)***

For problems 19 through 22, assume a FICA rate of 7.65% and a FICA maximum base of $50,400.

19. Teresa Ferrer earned a gross of $25,750 during one year. What were the total FICA withholdings for that year? *(5 points)*

20. A welder earns $9.47 per hour plus time and a half for overtime. During one week he worked 46 hours. How much was withheld for FICA? (Assume YTD is not close to maximum base.) *(10 points)*

21. A portion of an employee's payroll record is listed below. Fill in the blanks.

Month	YTD	Monthly Earnings	Monthly FICA
Sept	37,260.00	4,657.50	________
Oct	________	4,817.50	________
Nov	________	4,881.00	________
Dec	________	4,881.00	________

(15 points)

22. Kenneth Lanier receives a $5\frac{1}{2}$% straight commission on net sales. His January sales orders totaled $36,784, of which $2,346 were cancelled. How much was deducted from his monthly paycheck for FICA? *(10 points)*

23. What amount should be withheld from a monthly gross of $1,864.35 for a county tax of 0.25% of gross? *(5 points)*

24. The Schroeder Company deducts 7.2% from each weekly paycheck for its retirement plan, 1.6% for life insurance, and $21.55 for health insurance. If an employee's annual gross salary is $29,276, how much is deducted each week for these miscellaneous items? *(10 points)*

25. Charlene Abruzzi's gross earnings were $21,453.12 last year. In addition to weekly income tax and FICA withholdings totaling $66.23, $27.45 was deducted for health insurance, $26.82 for retirement, and $9.08 for union dues each week. What percent (to the nearest tenth of a percent) of Charlene's gross pay was take-home pay? *(15 points)*

Chapter 21 Sales and Property Taxes

OBJECTIVES

- Understand the role of retailers in sales tax collection and determine sales tax
- Understand property assessment value, calculate tax rates, and compute property tax

SECTIONS

INTRODUCTION

Taxes. You hear about them every day: the federal tax deficit, the state income tax increase, the local school referendum. Taxes generate conflict and debate. They also pay for the services that federal, state, and local governments provide.

Section 21.1 Sales Tax

KEY TERMS **Sales tax exemption** **Sales tax exclusion** **Sales tax tables**

Taxes, and Why They are Necessary

Governments levy taxes to generate the money that pays for the services they provide. Taxes increase as services increase or as the cost of services increases. See the brief summary of existing and proposed taxes.

Most states, as well as some counties and cities, impose a tax on retail sales. Retail businesses collect sales taxes from consumers and periodically remit the collected taxes to the taxing authorities. Some states allow a **sales tax exemption** for food and prescription drugs. Religious or nonprofit organizations may apply for a **sales tax exclusion** in states that allow these organizations to purchase any item without paying sales tax.

Use tax is a variation of sales tax. It allows a state to tax certain items its residents purchase out of state.

Types of Taxes

Type of Tax	Levied by	Taxpayers
Sales	States, lower units	Consumers
Property	Counties, municipalities	Businesses, individuals
Excise	Federal, states	Consumers of selected items
Income	Federal, states	Businesses, individuals
Payroll	Counties, municipalities	Employees
Type of Tax	**Proposed by**	**Taxpayers**
Tax on services (e.g. 5% tax on all advertising)	Certain states	Recipients of services
Value-Added Tax (VAT)	Economists propose a national sales tax imposed at all stages of manufacturing. In discussion stage. VAT is heavily relied upon in other countries.	Buyers of manufactured goods at each stage of production (ultimately the consumer).

The sales tax imposed by a state is a designated percent of the retail price, and may vary between one percent and seven percent. Lower governmental units, such as counties and municipalities, may collect additional percentage points.

State Tax:	5%
County Tax:	1%
City Tax:	1%

Many stores use computerized cash registers that compute sales tax. However, not all stores have computerized cash registers, and those that do occasionally experience malfunctions. So it is necessary to know how to calculate sales tax. Furthermore, you need to understand the distinction between the retail price and the price you actually pay.

States provide **sales tax tables** to retail businesses. Some tables show entire tax amounts up to a particular price. Other tables show the tax on amounts below $1. Some sales tax tables do not half-adjust. The brackets in table 21.1 are set up more for convenience than for accuracy. One state uses table 21.2. Its brackets half-adjust but are not easy to remember. Compute the tax mentally on whole dollars; use the table for the remainder.

Table 21.1

Sales Amount	*4% Tax*
.01–.10	.00
.11–.25	.01
.26–.50	.02
.51–.75	.03
.76–.99	.04

Tax on $7.83:
Tax on 7.00 = 0.28
Tax on 0.83 = 0.04
Tax on 7.83 = 0.32

Table 21.2

Sales Amount	*4% Tax*
.01–.12	.00
.13–.37	.01
.38–.62	.02
.63–.87	.03
.88–.99	.04

This 4% partial tax table half-adjusts.

Observe that $7.83 \times 0.04 = 0.3132$, half-adjusted to 0.31 (Half-adjusting means rounding up on 5 or above.)

Example 1 A customer bought taxable merchandise priced at $79.45. If there is a state sales tax of 4% and a county tax of 2%, what is the total tax on this merchandise?

Combined tax rate = 4% + 2% = 6%
$79.45 \times 0.06 = 4.767$

key: 79.45 [×] .06 [=]
display: 4.767
edited: $4.77

Example 2 Refer to table 21.2. What is the sales tax on an electric can opener priced at $11.69?

Tax on 11.00 = 11 × .04 = 0.44
Tax on 0.69 = 0.03
Tax on 11.69 = 0.47
($11.69 \times 0.04 = 0.4676 = 0.47$)

Example 3 At the end of the day a noncomputerized cash register contains $510.73, and the manager needs to calculate how much of the cash register's content is sales tax. All sales are subject to a 5% tax. The cash register amount is 105% (price + tax) of base (price), and the tax amount is 5% of base. Thus,

$$\text{Tax amount} = \frac{\text{cash register amount}}{1 + \text{rate}} \times \text{rate}$$

$$= \frac{510.73}{1.05} \times 0.05 = 24.32$$

key: 510.73 [÷] 1.05 [×] .05 [=]
display: 24.320476
edited: $24.32

PRACTICE ASSIGNMENT

Plan: Compute the sales tax on the total price of nonexempt items. Unless otherwise indicated, use your calculator and round the tax to the nearest cent. If there is one or more local tax in addition to the state tax, the combined rate applies. When using a partial sales tax table, compute the tax mentally on the whole dollars and use the table on the remainder; the tax is the sum of the two amounts. To solve other miscellaneous sales tax problems, recall that price is the base and tax the percentage.

1. Compute the 5% sales tax on items a–f by using table 21.3.

	Price					Tax
a.	0.72	______	+	______	=	______
b.	7.99	______	+	______	=	______
c.	5.64	______	+	______	=	______
d.	18.78	______	+	______	=	______
e.	20.50	______	+	______	=	______
f.	30.89	______	+	______	=	______

Table 21.3

Sales Amount	*Tax*
.01–.09	.00
.10–.29	.01
.30–.49	.02
.50–.69	.03
.70–.89	.04
.90–.99	.05

2. $2,757.36 in cash receipts include a 2% state sales tax and a 1% local sales tax.

a. How much must be forwarded to the state?

b. How much must be forwarded to the local authority?

3. If a resident in a 5% state + 2% county + 1% city sales tax area shops in the city, she must pay $349.36 for a gold wristwatch. How much would she save if she drove to a shopping mall outside the city and purchased the same watch at the same price?

4. How much would the consumer in problem 3 save if she drove 12.5 miles (each way) to a nontax county and purchased the watch at the same price?

5. Reconsider problem 4. Would the savings be worth the cost of the 25-mile trip at 25 cents per mile?
Cost in city: ______
Cost in nontax county: ______

Section 21.2 Property Tax

KEY TERMS **Assessed value** **Assessment rate** **Property tax rate** **Mill**

Taxes may be imposed on real estate (land and buildings) and personal property. Property taxes are levied on businesses as well as individuals, and provide a major source of revenue for counties and municipalities. Property taxes are usually imposed annually and may be paid annually, semiannually, or monthly.

The value of property for taxation purposes is called **assessed value,** and is derived by the tax assessor from fair market value. Some states set a uniform **assessment rate;** in states that do not, assessment rates vary from county to county or town to town.

$$\text{Tax rate} = \frac{\text{amount needed}}{\text{assessed values}}$$

Annual budget = 4,350,000
Assessed values = 201,700,000

$$\text{Tax rate} = \frac{4350000}{201700000}$$
$$= 0.0215666$$
$$= 2.157\%$$
$$= 21.57 \text{ mills}$$

The **property tax rate** for a given area is determined by the amount of money needed for government services and by the total assessment of taxable property in that area. It can be expressed as a decimal number, a percent, or in mills.

Mill denotes *per thousand,* and equals one-tenth of one percent. The decimal point is moved three places to the right rather than two places.

Example 1 Real property in an outlying suburb of Central City is assessed at 75% of fair market value, while real property within Central City limits is assessed at 95% of fair market value. How would an office building with fair market value of $350,000 be taxed in each area?

In the suburbs: 350000 × 0.75 × tax rate
= 262500 × tax rate

In Central City: 350000 × 0.95 × tax rate
= 332500 × tax rate

Example 2 In Grove City, the property tax rate is 21.57 mills, and business property is assessed at 82% of market value. What is the tax on a business site worth $272,500?

Assessed value = 272500 × 0.82 = 223450

21.57 mills = 2.157% = 0.02157

Tax = 223450 × 0.02157 = 4819.82

key: 272500 [×] .82 [×] .02157 [=]
display: 4819.8165
edited: $4,819.82

Example 3 If the tax on a property with an assessed value of $389,000 is $7,733.32, what is the tax rate in percent and in mills?
Assessed value is base. Tax is percentage.

$R = P \div B = 7733.32 \div 389000 = 0.01988$

key: 7733.32 [÷] 389000 [=]
display: 0.01988
edited: 1.988% or 19.88 mills

PRACTICE ASSIGNMENT

Plan: Assessed value is the percentage of market value determined by the assessment rate. The property tax rate is the revenue needed divided by the assessed amount available. It can be expressed in percent but is often expressed in mills. It may help to remember that one percent means one-hundredth, and one mill means one-thousandth. To convert from percent to mills remember that 1% = 10 mills. For example, 0.02931 = 2.931% = 29.31 mills. The tax amount is the percentage of the assessed value determined by the tax rate.

1. Determine the missing values.
Where necessary, round percent rates to three places and mills to two places.

	Assessed Value	Tax Rate in Percent	Tax Rate in Mills	Tax Amount
a.	82,500	2.061	________	________
b.	65,150	________	17.30	________
c.	58,000	________	________	1,642.32
d.	38,300	________	________	892.17
e.	________	1.88	________	2,840.68
f.	________	________	22.30	3,224.58

2. The annual budget of Clear Springs is $3,452,000; the total assessed value of all property is $234,750,000.

a. What is the tax rate in percent, rounded to three places?

b. What is the tax rate in mills, rounded to two places?

3. The building owned and occupied by the Rent-A-Tool Company has a current market value of $366,250. It is located in a county that assesses property at 90% of fair market value and taxes real estate at 23.45 mills. What is Rent-A-Tool's annual property tax?

4. A property owner's home was assessed at $85,000, one car at $4,500, and another car at $5,600, on the annual property tax bill.

a. What was the total property tax if the rate for real estate was 26.82 mills, and the rate for personal property was 22.5 mills?

b. How much was the bill if it was paid early at a 2% discount?

c. How much was the bill if it was paid late at a 6% penalty?

5. An apartment complex is assessed at 65% of market value. Five years ago, its market value was $8,750,000, and the annual property tax was $122,565.62. The market value is now $10,600,000; the tax and assessment rates remain the same.

a. What is the tax rate in mills?

b. What is the present tax amount?

Challenger

Consider the fiscal situation in Lake County, where the annual budget is $1,625,000 and the total assessed value of all property is $83,400,000 (at 75% of fair market value). This once rural area is gradually becoming suburban with a few remaining farms, new housing developments, and trailer parks.

Two large corporations recently moved their headquarters into the county after negotiating with government officials for initial five-year tax incentives. Following this five-year period, these corporate properties will be assessed at approximately twenty million dollars. In order to compensate residents for the impending loss of the rural character of the area, county officials have promised a 68% assessment rate and have assured homeowners that the value of their property will increase by 10% over the next five years. However, additional annual expenditures of 0.7 million dollars to service bonded indebtedness over the next twenty-five years for new public facilities are expected.

a. What is the tax rate now?

b. What will the estimated tax rate be after the five-year corporate tax incentive period?

c. On a home worth $120,000, what is the tax now, accurate to the nearest hundred dollars?

d. What will the tax be on the same home after the five-year period in accordance with the estimated new tax rate (accurate to the nearest hundred dollars)?

CHAPTER 21 SUMMARY

Concept	*Example*	*Procedure*	*Formula*
SECTION 21.1 Sales Tax			
Sales tax	Tax rate is 4% Retail price of nonexempt item = 10.35 10.35 × 0.04 = 0.41 Full price of item = 10.76	Sales tax rates imposed by states, counties, and municipalities may vary from 1% to 7%. To obtain the tax amount, multiply the retail price by the tax rate.	Tax = price × tax rate **price[×]taxrate[=]** price*taxrate
Sales tax tables	**Sales Amount** — **4% Tax** .01–.10 — .00 .11–.25 — .01 .26–.50 — .02 .51–.75 — .03 .76–.99 — .04 Tax on 10.35 = .40 + .02 0.42	Mentally compute tax on whole dollars. Use the table to find tax on remainder. Add the two figures.	
SECTION 21.2 Property Tax			
Assessment rate Assessed value	Market value = 100,000 Assessment rate = 75% Assessed value = 75,000	Assessment rate is 100% or less. Assessed value is a percentage of market value.	Assessment val. = market val. × assessment rate **marketval[×]assessmentrate[=]** marketval*assessmentrate
Property tax rate	County budget = 3,500,000 Total assessed value = 180,650,000 Property tax rate = $\frac{3500000}{180650000}$ = 0.0193744	Divide the amount needed by the total assessed value. This fraction is expressed as a decimal number, percent, or mills.	Rate = $\frac{\text{budget}}{\text{assessments}}$ **budget[÷]assessments[=]** budget/assessments
Mills	1.937% = 19.37 mills	One mill equals one-thousandth (one-tenth of one percent).	
Property tax	Assessed value = 84,750 Tax rate = 1.937% Tax = 84750 × 0.01937 = 1641.61	Property tax is a percentage of assessed value.	Tax = assessed val. × tax rate **assessedval[×]taxrate[=]** assessedval*taxrate

Chapter 21 Spreadsheet Exercise

The Effect of Sales Taxes on Consumers

Sales tax rates imposed by states may vary between one percent and seven percent. Some counties and municipalities levy their own sales taxes in addition to state taxes. A sample of sales tax rates are: three percent in Georgia, four percent in South Carolina, five percent in Kentucky, six percent in Pennsylvania, and seven percent in Nassau County, New York (six percent + one percent). Some of these rates may have increased since this writing.

Sales taxes on small purchases may amount to only a few cents, while the tax on a luxury car might be in excess of a thousand dollars. Use your chapter 21 template to illustrate how prices increase with larger tax rates.

A1: PR [W11] READY

	A	B	C	D	E	F	G
1		THE EFFECT OF SALES TAXES ON THE CONSUMER					
2	------------	------------	------------	------------	------------	------------	
3						Nassau	
4	Retail	GA	SC	KY	PA	County NY	
5	Price	3.00%	4.00%	5.00%	6.00%	7.00%	
6	------------	------------	------------	------------	------------	------------	
7	1.00						
8	5.00						
9	10.00						
10	50.00						
11	100.00						
12	250.00						
13	500.00						
14	750.00						
15	1,000.00						
16	2,500.00						
17	5,000.00						
18	7,500.00						
19	10,000.00						
20	15,000.00						

Column A contains retail prices ranging from $1–$20,000 in rows 7–21.

Row 5 contains tax rates ranging from 3%–7% in columns B–F.

Spreadsheet Exercise Steps

1. Compute the effective price from the retail price and tax rate for the various states.
   For example, if the tax rate is 6%, the effective price is retail price × (1 + .06).
   The retail price is in A7. The column address A must be absolute. The tax rate is in B5; the row address 5 must be absolute.
   The spreadsheet formula is
   $ ______ * (_____ + _____ $ _____).
   Complete the formula, store it in B7, and copy it from B7 to B7..F21.

2. What are the effective prices of a telephone costing $50.00?

Georgia	South Carolina	Kentucky	Pennsylvania	Nassau County NY
________	________	________	________	________

3. What are the effective prices of a desk costing $137.89? Change the retail price in row 11.

Georgia	South Carolina	Kentucky	Pennsylvania	Nassau County NY
________	________	________	________	________

4. What are the effective prices of a van costing $22,399? Change the retail price in row 21.

Georgia	South Carolina	Kentucky	Pennsylvania	Nassau County NY
________	________	________	________	________

NAME ______________________________

CLASS SECTION ______________ DATE ______________

SCORE __________ MAXIMUM 80 PERCENT __________

CHAPTER 21 ASSIGNMENT

1. A consumer purchased five nonexempt items at the following prices: $5.16, $7.88, $3.98, $12.50, $2.78. Compute sales tax of 3% on the total price and round to the nearest cent. *(5 points)*

2. A customer who purchased a sweater for $17.95 handed the cashier $20. If the sales tax was 4%, how much change did the customer receive? *(5 points)*

3. The property tax on a building is $2,753.78. If the tax rate is 1.835%, what is the assessed value? Round to the nearest dollar. *(5 points)*

4. The Dorseys own a home assessed at $62,700. If the property tax rate is 17.54 mills, how much annual property tax must they pay? *(5 points)*

5. If 6% sales tax on a new car amounted to $891, what was the price of the car? *(5 points)*

6. The price of a national brand battery at all National Tire Store locations is $62.85. At one location the customer paid $65.99 for the battery. What was the sales tax rate at that location? *(5 points)*

7. Gaston Township plans a budget of $982,830 for the next fiscal year. Total assessed valuation of all property is $33,400,000.
   a. What is the tax rate in percent, rounded to three places?
   b. What is the annual property tax bill for the owners of commercial property assessed at $384,500?
   (10 points)

8. At the end of the day, the cash register of a convenience store contained $8,007.63 in daily receipts. The monitor displayed food sales totalling $2,783.77. This store is located in an area with a 4% state sales tax on nonfood items. How much sales tax did the store manager remit to the state for that day's sales? *(10 points)*

9. A municipality assesses property at 80% of fair market value. If the tax rate is 2.532% and the tax on a particular property is $3,430.53, what is fair market value of that property to the nearest hundred dollars? *(10 points)*

10. Complete the following property tax bill. Observe that the millages for real estate and personal property differ. *(20 points)*

199X PROPERTY TAX BILL		Lloyd County		Commonwealth of Kentucky
class	mills	assessment	tax	Return notice whether paying in person or by mail. When paying by mail, include self-addressed stamped envelope for receipt.
real	21.3	85550		
personal	22.2	16412		
		total ➤		Make check payable to: Lloyd County
H & L Printing Co. 513 East Sixth Street Hendrick, KY 40641				Mail to: Sheriff John Miller County Court House Main Street Hendrick, KY 40641

TAXPAYER'S RECEIPT		
	– 2% discount if paid by Nov 30	
	if paid by Feb 28	
	+ 8% penalty if paid past Feb 28	
IMPORTANT!!! SEE NOTICE ON REVERSE SIDE	amount paid	date 3/15/199X

Chapter 22 Property Insurance

OBJECTIVES

- Set insurance premiums for annual and shorter policy terms
- Calculate indemnity

SECTIONS

INTRODUCTION

Property insurance protects property owners from loss, damage, or lawsuits due to injury. Insurance companies tailor a property insurance policy to the unique needs of each property owner, whether individual homeowner, landlord, small business owner, or large corporation.

Section 22.1 Property Insurance Premiums

KEY TERMS **Carrier** **Underwriter** **Face value**

Property insurance provides financial relief when losses to property occur. The **carrier** or **underwriter** is the insurance company that carries the policy. **Face value** is the maximum amount that the carrier will pay for a loss.

Businesses and homeowners purchase standard fire protection as a minimum coverage and may also purchase extended protection against theft, vandalism, water damage, etc. Comprehensive policies protect against lawsuits.

Property insurance premiums are rated according to risk factors: the type of building, its purpose, age, location, and proximity to fire fighting equipment. Property is therefore classified by risk. Table 22.1 shows sample classifications only; premiums tend to increase with inflation. Rates may also vary from one company to another.

Annual rates are usually quoted for each $100 of insurance, organized by rating classes.

$$\text{Premium} = \text{rate per hundred} \times \frac{\text{face value}}{100}$$

Table 22.1 Property Insurance Rates for $100 Face Value

Class	*Basic*	*Extended*	*Comprehensive*
A	0.85	1.06	1.18
B	0.92	1.23	1.35
C	0.98	1.19	1.31
D	1.06	1.27	1.39
E	1.12	1.33	1.45

Example 1 Mitchell's Shoe Store is in a class C building. The owner purchased extended coverage for property worth $75,000. What was the premium? Class C rate is 1.19 for a face value of $100.

$$\text{Premium} = \text{rate per hundred} \times \frac{\text{face value}}{100}$$

$$= 1.19 \times \frac{75000}{100} = 892.50$$

Most fire and accident policies cover a one-year period. The premium for a short-term policy (less than one year) may be proportional. If the policyholder cancels a policy before the end of the year, the refund is not ordinarily proportional. For example, if a property owner cancels a policy after six months, the refund will not be half of the annual premium because the carrier's selling, recording, and processing costs have not been reduced proportionally. On the other hand, rates covering more than one year may be lower. Short-term procedures vary and are often tabulated in "short-term tables." The rates in the box to the right are an example, not a standard. Many states require proportional refunds of premiums for policies cancelled by the carrier.

Owner cancellation rates	
Short Term	Penalty
0–239 days	10% of annual premium
240–299 days	5% of annual premium
300–329 days	3% of annual premium
330–360 days	0%

If the insured cancels a policy after 300 days, the proportional premium is 300/360 of the annual premium = $83\frac{1}{3}\%$. Premium paid = $83\frac{1}{3}\% + 3\%$.

If the carrier cancels a policy after 300 days, the premium paid is the proportional premium, that is $83\frac{1}{3}\%$.

Example 2 Bob's Moving Co. purchased an insurance policy worth $300,000 to cover its warehouse and contents for one year. The yearly rate per $100 was 1.35, and the cost for the policy was 1.35 × 3000 = 4050. After 274 days, Bob's went out of business, sold the warehouse, and cancelled the insurance policy. How much of the premium was refunded?

Premium paid = (insured days/360 + penalty rate) × annual premium
= (274/360 + 0.05) × 4050
= (0.76111 . . . + 0.05) × 4050
= 0.8111 . . . × 4050 = 3285

Premium refund = 4050 − 3285 = 765

key alg: 274 [÷] 360 [+] .05 [=] [×] 4050 [=]
key arith: 274 [÷] 360 [=] [+] .05 [+] [×] 4050 [=]
display: 3285. or 3284.9999
edited: $3,285

Example 3 The underwriter of the comprehensive policy of Jiffy Store cancelled the policy after 220 days because of the unusually large number of claims. The store is in a class E building and was insured for $92,500. How much of the premium was refunded?

Annual rate = 1.45 per $100

$$\text{Annual premium} = 1.45 \times \frac{92500}{100} = 1341.25$$

$$\text{Proportional rate} = \frac{220}{360} = \frac{11}{18}$$

$$\text{Premium paid} = \frac{11}{18} \times 1341.25 = 819.65$$

Premium refund = 1341.25 − 819.65 = 521.60

key: 11 [÷] 18 [×] 1341.25 [=]
display: 819.65278
edited: $819.65

PRACTICE ASSIGNMENT

Plan: To determine annual premium, use table 22.1. Annual premium = rate per hundred × face value ÷ 100. In cases of short terms that are not due to cancellation, assume proportional premiums: insured days/360. If the number of insured days is greater than 360, there is no short rate. If a policyholder cancels, the portion of the premium charged for short terms is:

(insured days/360 + penalty rate) × annual premium.

If the insurance company cancels, the premium charged is proportional. If time is given in months, assume each month has 30 days.

1. Compute the following premiums.

	Class	Coverage	Face Value	Annual Premium	Short Term	Premium if Insured Cancels Policy
a.	A	Comprehensive	75,000	________	70 days	________
b.	B	Basic	28,250	________	4 months	________
c.	E	Extended	62,750	________	350 days	________
d.	D	Basic	58,800	________	10 months	________
e.	A	Extended	185,000	________	6 months	________
f.	C	Comprehensive	97,300	________	11 months	________

2. Freddy's Sandwich Shop, located in a class D building, filed for bankruptcy and cancelled its extended property coverage on a $87,000 policy. If 50 days remained in the policy's year, how much of the premium was refunded?

3. A business insured its class B building for $85,000. The company moved to a class D building and upgraded coverage from basic to extended.

a. If the move occurred 104 days after the old policy became effective, how much of the premium was refunded on the old policy?

b. What was the premium for the remainder of the year on the new policy?

c. What was the difference between the two premiums for the next full year?

Section 22.2 Property Insurance Benefits

KEY TERMS **Indemnity** **Coinsurance**

The compensation paid by an insurance company to cover a loss is called **indemnity.** Most property losses are caused by fire, but fires do not always bring about complete destruction. Owners tend to insure their property for less than full value. Property valued at $80,000, for example, may be underinsured at $50,000 under the assumption that loss will not exceed that amount. Thus, the owner pays the premium based on a value of $50,000 for property worth $80,000.

A **coinsurance** clause discourages underinsurance and, if property is underinsured, reduces indemnity. The word *coinsurance* implies that the insured assumes some of the risk.

> Face value = $40,000
> Property value = $75,000
>
> A coinsurance clause of 80% means that the face value should be at least 80% of the property value.
>
> $40000 \div 75000 = 53.3\%$
> The coinsurance clause is not met.
>
> Required face value $= 75000 \times 0.8 = 60000$
>
> Loss is $20,000
>
> $$\text{Indemnity} = \frac{40000}{60000} \times 20000 = \$13{,}333.33$$

The coinsurance clause specifies that if coverage is not at least a certain percent of property value—often 80%—indemnity is reduced to the amount determined by the following calculation:

$$\frac{\text{carried face value}}{\text{required face value}} \times \text{loss}$$

If the face value is $60,000 for a property value of $75,000, the coinsurance clause is satisfied.

A loss of $20,000 would be fully compensated.

Indemnity is never greater than face value. In case of total loss, indemnity is not greater than $60,000.

Example 1 Maggie's Dollar Store, valued at $69,500, is insured for $56,000 with a coinsurance clause of 80 percent. A fire caused damage amounting to $15,000. What is the indemnity?

Required insurance = 80% of property value
= 69500 × 0.8 = 55600

The coinsurance clause is satisfied.

Indemnity = loss = $15,000

Example 2 A paint store with a market value of $87,500 is insured for $63,000. The policy contains an 80 percent coinsurance clause. If the store sustains damages of $25,000, how much indemnity will the insurance company pay?

Required insurance = 80% of property value
= 87500 × 0.8 = 70000
Indemnity is reduced to:

$$\frac{\text{carried face value}}{\text{required face value}} \times \text{loss} = \frac{63000}{70000} \times 25000 = 22500$$

key: 63 [÷] 70 [×] 25000 [=]
display: 22500.
edited: $22,500

PRACTICE ASSIGNMENT

Plan: To determine indemnity, consider the following:

A. If a loss is not total, indemnity pays for damage repair. If a loss is total, indemnity is market value or replacement value but not greater than face value.

B. If there is a coinsurance clause, determine whether face value is at least as great as the value required by the coinsurance clause. If so, indemnity is as in step A. If not, indemnity is reduced to:

$$\frac{\text{carried face value}}{\text{required face value}} \times \text{loss}$$

subject to the face value maximum.

1. Find the indemnity, assuming there is no coinsurance.

	Property Value	Face Value	Loss	Indemnity
a.	74,250	59,400	Total	______
b.	110,500	89,000	35,000	______
c.	38,500	25,000	Total	______
d.	64,300	64,000	50,000	______
e.	127,500	50,000	60,000	______

2. Find indemnity for each of the following cases.

	Property Value	Face Value	Loss	Coinsurance Clause	Indemnity
a.	64,500	55,000	Total	80%	________
b.	48,750	25,000	Total	80%	________
c.	110,000	87,500	25,000	80%	________
d.	93,000	80,000	35,000	85%	________
e.	57,000	43,500	31,000	75%	________
f.	300,000	245,000	150,000	85%	________

3. A business valued at $78,500 carried a $60,000 fire insurance policy for the building and its contents. Fire damage repairs totaled $15,000, and replacement furniture totaled $12,000. If the policy contained an 80% coinsurance clause, how much indemnity did the insurance company pay?

4. A drycleaning establishment with a market value of $55,250 is insured for $35,000 with a coinsurance clause of 80%. If a fire causes total destruction, how much indemnity will the carrier pay?

Challenger

An owner may insure a property with multiple carriers if a property value is too great to be covered by one company. The total indemnity, however, is subject to the limitations previously discussed. Each of the carriers compensates the insured in proportion to its policy's face value. For example, if property valued at $500,000 is insured for $300,000 with one company and $200,000 with another, indemnity payments will be in a ratio of 3 to 2.

The Lariat Hotel is 100% covered by property insurance. It is insured by Hartman Casualty for $1,500,000, by Vesuvius Underwriters for $1,000,000, by Wayfarer's Insurance Company for $750,000, and by Mutual of Warsaw for $500,000. A fire caused damages totaling $450,000. How much indemnity should each carrier pay?

CHAPTER 22 SUMMARY

Concept	*Example*	*Procedure*	*Formula*
SECTION 22.1 Property Insurance Premiums			
Rates	Rate per \$100 = 2.13 Face value = \$85,000 Premium = 2.13 × 850 = 1810.50 Premium = \$1,810.50	The premium rate is shown per \$100 of face value. To obtain the premium, multiply the rate by the face value and divide by 100.	$\text{Premium} = \text{rate per hundred} \times \frac{\text{face value}}{100}$ **rateperh[×]faceval[÷]100[=]** rateperh*faceval/100
Policy cancelled by policyholder	Annual premium = 1810.50 Premium for half a year = (50% + 10%) of annual premium = 60% of annual premium = 1086.30 Premium refund = \$724.20	If policy is cancelled by policyholder, premium is proportional to time with an added penalty.	$\text{Premium} = \left(\frac{d}{360} + \text{penalty}\right) \times \text{annual premium}$ **alg:** **d[÷]360[+]penalty[=][×]anprem[=]** **arith:** **d[÷]360[+]penalty[+][×]anprem[=]** (d/360+penalty)*anprem
Policy cancelled by carrier	Annual premium = 1810.50 Premium for half a year = 50% of annual premium = 905.25 Premium refund = \$905.25	Premium is proportional. No penalty applies. The same is true for short-term policies	$\text{Premium} = \frac{d}{360} \times \text{annual premium}$ **d[÷]360[×]anprem[=]** d/360*anprem
SECTION 22.2 Property Insurance Benefits			
Benefits	Property value = \$75,000 Face value = \$60,000 In case of total loss, indemnity = \$60,000 If cost of damage repair = \$20,000, indemnity = \$20,000	In case of total loss, indemnity equals property value, not to exceed face value. In case of damage, indemnity equals cost of damage repair.	
Coinsurance clause	Loss = \$40,000 Coinsurance clause = 80% Property value = \$100,000 Face value of policy = \$80,000 Coinsurance clause is satisfied. Indemnity = \$40,000 Face value = \$60,000 Coinsurance clause is not satisfied (60%). $\text{Indemnity} = \frac{60000}{80000} \times 40000 = 30000$	Face value must be a specified percent of property value (usually between 70% and 90%). If not, indemnity is reduced.	Required insurance = 0.8 × property value **.8[×]propval[=]** $\text{Indemnity} = \frac{\text{carried face value}}{\text{required face value}} \times \text{loss}$ **carried[÷]required[×]loss[=]** carried/(.8*propval)*loss

Chapter 22 Spreadsheet Exercise

Insurance Premium Lookup

Insurance agents use computer systems that enable them to look up a premium by entering input variables.

Consider table 22.1. The owner of a store in a class C location wants to purchase extended coverage for property worth $95,000.

The insurance agent enters:

C extended 95000

A computer program finds the correct rate for a face value of $100 (1.19) and performs the computation.

The screen displays:

class C extended face value $95,000 $1,130.50

Your chapter 22 template permits you to simulate a premium lookup system.

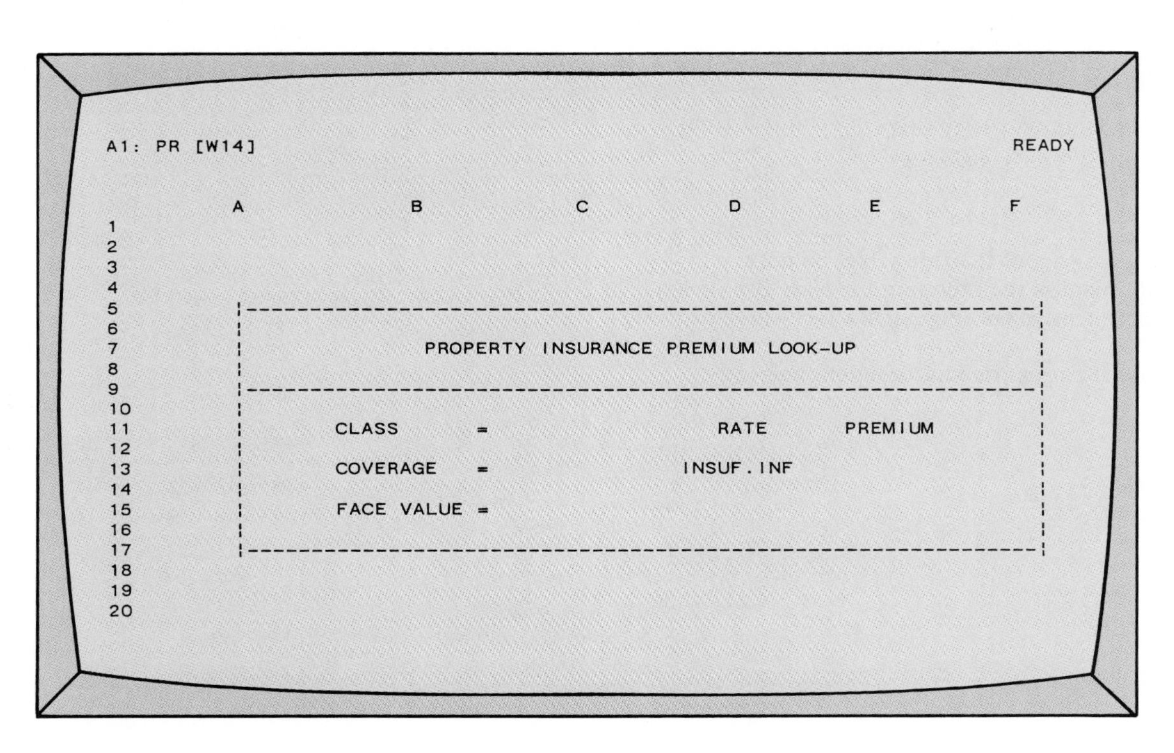

Spreadsheet Exercise Steps

1. Cell E13: Premium Calculation
   Have the program compute the premium from the face value and rate.
   The face value will be in C15, the rate in D13.
   If a spreadsheet formula begins with a cell address, it must be preceded by a sign (+).
   The spreadsheet formula is
   + _______ / _______ * _______.
   Complete the formula and store it in E13.
2. Try the system by inputting some data.
   Enter a class (a, b, c, d, or e) into C11.
   Enter a coverage (basic, extended, or comprehensive) into C13. (Be sure the entry is spelled correctly!)
   Enter a face value (pure numeric form: no dollar sign, no commas) into C15.
   If you have entered the data correctly, the rate and the premium will display; if not, INSUF.INF will display to indicate "insufficient information."

3. What is the premium for class D property with extended coverage and a face value of $175,000? ____________
4. What is the premium for class B property with comprehensive coverage and a face value of $110,000? __________
5. This is how the program finds the rates: The table entries are stored in columns G, H, and I, but these columns are hidden. Data in hidden columns are used in calculations, but their display is suppressed. If you want to change the rates, you must "unhide" these columns.

   /
   Worksheet[enter]
   Column[enter]
   Display[enter]
   prompt for columns to "unhide"
   G1..I1[enter]

   or:

   /WCD
   prompt for columns to "unhide"
   G1..I1[enter]

   Change the comprehensive rates in column I from

   1.18 1.35 1.31 1.39 1.45
   to
   2.18 2.35 2.32 2.39 2.45

   Row 16 is not part of the rate table; do not try to change it. What is the premium for class B property with comprehensive coverage and a face value of $110,000? __________
   Verify that the program still functions correctly.

NAME ____________________

CLASS SECTION ____________ DATE ____________

SCORE ________ MAXIMUM 85 PERCENT ________

CHAPTER 22 ASSIGNMENT

Use table 22.1 and penalty rates shown on page 298.

1. A professional building is located in a class A area and receives basic fire protection. If the building is insured for $425,000, what is the annual premium? *(5 points)*

2. The new King Hardware Store building in Long Branch has a market value of $185,000. Merchandise is worth about $50,000. The store's location near a firehouse and a police station qualifies it for class B fire and casualty insurance.

a. How much will an extended policy cost?

b. How much will a comprehensive policy cost? *(10 points)*

3. Slick's Formal Rental is insured for $25,000. A fire caused loss of stock merchandise totaling $15,000 and structural damages totaling $14,000. How much indemnity will the carrier pay? *(5 points)*

4. Sans Souci, a photography studio, was insured for $46,000 with a coinsurance clause of 90%. Actual market value was $51,000. If the studio sustained damages totaling $27,000, how much indemnity would the carrier pay? *(5 points)*

5. If the Haverford Fire and Casualty Insurance Company cancels a basic class D policy with a face value of $50,000 after 128 days, what is the premium refund? *(5 points)*

6. A fire at a furniture plant caused a total loss. The plant, valued at $375,000, was insured for $300,000 with a coinsurance clause of 80%. How much indemnity will the plant receive? *(5 points)*

7. A policy with face value of $243,500 is purchased at a rate of 2.01 per $100.

a. What is the annual premium?

b. What is the refund if the policyholder cancels after 175 days?
What percent of the premium is kept by the carrier?

c. What is the refund if the insurer carrier cancels after 175 days?
What percent of the premium is kept by the carrier? *(15 points)*

8. Romberg Optical Wear carries $40,000 in property insurance on property valued at $65,000. The insurance policy contains an 80% coinsurance clause. How much indemnity will the carrier pay on damages totaling $22,500? *(5 points)*

9. A chemical supplier is covered by two insurance policies, one for its merchandise and the other for its building and equipment. Each policy carries an 80% coinsurance clause. The merchandise, valued at $67,500, is insured for $55,000; and the building and equipment, worth $118,000, are insured for $92,000. If a fire destroys merchandise worth $40,000 and equipment worth $45,000, how much indemnity will the policies pay? *(10 points)*

10. When Barton's Drug Store, valued at $158,000, moved to a new location, owners cancelled the basic coverage policy for a class C building and purchased an extended policy for a class B building. The move was made 293 days after the beginning of the year.

a. What was the annual cost of the basic class C policy?

b. How much of the old policy premium was refunded?

c. What was the annual premium for the new extended class B policy?

d. What was the premium for the remainder of the year? *(20 points)*

PART 6 ACCOUNTING

Chapter 23 Inventory

OBJECTIVES

- Evaluate inventory with different methods
- Calculate inventory turnover

SECTIONS

INTRODUCTION

Inventory is the merchandise a business keeps in stock. Keeping inventory can be tricky. A company that runs short of goods may lose sales. On the other hand, merchandise that sits on the shelf costs a company money. Because inventory impacts profitability, a business must have an efficient and accurate method of inventory management and control.

Section 23.1 Inventory Evaluation

KEY TERMS **Specific identification method** **Average cost method**

What Methods Are Used to Evaluate Inventory?

A perpetual inventory system records all movements into or out of inventory—sales, purchases, and returns—so that balances are perpetually adjusted. Many manual systems that use inventory cards have been replaced by on-line computer systems that record the data as transactions occur.

A periodic inventory system determines accurate balance data only after each periodic inventory verification. In between verifications, transactions are recorded, and the quantity of available goods is estimated through calculations: Available goods = last balance + incoming goods.

With either system businesses must periodically determine or estimate the value of current inventory.

A company that keeps a small number of items in stock can adequately manage inventory with the **specific identification method.** As the name suggests, each item in stock is identified, and the sum of the costs of all items in stock is the inventory value.

Date Received	Quantity Purchased	Unit Cost	Balance 12/31
3/21	50	8.76	1
4/12	100	7.70	21
8/16	125	8.12	34
11/24	50	8.98	9

$$\text{Inventory value} = 8.76 + 21 \times 7.70 + 34 \times 8.12 + 9 \times 8.98 = 527.36$$

Companies that keep large inventories, however, may find the **average cost method** more practical. This method, recognized by the accounting profession, results in a reasonable estimate of actual inventory value. Average cost is calculated, and the inventory value is computed from this figure.

Date Received	Quantity Purchased	Unit Cost	Extension
3/21	50	8.76	438.00
4/12	100	7.70	770.00
8/16	125	8.12	1,015.00
11/24	50	8.98	449.00
Total	325		2,672.00

Balance 12/31: 65

$$\text{Average cost} = \frac{\text{total cost}}{\text{items purchased}} = \frac{2672}{325} = 8.22$$

$$\text{Inventory value} = 65 \times 8.22 = 534.30$$

PRACTICE ASSIGNMENT

Plan: Determine inventory value by either the specific identification method or the average cost method.
For the specific identification method, the cost of each remaining item in stock is known, and inventory value is:
sum of (balance $\times$ unit cost).
For the average cost method, inventory value is:
average cost $\times$ total number of items in stock
where
average cost = total cost $\div$ number of items purchased.

1. Lathrop's Clothing Store keeps track of inventory balances on individual purchases. The following chart records purchases of blazers. Fill in the blanks on the chart.

Date Received	Quantity Purchased	Unit Cost	Balance 6/30	Value
3/10	15	27.25	3	______
3/25	18	28.05	7	______
4/05	20	28.55	11	______
4/30	12	28.75	9	______

What is the inventory value, using the specific identification method?

2. A sporting goods store received several shipments of baseballs. On August 31, 38 baseballs were left in stock. Fill in the blanks in the following purchase record.

Date Received	Quantity Purchased	Unit Cost	Extension
3/15	35	3.45	______
4/01	50	3.25	______
4/25	60	3.30	______
5/20	75	3.20	______
6/18	40	3.65	______
Total	______		______

What was the average cost of each baseball?
What was the value of the remaining inventory, using the average cost method?

Section 23.2 Inventory Evaluations Using FIFO and LIFO

KEY TERMS **FIFO (first-in first-out)** **LIFO (last-in first-out)**

What Other Methods Are Used to Evaluate Inventory?

The **FIFO (first-in first-out)** evaluation method assumes that the first goods bought were the first ones sold, and that any stock remaining in inventory came from the latest purchases. FIFO is an accounting procedure used to determine inventory value, however, and does not dictate the order in which merchandise is actually sold.

Date Received	FIFO Quantity Purchased	Unit Cost
3/21	50	8.76
4/12	100	7.70
8/16	125	8.12
11/24	50	8.98

Balance 12/31: 65
Assume the balance of 65 remains from the latest purchases.

$$\text{Inventory value} = 50 \times 8.98 + 15 \times 8.12 = 570.80$$

Example 1 On December 31, Sundries Outlet had 215 boxes of disposable diapers in stock. Purchases of the diapers are shown in the following table. What is the inventory value using FIFO?

Date Received	Quantity Purchased	Unit Cost
2/18	95	8.76
5/23	100	7.70
7/15	125	8.12
10/30	75	8.98

FIFO assumption:

75	units were purchased at 8.98	215 − 75 = 140
125	units were purchased at 8.12	140 − 125 = 15
15	units were purchased at 7.70	
215		

$$\text{Inventory value} = 75 \times 8.98 + 125 \times 8.12 + 15 \times 7.70 = 1804.00$$

key: 75 [×] 8.98 [=] [M+]
125 [×] 8.12 [=] [M+]
15 [×] 7.7 [=] [M+] [MR]
display: 1804.
edited: $1,804

PRACTICE ASSIGNMENT 1

Plan: Use FIFO to determine the value of inventory. Under FIFO, evaluate the balance from the bottom up when purchases are listed vertically in chronological order:

A. Begin at the bottom of the purchasing series.
B. Compare the balance to the quantity purchased.
C. Take the smaller of the balance or quantity purchased and multiply it by the unit price.
If the balance is smaller than (or equal to) the quantity purchased, finish the process with step D.
If the balance is larger, reduce it by the quantity purchased, move up to the next earlier purchase, and return to step B.
D. Compute the sum of the products obtained in step C.

1. Value-Town has 52 folding chairs left in stock. Refer to the following record of purchases and use FIFO to determine the value of the remaining chairs.

Date Received	Quantity Purchased	Unit Cost
4/21	60	18.85
6/17	100	18.95
8/05	40	19.10
10/04	88	19.15
11/15	60	19.25

Inventory value:

$52 \times 19.25 =$

2. Purchases of 125-watt stereo systems by a mail-order house are listed in the following table.

Date Received	Quantity Purchased	Unit Cost
1/02	22	472.50
3/12	35	480.00
5/31	40	480.00
8/07	25	482.50
9/18	20	485.25
11/05	15	488.00

Determine the inventory value of the 62 stereo systems left in stock using FIFO:

The **LIFO (last-in first-out)** evaluation method assumes that the last goods bought were the first ones sold, and that any remaining stock came from the earliest purchases. As with FIFO, LIFO is an accounting procedure and does not dictate the order in which merchandise is actually sold.

Date Received	LIFO Quantity Purchased	Unit Cost
3/21	50	8.76
4/12	100	7.70
8/16	125	8.12
11/24	50	8.98

Balance 12/31: 65

Assume the balance remains from the earliest purchases.

$$\text{Inventory value} = 50 \times 8.76 + 15 \times 7.70 = 553.50$$

How Does a Company Decide Whether to Use FIFO or LIFO?

While FIFO may seem to be the more logical choice, there are times when LIFO results in tax advantages that save money. During periods of rising prices, LIFO generates lower inventory values and lower gross profits that result in lower income taxes.

Example 2 Reconsider the 215 boxes of disposable diapers Sundries Outlet has in stock (example 1). What is the inventory value using LIFO?

Date Received	Quantity Purchased	Unit Cost
2/18	95	8.76
5/23	100	7.70
7/15	125	8.12
10/30	75	8.98

LIFO assumption:

95 units were purchased at $8.76   215 − 95 = 120
100 units were purchased at $7.70   120 − 100 = 20
20 units were purchased at $8.12
215

$$\text{Inventory value} = 95 \times 8.76 + 100 \times 7.70 + 20 \times 8.12 = 1764.60$$

(Compare to the FIFO evaluation in example 1 of $1,804.)

key: 95 [×] 8.76 [=] [M+]
100 [×] 7.70 [=] [M+]
20 [×] 8.12 [=] [M+] [MR]
display: 1764.6
edited: $1,764.60

PRACTICE ASSIGNMENT 2

Plan: Use LIFO to determine the value of inventory. Under LIFO, evaluate the balance from the top down when the purchases are listed vertically in chronological order.

A. Begin at the top of the purchasing series.

B. Compare the balance to the quantity purchased.

C. Take the smaller of the balance or quantity purchased and multiply it by the unit price.
If the balance is smaller than (or equal to) the quantity purchased, finish the process with step D.
If the balance is larger, reduce it by the quantity purchased, move down to the next purchase, and return to step B.

D. Compute the sum of the products obtained in step C.

1. A campus bookstore made the following purchases of one style of ring binder:

Date Received	Quantity Purchased	Unit Cost
8/10	150	3.75
9/04	175	3.95
10/17	144	4.10
11/09	75	4.15

By mid-December 138 binders were left in stock. What was the LIFO value of this stock?

2. Timberline is a home improvement store that uses the LIFO method of evaluating inventory. The following record lists plywood purchases:

Date Received	No. of Sheets Purchased	Cost per Sheet
3/04	45	5.17
4/17	50	5.19
5/06	72	5.25
5/28	85	5.33
6/30	60	5.35
7/21	48	5.39

What was the value of the plywood stock at the end of July if there were 106 sheets in inventory?

Section 23.3 Inventory Turnover

KEY TERMS **Turnover ratio** **Cost of goods sold** **Average inventory**

Careful inventory management requires detailed records and accurate statistics. **Turnover ratio** is a useful statistic that measures the number of times an inventory is purchased and sold (turned over) during a specific time period; it is also used to indicate how long an inventory can be expected to last.

In this text, assume that the turnover ratio is based on cost. There are several methods for finding the **cost of goods sold.** Add the cost of beginning inventory to the cost of purchases; the sum is the cost of goods available for sale. Then subtract the cost of ending inventory to yield the cost of goods sold.

Turnover ratios are meaningful when compared to standards published by trade or research organizations. High turnover ratios indicate that goods are moving quickly. Extremely high ratios, however, may signal excessive costs.

$$\text{Turnover ratio (at cost)} = \frac{\text{cost of goods sold}}{\text{average cost of inventory}}$$

Beginning inventory
+ Purchases
− Ending inventory
= Cost of goods sold

Example 1 At the beginning of the year, Brock's Furniture Mart held inventory valued at $89,758. Net yearly purchases amounted to $492,550. Inventory at the end of the year was valued at $94,462. What was the cost of goods sold during that year?

Cost of goods sold
= beginning inventory + purchases − ending inventory
= 89,758 + 492,550 − 94,462 = 487,846
Cost of goods sold = $487,846.

Example 2

Auto Parts Inventories

January	27,400
April	26,300
July	23,800
October	28,500
December	26,200

Average inventory is the sum divided by 5.

key algebraic: 27400 [+] 26300 [+] 23800 [+] 28500 [+] 26200 [=] [÷] 5 [=]
key arithmetic: 27400 [+] 26300 [+] 23800 [+] 28500 [+] 26200 [+] [÷] 5 [=]
display: 26440
edited: $26,440

Suppose the cost of goods sold during the year was $58,000.

$$\text{Turnover ratio} = \frac{\text{cost of goods sold}}{\text{average inventory}} = \frac{58000}{26440} = 2.2$$

The sales period is one year. A turnover ratio of 2.2 means that, on the average, goods were purchased and sold (turned over) slightly more often than twice per year. This result could be compared to the standard turnover ratio for the auto parts business.

In a sales period of 12 months, the calculation

$$\frac{12}{\text{turnover ratio}}$$

indicates the average number of months for which the inventory was sufficient.

$$12 \div 2.2 = 5.45.$$

The auto parts inventory lasts approximately $5\frac{1}{2}$ months.

PRACTICE ASSIGNMENT

Plan: To find the cost of goods sold from inventory data:

Cost of goods sold = beginning inventory + purchases − ending inventory
Average inventory = sum of the given inventories ÷ number of inventories

The time period used for computing the average inventory need not be the same as the time period used for computing the cost of goods sold.

Turnover ratio = cost of goods sold ÷ average inventory

Round the turnover ratio to the nearest tenth.
If the cost of goods sold is calculated for a one-year period, the result represents the average number of times in a year that goods were purchased and sold. The result of

12 ÷ turnover ratio

represents the average number of months for which the inventory was sufficient.

1. Use the following data to find the cost of goods sold.

Company	Beginning Inventory	Purchases	Ending Inventory	Cost of Goods Sold
a. Acme	23,470	49,500	15,700	________
b. Craft	114,000	233,000	142,500	________
c. Flare	78,800	118,630	51,700	________
d. Morgan	492,650	907,500	388,900	________
e. Selcon	91,150	325,800	102,600	________

2. Compute the average inventories.

Company	Inventories Jan 1	May 1	Sept 1	Dec 1	Average
a. Acme	23,470	20,700	14,700	25,000	________
b. Craft	114,000	113,700	105,000	117,500	________
c. Flare	78,800	74,000	68,000	80,000	________
d. Morgan	492,650	458,000	391,000	447,500	________
e. Selcon	91,150	92,050	99,500	103,750	________

3. Use the information obtained in problems 1 and 2 to compute the yearly turnover ratios for the five companies. Round the ratios to the nearest tenth.

Company	Cost of Goods Sold	Average Inventory	Turnover Ratio
a. Acme	________	________	______
b. Craft	________	________	______
c. Flare	________	________	______
d. Morgan	________	________	______
e. Selcon	________	________	______

4. Interpret the turnover ratios for the five companies in problem 3. Round all results to whole numbers.

a. Acme
On the average, goods were purchased and sold _____ times per year.
On the average, inventory was adequate for _____ months.

b. Craft
On the average, goods were purchased and sold _____ times per year.
On the average, inventory was adequate for _____ months.

c. Flare
On the average, goods were purchased and sold _____ times per year.
On the average, inventory was adequate for _____ months.

d. Morgan
On the average, goods were purchased and sold _____ times per year.
On the average, inventory was adequate for _____ months.

e. Selcon
On the average, goods were purchased and sold _____ times per year.
On the average, inventory was adequate for _____ months.

Challenger

When it is impractical to take a physical inventory, an estimated inventory may be calculated with the following formula:

$$\text{Ending inventory} = \text{beginning inventory} + \text{purchases} - \text{cost of goods sold}$$

If the average markup rate based on cost is known, the approximate cost of goods sold can be derived from the markup equation:

$$\text{Cost} \times (1 + \text{markup rate based on cost}) = \text{sales}$$

Use the following data to estimate the inventory value. Assume an average markup rate of 18% based on cost.

Beginning inventory = 35,400
Purchases = 12,600
Sales = 14,250

CHAPTER 23 SUMMARY

Concept	*Example*	*Procedure*	*Formula*
SECTION 23.1 Inventory Evaluation			
Periodic inventory evaluation	**Date Received / Quantity Purchased / Unit Cost / Balance 12/21** 3/26 50 8.76 9 6/14 100 8.80 12 11/30 100 8.85 14 Totals 250 35	Inventory balances are determined periodically.	
Inventory evaluation by specific identification	9 × 8.76 + 12 × 8.80 + 14 × 8.85 = 308.34	Each item purchased is identified and recorded with its cost. The remaining balance of each purchase is multiplied by its unit cost, and the products are added.	
Inventory evaluation by average cost	For purchase data shown above: Total cost = 50 × 8.76 + 100 × 8.80 + 100 × 8.85 = 2203.00 Average cost = 2203 ÷ 250 = 8.81 Inventory = 35 × 8.81 = 308.35	To find the total cost, multiply each quantity purchased by its unit cost and add the results. Divide this sum by the total quantity purchased to obtain the average unit cost. Multiply this average by the remaining balance.	$\text{Average cost} = \frac{\text{total cost}}{\text{quantity purchased}}$ totcost[÷]quant[=] Inventory = balance × average cost bal[×]avcost[=] totcost/quant*bal
SECTION 23.2 Inventory Evaluations Using FIFO and LIFO			
Inventory evaluation by FIFO (first-in first-out)	For purchase data shown above: Balance = 35 Last purchase = 100 Last purchase price = 8.85 35 × 8.85 = 309.75	Assume that the balance remained from the latest purchases. Moving up from the bottom, summarize the following products: smaller of total balance or quantity purchased multiplied by unit cost. Reduce the total balance at each step until it has all been evaluated.	
Inventory evaluation by LIFO (last-in first-out)	For purchase data shown above: Balance = 35 First purchase = 50 First purchase price = 8.76 35 × 8.76 = 306.60	Assume that the balance remained from the earliest purchases. Moving down from the top, summarize the following products: smaller of total balance or quantity purchased multiplied by unit cost. Reduce the total balance at each step until it has all been evaluated.	
SECTION 23.3 Inventory Turnover			
Cost of goods sold	Beginning inventory = 90,000 Purchases = 250,000 Ending inventory = 95,000 Cost of goods sold = 90,000 + 250,000 − 95,000 = 245,000	To obtain cost of goods sold, add purchases to beginning inventory. Subtract ending inventory from this sum.	Cost of goods sold = beginning inventory + purchases − ending inventory
Average inventory	Jan inventory = 17,000 June inventory = 18,500 Sept inventory = 16,400 Average inventory = (17,000 + 18,500 + 16,400) ÷ 3 = 17,300	Average inventory is the average of periodic inventories.	Average inventory $= \frac{\text{sum of periodic inventories}}{\text{number of inventories}}$
Turnover ratio	Cost of goods sold = 245,000 Average inventory = 17,300 $\text{Turnover ratio} = \frac{245{,}000}{17{,}300} = 14.2$ 12 ÷ 14.2 = 0.85 (slightly less than a month's supply)	The turnover ratio, which indicates how often stock must be replaced, is obtained by dividing the cost of goods sold by the average inventory. If the sales period is one year, divide 12 by the turnover ratio to obtain the number of months for which the stock is adequate.	$\text{Turnover ratio} = \frac{\text{cost of goods sold}}{\text{average inventory}}$ costofgoodssold[÷]avinv[=] (beginv+purch−endinv)/(SUM(inv)/invcount) $\text{Number of months} = \frac{12}{\text{turnover ratio}}$ 12[÷]turnover[=] 12/turnover

Chapter 23 **Spreadsheet Exercise**

Inventory Evaluations

The chapter 23 spreadsheet produces inventory evaluations by specific identification, average cost, LIFO, and FIFO, which enables you to compare the different methods. The template appears as follows:

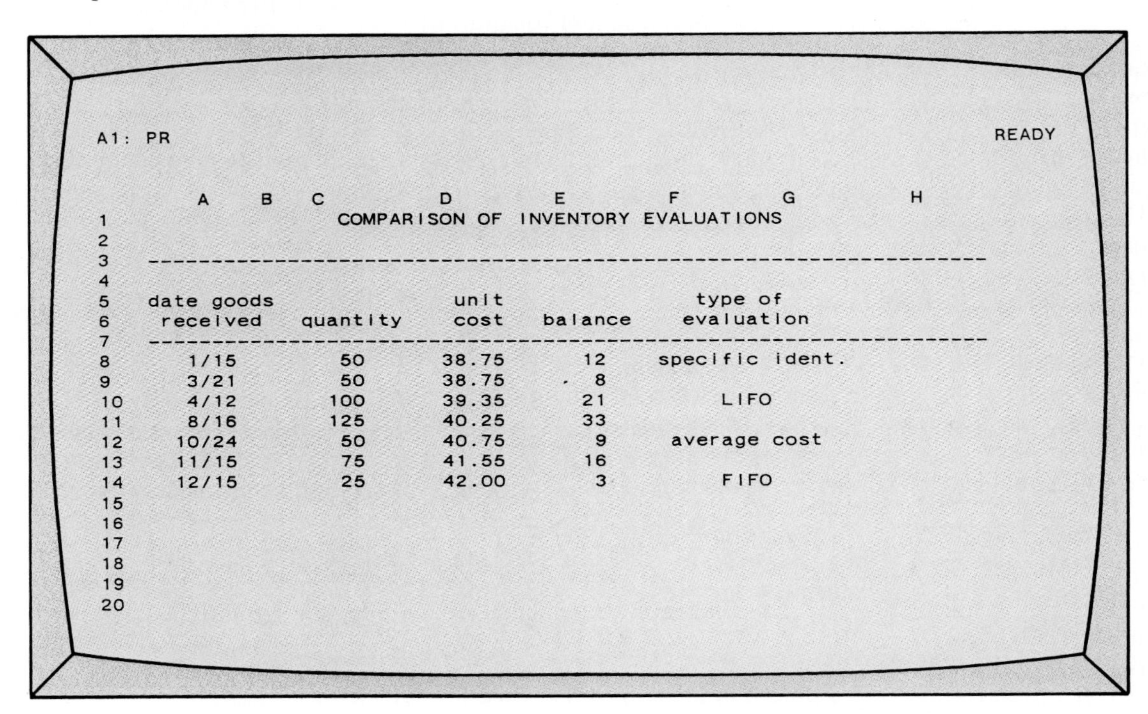

A1: PR READY

	A B	C	D	E	F G	H
1		COMPARISON OF INVENTORY EVALUATIONS				
2						
3	-----------	-----------	-----------	-----------	-----------	
4						
5	date goods		unit		type of	
6	received	quantity	cost	balance	evaluation	
7	-----------	-----------	-----------	-----------	-----------	
8	1/15	50	38.75	12	specific ident.	
9	3/21	50	38.75	8		
10	4/12	100	39.35	21	LIFO	
11	8/16	125	40.25	33		
12	10/24	50	40.75	9	average cost	
13	11/15	75	41.55	16		
14	12/15	25	42.00	3	FIFO	
15						
16						
17						
18						
19						
20						

The template displays inventory data in rows 8–14: dates when goods were received in column A, quantities in column C, unit costs in column D, and balances in column E.

Spreadsheet Exercise Steps

1. Cell H8 (and associated column I): specific identification
The specific identification inventory value is the sum of the balances multiplied by their unit costs: sum of (balance × cost).
The balance is in E8; the cost in D8.
If the spreadsheet formula begins with a cell address, it must be preceded by a sign (+).
The spreadsheet formula for balance × cost is + ______ * ______ .
Complete the formula, store it in I8, and copy it from I8 to I9..I14.
Since the (balance × cost) products are now stored in column I, the formula for the inventory value is @SUM(______ .. _______).
Complete the formula and store it in H8.

2. Cell H12 (and associated column J): Average Cost
The average cost inventory value is average cost × total balance.

$$\text{Average cost} = \frac{\text{sum of (quantity} \times \text{cost)}}{\text{sum of (quantities)}}$$

$$\text{Total balance} = \text{sum of balances}$$

Therefore,

$$\text{average cost inventory} = \frac{\text{sum of (quantity} \times \text{cost)}}{\text{sum of (quantities)}} \times \text{sum of (balances)}$$

The quantity is in C8; the cost in D8.
The spreadsheet formula for quantity × cost is + ______ * ______ .
Complete the formula, store it in J8, and copy it from J8 to J9..J14.
Since the (quantity × cost) products are now stored in column J, the quantities are in column C and the balances are in column E, the formula for the inventory value is
@SUM(______ .. ______)/@SUM(______ .. ______)*@SUM(______ .. ______)
Complete the formula and store it in H12.

3. Cell H14 (and associated columns K and M): FIFO
The FIFO inventory value is the sum of the balances accumulated from the bottom up multiplied by their costs.
Cost is in D8.
The FIFO balances have already been stored in column K (beginning with K8).
Complete + ______ * ______ , store it in M8, and copy it from M8 to M9..M14.
(There are two hidden columns, L and O, which appear in some versions of Lotus 1-2-3 while you copy. Ignore them; they will disappear.)
Since the FIFO products are now stored in column M, the formula for the inventory value is
@SUM(______ .. ______).
Complete the formula and store it in H14.

4. Cell H10 (and associated columns N and P): LIFO
The LIFO inventory value is the sum of the balances accumulated from the top down multiplied by their costs.
Cost is in D8.
The LIFO balances have already been stored in column N (beginning with N8).
Complete + ______ * ______ , store it in P8, and copy it from P8 to P9..P14.
Since the LIFO products are now stored in column P, the formula for the inventory value is
@SUM(______ .. ______).
Complete the formula and store it in H10.

5. Specific identification produces the most correct value.
Average cost provides a close estimate that falls between the lowest and highest possible values.
When costs are rising—and in periods of creeping inflation they usually are—LIFO yields a low estimate, because it is assumed that the remaining goods were purchased early in the inventory evaluation period. FIFO yields a high estimate, because it is assumed that the remaining goods were purchased late in the inventory evaluation period.
Verify that the three estimates are close and in ascending order.

	Exact	**Rounded to Nearest Hundred**
Specific identification value:	______	______
LIFO estimate:	______	______
Average cost estimate:	______	______
FIFO estimate:	______	______

NAME ______________________________

CLASS SECTION ______________ DATE ______________

SCORE ____________ MAXIMUM 200 PERCENT ____________

CHAPTER 23 ASSIGNMENT

1. Kazan's Office Equipment recorded the following purchases of Contax model 350 fax machines. Fill in the missing values.

Date Received	Quantity	Unit Cost	Balance 9/30	Value
1/03	28	925.50	7	______
2/18	22	929.50	11	______
4/12	35	932.00	13	______
6/05	25	934.50	18	______
8/14	40	939.50	28	______
Total balance			______	

What is the value of the fax machines in stock on September 30? *(25 points)*

2. Using the information in problem 1, evaluate the inventory with the average cost method.

Date Received	Quantity		Unit Cost		Extension
1/03	28	×	925.50	=	______
2/18	22	×	929.50	=	______
4/12	35	×	932.00	=	______
6/05	25	×	934.50	=	______
8/14	40	×	939.50	=	______
Total	______				______

Average cost of each fax machine: ______

Inventory value 9/30: ______ *(35 points)*

3. Use FIFO to determine the value of 128 bags of plant food remaining in stock for the following purchases.

Date Received	Quantity	Unit Cost
2/02	60	11.75
5/12	75	12.78
8/05	90	14.12
10/14	75	14.98
12/03	50	15.06

Inventory value: ______ *(15 points)*

4. On January 31, a department store inventory count indicated 79 linen sets in stock from the following purchases.

Date Received	Quantity	Unit Cost
7/06	60	7.29
7/29	96	7.39
8/15	48	7.45
9/11	36	7.50
10/31	48	7.67

Use LIFO to determine the inventory value of linen sets left in stock: ______ *(15 points)*

5. On August 10, Top-Notch Hardware received a shipment of 48 boxes of 1-inch nails at $1.85 per box, 72 boxes of $1\frac{1}{2}$-inch nails at $2.05 per box, and 64 boxes of 2-inch nails at $2.48 per box. On November 1, there were 13 boxes of 1-inch nails, 16 boxes of $1\frac{1}{2}$-inch nails, and 21 boxes of 2-inch nails remaining from this shipment. What was the inventory value of the remaining nails using the specific identification method? *(10 points)*

6. On September 28, a department store received 37 Turnip Seed dolls costing $12.75 each. The store purchased 44 more of these dolls at a price of $13.47 on October 15 and an additional 65 dolls at $13.95 each on November 5. On December 27, 19 dolls were left in stock. Determine inventory value, using the average cost method.
Total cost of purchases: ___________
Average cost of doll: ___________
Inventory value on 12/27: ___________ *(15 points)*

7. R & L Department Store purchased 48 umbrellas at $6.75 each on February 8. On April 14, the store purchased 65 umbrellas at $7.15 each. The last purchase of 15 umbrellas at $7.30 each was made on May 27. Inventory taken in June showed 23 umbrellas remaining in stock. What is this inventory worth according to the FIFO method?
What is it worth using average cost?
(30 points)

8. Hal's Electronic Supplies purchased 18 Ultravox #6000 camcorders costing $892.50 each on March 28. On May 8, Hal's purchased 25 camcorders at $909 each. Another 20 were bought at $919.50 each on September 8. At the end of the year, there were 21 camcorders in stock. What was the LIFO value of this inventory?
What was the FIFO value? *(30 points)*

9. Marlowe's Music House began the year with an inventory valued at $97,482. During the year, quarterly inventories were recorded at $84,900, $118,250, $98,277, and $86,745. The last figure represents the ending inventory. Purchases for the year totaled $387,500.

a. What was the cost of goods sold?

b. What was the average inventory?

c. What was the turnover ratio to the nearest tenth? *(15 points)*

10. Walk-More's shoe inventory was valued at $11,921 at the beginning of the year. At the end of the year, inventory was valued at $12,063. Purchases of $38,665 had been made during the year.

a. What was the turnover ratio to the nearest tenth?

b. On the average, for how many months did the stock last? *(10 points)*

Chapter 24 Depreciation

OBJECTIVES

- Analyze different depreciation methods
- Compute annual depreciation amounts and write depreciation schedules

SECTIONS

INTRODUCTION

In the course of establishing and running a business, companies acquire assets, such as buildings, machinery, vehicles, and other equipment. As time passes, machinery wears out, vehicles get old, and equipment becomes obsolete, causing these assets to depreciate or decline in value. By deducting depreciation from revenues, businesses partially recover the costs of their assets (in the form of tax write-offs). This chapter covers three classical methods of depreciation—straight-line, sum-of-the-years-digits, and declining-balance—along with the depreciation systems implemented by the federal government—the Accelerated Cost Recovery System (ACRS) and the Modified Accelerated Cost Recovery System (MACRS).

Section 24.1 Straight-Line Depreciation

KEY TERMS **Total depreciation** **Scrap value** **Annual depreciation** **Depreciation rate** **Book value**

The straight-line method of depreciation is fundamental. To calculate **total depreciation,** subtract scrap value from cost. **Scrap value,** also called salvage value, is the estimated price for which the asset can be sold at the end of its service life. Service life or useful life is the number of years an asset is expected to be useful. To calculate the **annual depreciation** amount, divide total depreciation by the number of years in service. The annual depreciation amount is the same each year.

The annual **depreciation rate** is 1 divided by the number of years in service and may be expressed as a fraction or in percent form. Total depreciation multiplied by depreciation rate results in the annual depreciation.

All of the annual depreciation amounts plotted on a line graph form a straight line; thus *straight-line* method.

(Chapter 30 displays depreciations on a multi-line graph.)

Depreciation of a Lathe

Cost of a new lathe is $15,000. Useful life is estimated to be 5 years with a scrap value of $1,000.

Total depreciation amount = 15000 − 1000 = 14000

Annual depreciation amount = 14000 ÷ 5 = 2800

Rate of depreciation is $\frac{1}{5}$ or 20% per year.

Each year, the depreciation is 20% of the total depreciation.

Example 1 A vending machine has an estimated useful life of eight years. What is the annual rate of depreciation?

$$\frac{1}{8} = 0.125 = 12\frac{1}{2}\%$$

Example 2 A car telephone system costing $1,100 has an estimated service life of four years and a salvage value of $50. What is the depreciation each year?

Total depreciation = cost − scrap value
= 1100 − 50 = 1050

Annual depreciation = 1050 ÷ 4 = 262.50
(25% of 1050)

A depreciation schedule shows service years, annual depreciation, accumulated depreciation, and remaining book values. To calculate **book value,** subtract accumulated depreciation from cost. Refer to the lathe on p. 321 and study the depreciation schedule to the right:

At the end of service life, accumulated depreciation equals total depreciation, and book value equals scrap value.

Straight-Line Depreciation Schedule

End of Service Year	Annual Depreciation	Accumulated Depreciation	Book Value
1	$2,800	$ 2,800	$12,200
2	2,800	5,600	9,400
3	2,800	8,400	6,600
4	2,800	11,200	3,800
5	2,800	14,000	1,000

Example 3 A movie projector costing $875 was expected to last five years with no scrap value. What were accumulated depreciation and book value after three years?

Depreciation rate = 1/5 = 20%
Annual depreciation = 875 ÷ 5 = 175 (20% of 875)
Accumulated depreciation = 175 × 3 = 525
Book value = 875 − 525 = 350

PRACTICE ASSIGNMENT

Plan: Total depreciation is cost − scrap value. Annual depreciation is

$$\frac{\text{cost} - \text{scrap value}}{\text{number of years}} = \text{total depreciation} \times \text{depreciation rate.}$$

The rate of depreciation for each service year is

$$\frac{1}{\text{number of years}}$$

To find the accumulated depreciation for a given service year, multiply the annual depreciation by the number of that service year.

In the depreciation schedule in problem 3:

A. Column 1 shows the service year ending.

B. Column 2 shows the annual depreciation.

C. Column 3 shows the accumulated depreciation:
For year 1, it is the current depreciation.
For each of the following years, add the current depreciation to the previous accumulated depreciation.

D. Column 4 shows the book value:

cost − accumulated depreciation.

1. Show the straight-line depreciation rate for each of the following assets in percent form, accurate to the nearest tenth of a percent.

	Service Life	Depreciation Rate
a.	4 years	________
b.	5 years	________
c.	10 years	________
d.	15 years	________
e.	6 years	________
f.	20 years	________

2. Find the total depreciation and annual depreciation amounts. Divide total depreciation by the number of years in service to obtain annual depreciation.

	Cost of Asset	Scrap Value	Service Life in Years	Total Depreciation Amount	Annual Depreciation Amount
a.	5,000	0	4	________	________
b.	12,000	300	5	________	________
c.	45,000	5,000	10	________	________
d.	16,900	400	6	________	________
e.	52,500	3,500	15	________	________
f.	120,000	12,000	20	________	________

3. Write the depreciation schedule for problem 2d.

Service Year	Annual Depreciation	Accumulated Depreciation	Book Value
1	________	________	________
2	________	________	________
3	________	________	________
4	________	________	________
5	________	________	________
6	________	________	________

In problems 4 and 5, divide total depreciation by the number of service years to obtain annual depreciation. Round your result to the nearest dollar.

4. Lambert and Stagg Agency purchased office furniture at a cost of $7,200. The furniture has a salvage value of $500 and is expected to have a useful life of twelve years. How much will the furniture depreciate after seven years?

5. Amy McClanahan purchased a car for $11,750. The car was expected to last seven years with a scrap value of $350. What was book value after four years?

Section 24.2 Sum-of-the-Years-Digits Method

KEY TERM **Accelerated depreciation**

The straight-line method produces depreciation amounts that are equal for each year of service life. The sum-of-the-years-digits method produces depreciation amounts that decline during service life, so that depreciation is greater in the early years of service when an asset is worth more. **Accelerated depreciation** reduces profits and taxes during the early years of asset ownership and thereby encourages businesses to invest in new equipment.

Cost = $15,000.
Estimated service life = 5 years.
Scrap value = $1,000.

Total depreciation amount = 15000 − 1000 = 14000

Annual depreciation amounts rounded to the nearest dollar:

5/15 × 14000	=	4667
4/15 × 14000	=	3733
3/15 × 14000	=	2800
2/15 × 14000	=	1867
1/15 × 14000	=	933
15/15 Total depreciation	=	14000

The sum-of-the-years-digits depreciation rate is a fraction whose numerator is the number of service years in descending order (5, 4, 3, 2, 1 for 5 years) and whose denominator is the sum of the digits of the number of service years (1 + 2 + 3 + 4 + 5 = 15). The digits can be summed with the formula

$$\frac{n(n+1)}{2}$$

(5 × 6 ÷ 2 = 15).

Sum-of-the-Years-Digits — Depreciation Schedule

End of Service Year	Annual Depreciation	Accumulated Depreciation	Book Value
1	$4,667	$ 4,667	$10,333
2	3,733	8,400	6,600
3	2,800	11,200	3,800
4	1,867	13,067	1,933
5	933	14,000	1,000

Example 1 The estimated useful life of a forklift is seven years. By what percent is it depreciated during the first three service years, accurate to the nearest tenth of a percent?

The numerator is 7 for the first year, 6 for the second year, and 5 for the third year. The denominator is 7 × (7 + 1) ÷ 2 = 7 × 8 ÷ 2 = 28.

Year 1: $\frac{7}{28} = 25\%$ Year 2: $\frac{6}{28} = 21.4\%$ Year 3: $\frac{5}{28} = 17.9\%$

Example 2 New-Life Fitness Center purchased exercise equipment costing $4,500. The equipment is expected to last six years and has a salvage value of $400. How much of the cost can the fitness center write off in the second year?

Depreciated amount = 4500 − 400 = 4100

Counting backwards, numerator for year 2 is 5

Denominator = 6 × (6 + 1) ÷ 2 = 6 × 7 ÷ 2 = 21

$$\frac{5}{21} = 23.8\%$$

Calculation 1: $\frac{5 \times 4100}{21} = 976.19$

Calculation 2: 4100 × 0.238 = 975.80

Depreciation for year 2 rounded to the nearest dollar: $976

Example 3 Continuing with example 2, what is the accumulated depreciation and book value after the second year?

Fraction for year 1: $\frac{6}{21}$ Fraction for year 2: $\frac{5}{21}$ $\frac{6}{21} + \frac{5}{21} = \frac{11}{21} = 52.4\%$

Calculation 1: $\frac{11 \times 4100}{21} = 2147.62$

Calculation 2: $4100 \times 0.524 = 2148.40$

Accumulated depreciation rounded to the nearest dollar: $2,148

Book value = 4500 − 2148 = 2352

PRACTICE ASSIGNMENT

Plan: To obtain the numerator of each depreciation fraction, start with the number of service years and count backwards.

Calculate the denominator with the formula $n(n + 1) \div 2$, where n is the number of service years. Each depreciation fraction can be converted to percent form. Again, the total amount to be depreciated is cost − scrap value.

The annual depreciation amount is total depreciation × depreciation rate for the year.

Round all annual depreciations to the nearest dollar.

To obtain the accumulated depreciation, add the fractions up to and including the current year and multiply the total depreciated amount by this result.

Again, book value is cost − accumulated depreciation.

The depreciation schedule again shows annual depreciation, accumulated depreciation over the years of service life, and remaining book value.

1. Show the yearly depreciation rates in fraction form and percent form, accurate to the nearest tenth of a percent.

a. 6 Service Years

Service Year	Fraction	%
1	_______	_______
2	_______	_______
3	_______	_______
4	_______	_______
5	_______	_______
6	_______	_______

b. 8 Service Years

Service Year	Fraction	%
1	_______	_______
2	_______	_______
3	_______	_______
4	_______	_______
5	_______	_______
6	_______	_______
7	_______	_______
8	_______	_______

c. 10 Service Years

Service Year	Fraction	%
1	_______	_______
2	_______	_______
3	_______	_______
4	_______	_______
5	_______	_______
6	_______	_______
7	_______	_______
8	_______	_______
9	_______	_______
10	_______	_______

2. Compute the annual depreciations and total depreciation.

Cost of Asset	Scrap Value	Service Life	Total Depreciation
26,400	500	6 years	$25,900

Service Year	Annual Depreciation
1	______
2	______
3	______
4	______
5	______
6	______

3. Write the depreciation schedule for problem 2.

Service Year	Annual Depreciation	Accumulated Depreciation	Book Value
1	______	______	______
2	______	______	______
3	______	______	______
4	______	______	______
5	______	______	______
6	______	______	______

For problems 4 and 5, use the accumulated depreciation rate in fraction form.

4. Easy-Shop installed a security system costing $18,000. The system has a useful life of eight years and a salvage value of $1,500. What will accumulated depreciation be after four years?

5. Adam Nitti purchased a van for $31,000. The van is expected to last seven years with a scrap value of $1,700. How much will it be worth after three years?

Section **24.3 Declining-Balance Method**

KEY TERMS **Declining-balance factor** **Double-declining-balance method**

In the straight-line method, total depreciation is multiplied by the same rate each year, resulting in equal annual depreciation amounts.

In the sum-of-the-years-digits method, total depreciation is multiplied by a declining rate each year, resulting in decreasing annual depreciation amounts.

The declining-balance method produces a decreasing annual depreciation by another technique: the declining book value is multiplied by the same depreciation rate each year.
The depreciation rate is $1/n \times$ declining balance factor, where n is the number of service years.
The **declining-balance factor** is usually a number between 1.25 and 2. When the factor is 2, the procedure is called the **double-declining-balance method.**

Cost = $15,000
Service life = 5 years
Depreciation rate = $1/5 \times 2 = 2/5$
$= 0.4 = 40\%$
The initial book value is the cost.

1st depreciation = $15000 \times 0.4 = 6000$
Book value = 9000
2nd depreciation = $9000 \times 0.4 = 3600$
Book value = 5400
3rd depreciation = $5400 \times 0.4 = 2160$
Book value = 3240

Example 1 A chain saw has an estimated useful life of seven years. What is its annual depreciation rate with the double-declining-balance method? Double-declining implies that the declining balance factor is 2.

$$1/n \times 2 = 1/7 \times 2 = 2/7 = 28.6\%$$

Declining-balance depreciation does not terminate with scrap value; at the end of service life, an adjustment is required. Accounting practice allows for several methods of adjustment. In most of this text, you obtain the scrap value by adjusting the final depreciation amount. Observe the following depreciation schedule that corresponds to example 1 (scrap value = $1,000).

Annual depreciation
= depreciation rate × previous book value

Double-Declining-Balance Depreciation Schedule

Service Year	Annual Depreciation	Accumulated Depreciation	Book Value
0			$15,000
1	$6,000	$ 6,000	9,000
2	3,600	9,600	5,400
3	2,160	11,760	3,240
4	1,296	13,056	1,944
5	944*	14,000	1,000

*1,944 × 0.4 = 778, would result in a final value greater than the scrap value. Therefore, the last depreciation is increased to 944.

PRACTICE ASSIGNMENT

Plan: Depreciation rate = $1/n$ × declining balance factor.
For the double-declining-balance method, the factor is 2, and the depreciation rate is $2/n$.
Round all annual depreciations to the nearest dollar.
Again, the depreciation schedule shows annual depreciation, accumulated depreciation over the years of service life, and remaining book value. Remember that you may need to adjust the last depreciation amount to allow for scrap value.

A. The cost of the asset is the first book value.
B. Compute the annual depreciation as book value × depreciation rate.
C. The next book value is previous book value − previous annual depreciation.
D. Repeat steps B and C, but adjust the last depreciation amount in order to obtain the scrap value.

1. Show the rate of depreciation in percent form, accurate to the nearest tenth of a percent.

	Service Life	Declining-Balance Factor	Depreciation Rate (%)
a.	5 years	1.25	________
b.	6 years	1.5	________
c.	8 years	2	________
d.	4 years	1.75	________
e.	3 years	1.5	________
f.	15 years	1.9	________

In problems 2 through 4, apply the percent form of the double-declining-balance depreciation rate, accurate to the nearest tenth of a percent.

2.

Cost of Asset	Service Life	Scrap Value	Service Year	Annual Depreciation	Accumulated Depreciation	Book Value
$44,000	8 years	$4,000	0			$________
			1	$________	$________	________
			2	________	________	________
			3	________	________	________
			4	________	________	________
			5	________	________	________
			6	________	________	________
			7	________	________	________
			8	________	________	________

3. A supermarket purchased freezers that are expected to last five years and have a salvage value of $500. If book value is $9,500 after the first year, how much should be depreciated the second year?

4. The Rendezvous Lounge has lighting fixtures with an estimated life of six years and scrap value of $50. The book value of these fixtures was $158 after four years. How much must be depreciated for the fifth year? The sixth year?

Section 24.4 Accelerated and Modified Accelerated Cost Recovery Systems

KEY TERMS **Real property** **Tangible property** **Recovery period** **Life class** **Partial depreciation**

The Accelerated Cost Recovery System (ACRS) was established in 1981 as part of the Economic Recovery Tax Act, enacted by the United States Congress to offer federal tax incentives that would encourage business expansion. Assets placed into service prior to 1981 are subject to depreciation by one of the classical methods. Assets placed into service on or after January 1, 1981, may be depreciated by ACRS.

Property is divided into two primary categories, real and tangible. **Real property** is anything erected on, attached to, or growing on land; the land itself, however, is not depreciable. **Tangible property** includes vehicles, machinery, and equipment (not real estate).

Under ACRS, the Internal Revenue Service (IRS) assigns each item of property a **recovery period;** that time during which the property is depreciated. The IRS provides a table that lists life classes and the depreciation rate for each year. The entire cost of an asset is depreciated with no regard to scrap value.

ACRS Life Classes			
Three-year Assets	Five-year Assets	Ten-year Assets	Fifteen-year Assets
Automobiles Light trucks Some tools	Heavy trucks Machinery Office equipment Airplanes	Mobile homes Buses Office furniture Some real properties*	Locomotives Heavy construction equipment Public utility equipment

*Other real properties are 15-, 18-, or 19-year assets.

A **life class** is a group of assets with a particular recovery period. The list here is necessarily brief. Tax professionals keep complete listings.

Table 24.1 ACRS Tangible Property

	Life Class			
	3-year	*5-year*	*10-year*	*15-year*
Year	**Depreciation Rates in Percent**			
1	25	15	8	5
2	38	22	14	10
3	37	21	12	9
4		21	10	8
5		21	10	7
6			10	7
7			9	6
8			9	6
9			9	6
10			9	6
11–15				6

Example 1 Depreciate a heavy truck purchased for $43,500. Heavy-duty trucks are classified as five-year assets. The depreciation rates are 15%, 22%, 21%, 21%, and 21%.

The depreciation amounts are:

Year 1 = 43500 × 0.15 = 6525
Year 2 = 43500 × 0.22 = 9570
Year 3 = 43500 × 0.21 = 9135
Year 4 = 43500 × 0.21 = 9135
Year 5 = 43500 × 0.21 = 9135

PRACTICE ASSIGNMENT 1

Plan: In this assignment, use ACRS to depreciate for federal tax purposes tangible property that was placed into service between 1981 and 1986. Disregard scrap value. Determine whether the life is 3, 5, 10, or 15 years. Each annual depreciation rate is shown in table 24.1. Round amounts to the nearest dollar.

1. Compute the annual depreciation for the first three years.

Type of Asset	Cost	Year 1	Year 2	Year 3
a. Production machinery	24,500	______	______	______
b. Office furniture	17,200	______	______	______
c. Heavy truck	55,500	______	______	______
d. Computer system	230,000	______	______	______
e. Lathe	15,000	______	______	______

2. Show the depreciation schedule for problem 1b.

Service Year	Annual Depreciation	Accumulated Depreciation	Book Value
1	$______	$______	$______
2	______	______	______
3	______	______	______
4	______	______	______
5	______	______	______

Service Year	Annual Depreciation	Accumulated Depreciation	Book Value
6	______	______	______
7	______	______	______
8	______	______	______
9	______	______	______
10	______	______	______

The Tax Reform Act of 1986 altered the Accelerated Cost Recovery System, which became the Modified Accelerated Cost Recovery System (MACRS). The MACRS can be used for federal tax purposes for property placed into service on or after January 1, 1987. Property placed into service between January 1, 1981 and December 31, 1986 is subject to the ACRS.

Under the MACRS, all real property is depreciated with the straight-line method. Residential property has a life of $27\frac{1}{2}$ years, non-residential property a life of $31\frac{1}{2}$ years. The six life classes for tangible property are briefly listed here. Complete listings are available from tax professionals. (The IRS has begun recently to designate MACRS as General Depreciation System or GDS.)

MACRS Life Classes		
Three-year Assets	Five-year Assets	Seven-year Assets
Over-the-road tractors Race horses Other horses over 12 yrs. old	Automobiles Trucks Airplanes Computers Office machinery Research equipment	Office furniture Buses Single purpose agricultural equipment
Ten-year Assets	Fifteen-year Assets	Twenty-year Assets
Mobile homes Assets used for entertainment Water transportation equipment	Municipal waste-water treatment plants	Farm property Municipal sewers

MACRS uses the 200% declining-balance method (double-declining) for three-, five-, seven-, and ten-year assets; it uses the 150% declining-balance method (factor = 1.5) for fifteen- and twenty-year assets.

Because the MACRS assumes that all tangible assets are owned for one-half of the year in which they were purchased, half-year **partial depreciations** are applied.

Example 2 In 1989, a company purchased a minicomputer system for $155,000. What is the MACRS depreciation schedule? Computers are classified as five-year assets and are depreciated with the double-declining-balance method (rate = $\frac{1}{5} \times 2 = 40\%$). The half-year depreciation for year one is one-half of the annual depreciation ($\frac{1}{2}$ of 40% of book value). The depreciation for the last six months has been adjusted to reach zero book value.

Recovery Period	Depreciation	Accumulated Depreciation	Book Value
0	0	0	155,000
6 months in year 1	31,000	31,000	124,000
year 2	49,600	80,600	74,400
year 3	29,760	110,360	44,640
year 4	17,856	128,216	26,784
year 5	10,714	138,930	16,070
6 months in year 6	16,070	155,000	0

You will notice that the declining-balance method does not reach a desired book value, zero or otherwise, without an adjustment. When declining-balance depreciation drops below straight-line depreciation, switch to the straight-line method. This procedure is permitted under MACRS and is often beneficial.

To find the straight-line depreciation for a remaining year, divide the remaining life into the remaining book value.

Example 3 Recompute the depreciation schedule for example 2; switch to the straight-line method when declining-balance depreciation drops below straight-line depreciation.

Recovery Period	Depreciation	Accumulated Depreciation	Book Value
0	0	0	155,000
6 months in year 1	31,000	31,000	124,000
year 2	49,600	80,600	74,400
year 3	29,760	110,360	44,640
year 4	17,856	128,216	26,784
year 5*	17,856	146,072	8,928
6 months in year 6	8,928	155,000	0

*In year 5, the double-declining-balance depreciation would have been $26784 \times 0.4 = 10714$, but the straight-line depreciation is $26784 \div 1.5 = 17856$.

PRACTICE ASSIGNMENT 2

Plan: In this assignment, use MACRS to depreciate tangible property that was placed into service after 1986. Disregard scrap value.
Determine whether the life is 3, 5, 7, 10, 15, or 20 years.
Use the 200% declining-balance method for 3-, 5-, 7-, and 10-year assets; use the 150 percent declining-balance method for 15- and 20-year assets. Use half-year partial depreciation. As indicated in the problems, either adjust to reach zero book value or switch to the straight-line method when that depreciation becomes larger. Round amounts to the nearest dollar.

1. An agricultural corporation purchased a fruit sorter costing $185,000. What is book value after the first $3\frac{1}{2}$ years.
$1/n \times$ declining balance factor = ______ $\times$ ____ = ______
= ______

Recovery Period	Depreciation	Accumulated Depreciation	Book Value
0	0	0	______
6 months in year 1	______	______	______
Year 2	______	______	______
Year 3	______	______	______
Year 4	______	______	______

2. Complete the MACRS depreciation schedule for office desks costing $3,600. Adjust the last depreciation in order to reach zero book value.
$1/n \times$ declining balance factor = ______ $\times$ ____ = ______
= ______

Recovery Period	Depreciation	Accumulated Depreciation	Book Value
0	0	0	______
6 months in year 1	______	______	______
Year 2	______	______	______
Year 3	______	______	______
Year 4	______	______	______
Year 5	______	______	______
Year 6	______	______	______
Year 7	______	______	______
6 months in year 8	______	______	______

3. Recompute problem 2; switch to the straight-line method when appropriate.

Recovery Period	Depreciation	Accumulated Depreciation	Book Value
0	0	0	______
6 months in year 1	______	______	______
Year 2	______	______	______
Year 3	______	______	______
Year 4	______	______	______
Year 5	______	______	______
Year 6	______	______	______
Year 7	______	______	______
6 months in year 8	______	______	______

Challenger

A depreciation method not covered in this chapter is units-of-production. The units-of-production method depreciates an asset on the basis of use—pieces produced, hours operated, or miles traveled. The yearly depreciation amounts are determined by how much the asset was used during the year.

The estimated life in units must be known.

Total depreciation remains cost − scrap value.

$$\text{Depreciation per unit} = \frac{\text{cost} - \text{scrap value}}{\text{life in units}}$$

$$\text{Annual depreciation} = \text{unit depreciation} \times \text{units produced}$$

A machine with an estimated production potential of 550,000 pieces in five years was purchased for $25,000 and may be sold for a scrap value of $1,500.

a. What is the depreciation per unit?

b. What is the depreciation schedule for the following production?

Service Year	Planned Production in Pieces
1	75,000
2	120,000
3	140,000
4	120,000
5	95,000

CHAPTER 24 SUMMARY

Concept	*Example*	*Procedure*	*Formula*

SECTION 24.1 Straight-Line Depreciation

Concept: Straight-line depreciation

Example:

Cost of asset = $6,500
Service life = 3 years
Scrap value = $500

Procedure: Subtract the scrap value from the cost to obtain the total depreciation amount.

Formula:

Total depreciation = cost − scrap value

alg: cost[−]scrap[=]
arith: cost[+]scrap[−]

cost−scrap

Example:

Total depreciation = 6500 − 500
= 6000

Depreciation rate = $\frac{1}{3} = 33\frac{1}{3}\%$

Annual depreciation = $33\frac{1}{3}\%$ of 6000
= 2000

Procedure: The annual depreciation rate is 1 divided by the number of service years. Multiply the total depreciation by this rate to derive the annual depreciation.

Formula:

$$\text{Rate} = \frac{1}{\text{number of years}}$$

1[÷]n[=]

1/n

Example:

Service Year	Depr.	Acc. Depr.	Book Value
0			6,500
1	2,000	2,000	4,500
2	2,000	4,000	2,500
3	2,000	6,000	500

Procedure: The rate and depreciation amount are the same for each service year.

Book value is cost minus accumulated depreciation.

Formula:

$$\text{Annual depreciation} = \frac{\text{cost} - \text{scrap}}{\text{number of years}}$$

alg: cost[−]scrap[=][÷]n[=]
arith: cost[+]scrap[−][÷]n[=]

(cost−scrap)/n

SECTION 24.2 Sum-of-the-Years-Digits Method

Concept: Sum-of-the-years-digits method

Example:

Cost, service life, and scrap value are the same as the straight-line example.

Total depreciation = 6,000

Depreciation rates are:

$\frac{3}{6}\ \frac{2}{6}\ \frac{1}{6}$

Service Year	Depr.	Acc. Depr.	Book Value
0			6,500
1	3,000	3,000	3,500
2	2,000	5,000	1,500
3	1,000	6,000	500

Procedure: Total depreciation is the same as straight-line.

The depreciation rate is different each service year.
The numerator of the rate fraction is the number of the service year in descending order. The denominator is the sum of each year's number or $\frac{n(n+1)}{2}$

Multiply total depreciation by each year's rate.

Formula:

$$\text{Rate} = \frac{\text{reverse year}}{n(n+1)/2}$$

Annual depreciation = (cost − scrap) × annual rate

alg: cost[−]scrap[=][×]rate[=]
arith: cost[+]scrap[−][×]
rate[=]

(cost−scrap)*revyr/(n*(n+1)/2)

SECTION 24.3 Declining-Balance Method

Concept: Declining-balance method

Example:

Cost, service life, and scrap value are the same as the straight-line example.

Total depreciation = 6,000

In the double-declining-balance method, declining balance factor = 2.

Rate = $\frac{1}{3} \times 2 = \frac{2}{3}$ = 66.7%

Procedure: Total depreciation is the same as straight-line.

The depreciation rate is the same each year. It is the declining balance factor (ranging between 1.25 and 2) divided by the number of years.
The base is the book value (beginning with cost), which declines each year.

Formula:

Rate = $\frac{1}{n} \times$ factor

factor[÷]n[=]

factor/n

Example:

Service Year	Depr.	Acc. Depr.	Book Value
0			6,500
1	4,333	4,333	2,167
2	1,445	5,778	722
3	222	6,000	500

Procedure: Multiply each year's book value by the depreciation rate.
Adjust the final depreciation to obtain scrap value.

Formula:

Annual depreciation = last book value × rate

lastbook[×]rate[=]

lastbook*rate

CHAPTER 24 SUMMARY—*(Continued)*

Concept	*Example*	*Procedure*	*Formula*

SECTION 24.4 Accelerated and Modified Accelerated Cost Recovery Systems

Accelerated cost recovery system (ACRS)—1981

Example: An automobile is a 3-year asset. The rates are: 25%, 38%, 37%.

Cost is $6,500.

Service Year	Depr.	Acc. Depr.	Book Value
0			6,500
1	1,625	1,625	4,875
2	2,470	4,095	2,405
3	2,405	6,500	0

Procedure: Annual rates (which may vary from year to year) are given in tables for three-, five-, ten-, and fifteen-year assets. Scrap value is ignored.

Multiply the cost by the annual rate.

Formula: Annual depreciation = cost × annual rate

cost[×]rate[=]

cost*rate

Modified accelerated cost recovery system (MACRS)—1986

Example: A truck is a 5-year asset. Cost = $22,500.

Service Year	Depr.	Acc. Depr.	Book Value
0			22,500
$\frac{1}{2}$ of year 1	4,500	4,500	18,000
2	7,200	11,700	10,800
3	4,320	16,020	6,480
4	2,592	18,612	3,888
5	1,555	20,167	2,333
$\frac{1}{2}$ of year 6	2,333	22,500	0

Procedure: Real property is depreciated with the straight-line method.
In the case of tangible assets, the double-declining-balance method is used for three-, five-, seven-, and ten-year assets; the 150% declining-balance method is used for fifteen- or twenty-year assets. You may switch to straight-line for the remainder of the asset's life when this method produces greater depreciation than the declining-balance method. Assume the asset was purchased halfway through the year.

Chapter 24 **Spreadsheet Exercise**

Three Depreciation Methods

This exercise introduces powerful financial functions available in spreadsheet software for the traditional depreciation methods: SLN (straight-line), SYD (sum-of-the-years-digits) and DDB (double-declining-balance).

Your chapter 24 template will appear as follows:

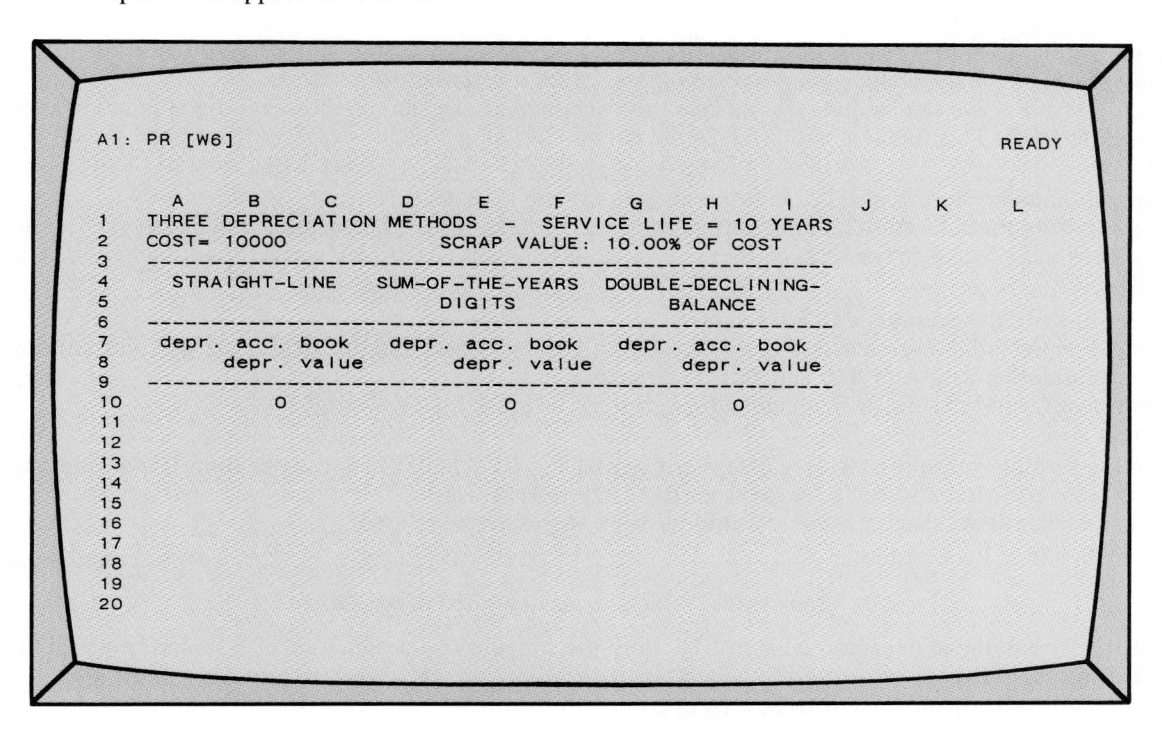

Columns A, B, and C are reserved for straight-line; D, E, and F for sum-of-the-years-digits; and G, H, and I for double-declining-balance.

A service life of ten years will be shown in rows 10–20. All figures are to the nearest dollar. The cost is variable (up to 99999) and is contained in B2. The scrap value is a percent of the cost; the variable number of percent is in G2.

Spreadsheet Exercise Steps

1. Column A: Straight-Line Depreciation
The general form of the straight-line depreciation function is
@SLN(cost, scrap value, service life).
Cost is in B2. Scrap value is a percent of cost; the number of percent is in G2. Service life is set at 10 years.
The row designation in B2 and G2 must be absolute.
The function is
@ SLN(_____ $ _____ , _____ $ _____ * _____ $ _____ , _____).
Complete this function, which produces the straight-line depreciation for one year. Store the function in A11 and copy it from A11 to A12..A20.

2. Column D: Sum-of-the-Years-Digits Depreciation
The general form of the sum-of-the-years-digits depreciation function is
@SYD(cost, scrap value, service life, year).
Cost, scrap value, and service life are coded as in step 1. There is no column containing the years 1, 2, . . . 10.
However, the year can be extracted from the row number since year = row − 10.
The CELL function @CELL("row",A11..A11) extracts the row number from the cell location. (To learn more about this and other logical functions, consult a spreadsheet text or software manual.)
The entire coding is
@SYD(_____ $ _____ , _____ $ _____ * _____ $ _____ , _____ , @CELL("row",A11..A11)−10).
Complete this function, which produces the sum-of-the-years-digits depreciation for one year. Store the function in D11 and copy it from D11 to D12..D20.

3. Column G: Double-Declining-Balance Depreciation
The general form of the double-declining-balance depreciation function is
@DDB(cost, scrap value, service life, year). Code this function like the sum-of-the-years-digits function except replace @SYD with @DDB. This function will produce the double-declining-balance depreciation for one year.
@DDB(_____ $ _____ , _____ $ _____ * _____ $ _____ , _____ , @CELL("row",A11..A11)−10).
Complete the function, store it in G11, and copy it from G11 to G12..G19.
Enter the following formula into G20, the bottom cell: +I19−G$2*B$2. This formula will adjust the last depreciation to reach a book value equal to the scrap value. The value in G20 will not display correctly until all book values are computed.

4. Columns B, E, and H: Accumulated Depreciation
Accumulated depreciation begins with 0 and proceeds with previous accumulated depreciation + current depreciation. Previous accumulation will be in B10 and current depreciation in A11.
If a spreadsheet formula begins with a cell address, it must be preceded by a sign (+).
The formula is + _______ + _______ .
Complete the formula, store it in B11, and copy it from B11 to B12..B20. Then, copy column B into columns E and H by copying from B11..B20 to E11..E20 and from B11..B20 to H11..H20.
The value in H20 will not display correctly until all book values are computed.

5. Columns C, F, and I: Book Value

Book value = cost − accumulated depreciation

Cost is in B2. Accumulated depreciation is in B10. Both column and row designation of B2 must be absolute. The spreadsheet formula for book value is $ _____ $ _____ − _______ .
Store the formula in C10 and copy it from C10 to C11..C20.
Then, copy all of column C into all of columns F and I by copying from C10..C20 to F10..F20 and from C10..C20 to I10..I20.

6. Verify with your calculator each depreciation for year 5 (row 15). Cost = 10000, scrap value = 10% of cost = 1000, total depreciation = 9000.

SLN	**SYD**	**DDB**
$1/n \times$ total deprec.	Deprec. fraction = _______	$2/n \times$ prev. book
= 1/ _____ × _______	Depreciation	= 2/ _____ × _______
= _______	= _______ × total deprec.	= _______
	= _______ × _______	
	= _______	
Spreadsheet Value	**Spreadsheet Value**	**Spreadsheet Value**
_______	_______	_______

7. Change the scrap value to 15% of cost (enter as .15). What are the book values at the end of the eighth year?
_______ _______ _______

8. In which year does the straight-line depreciation begin to exceed the other two depreciations? year:_____
Vary the cost. Does your answer seem to be correct for any cost? Yes:_____ No_____

NAME ____________________

CLASS SECTION ______________ DATE ______________

SCORE __________ MAXIMUM 200 PERCENT __________

CHAPTER 24 ASSIGNMENT

Round all annual depreciations to the nearest dollar where necessary.

1. Write the straight-line depreciation schedule for an asset with a cost of $34,000, scrap value of $3,000, and service life of 8 years.

Service Year	Annual Depreciation	Accumulated Depreciation	Book Value
1			
2			
3			
4			
5			
6			
7			
8			

(25 points)

2. Compute annual depreciations and total depreciation by the sum-of-the-years-digits method.

Cost of Asset	Scrap Value	Service Life	Total Depreciation
44,000	4000	8	40,000

Service Year	Annual Depreciation
1	
2	
3	
4	
5	
6	
7	
8	

(10 points)

3. Write the depreciation schedule for problem 2.

Service Year	Annual Depreciation	Accumulated Depreciation	Book Value
1			
2			
3			
4			
5			
6			
7			
8			

(25 points)

4. Complete the following depreciation schedule by the double-declining-balance method.

Cost of Asset	Service Life	Scrap Value
$55,000	10 years	$5,000

Service Year	Annual Depreciation	Accumulated Depreciation	Book Value
0			
1			
2			
3			
4			
5			
6			
7			
8			
9			
10			

(25 points)

5. A dentist used the straight-line method to depreciate a $21,500 drill over a period of eight years with no salvage value. What was the accumulated depreciation after five years? *(5 points)*

6. Sunflower Movers purchased a truck with an estimated useful life of eight years and a salvage value of $1,000. Book value after three years was $17,800. What will the truck be worth after four years according to the double-declining-balance method? *(5 points)*

7. A tractor costs $27,500 and is expected to last five years. Its salvage value is $1,200. Calculate accumulated depreciation and book value after three years, using the straight-line method.

a. Accumulated depreciation: __________

b. Book value: __________ *(10 points)*

8. A radio transmitter purchased four years ago has a current book value of $7,250 and is expected to last another three years. There is no salvage value. How much should be depreciated next year by the double-declining-balance method? What will book value be then?

a. Fifth-year depreciation amount: ________

b. Book value: ________ *(10 points)*

9. Use the sum-of-the-years-digits method in this problem. Community Hospital purchased cardiac monitors costing $30,000, which were expected to have a useful life of five years and a salvage value of $1,000. What amount did the hospital write off during the third year? At the end of the third year, what was the accumulated depreciation and what was book value?

a. Third-year depreciation: ________

b. Accumulated depreciation: ________

c. Book value: ________ *(15 points)*

10. Show the ACRS depreciation schedule for an office computer costing $24,500. (Use table 24.1 on page 328.)

Service Year	Annual Depreciation	Accumulated Depreciation	Book Value
1	________	________	________
2	________	________	________
3	________	________	________
4	________	________	________
5	________	________	________

(20 points)

11. Show the MACRS depreciation schedule for a mobile home costing $31,500. Adjust to reach zero book value.

$1/n \times$ declining balance factor

$=$ ________ $\times$ ______ $=$ ________

$=$ ________

Recovery Period	Depreciation	Accumulated Depreciation	Book Value
0	0	0	________
6 months in year 1	________	________	________
year 2	________	________	________
year 3	________	________	________
year 4	________	________	________
year 5	________	________	________
year 6	________	________	________
year 7	________	________	________
year 8	________	________	________
year 9	________	________	________
year 10	________	________	________
6 months in year 11	________	________	________

(25 points)

12. Recompute problem 11; switch to the straight-line method when appropriate.

Recovery Period	Depreciation	Accumulated Depreciation	Book Value
0	0	0	________
6 months in year 1	________	________	________
year 2	________	________	________
year 3	________	________	________
year 4	________	________	________
year 5	________	________	________
year 6	________	________	________
year 7	________	________	________
year 8	________	________	________
year 9	________	________	________
year 10	________	________	________
6 months in year 11	________	________	________

(25 points)

Chapter 25 Financial Statements

OBJECTIVES

- Complete an income statement and a balance sheet
- Perform vertical and horizontal analyses of financial statements
- Calculate financial ratios

SECTIONS

INTRODUCTION

How do businesses track their progress? Have they made a profit or lost money? Are expenses up or down? Can they handle additional debt? The income statement and the balance sheet are two tools businesses use to assess financial health.

Section 25.1 The Income Statement

KEY TERMS **Net sales** **Net purchases** **Gross profit** **Operating income** **Operating expenses** **Net income before taxes** **Net income after taxes**

Did we make a profit? This is an important question, and the answer is literally the bottom line of the income statement. The income statement is a summary of a company's activities during an accounting period, usually one year.

Figure 25.1 shows an example of an income statement for a business that sells merchandise. You may recognize such terms as *cost of goods sold* and *depreciation* from previous chapters. Formats may vary among companies and often depend upon the type of business.

Figure 25.1
Income Statement

BREMMER HOME IMPROVEMENTS
Income Statement
For the Year Ended December 31, 199X

Gross Sales		1,454,000	
Less Returns and Allowances		16,300	
Net Sales			$1,437,700
Inventory, January 1, 199X		372,000	
Purchases	584,000		
Less Returns and Allowances	15,700		
Net Purchases		568,300	
Goods Available for Sale		940,300	
Less Inventory, December 31, 199X		312,000	
Cost of Goods Sold			628,300
Gross Profit on Sales			809,400
Payroll		214,600	
Utilities		25,000	
Advertising		5,800	
Supplies Expense		1,500	
Real Estate Tax		17,300	
Insurance		4,000	
Depreciation		11,200	
Total Operating Expenses			279,400
Operating Income			530,000
Interest Expense			7,500
Net Income before Income Taxes			522,500
Income Taxes			80,900
Net Income after Income Taxes			$ 441,600

Income Statement Steps

Step		Example	
1. Compute **net sales:** Gross sales − Returns and allowances		Gross sales − Returns and allowances = **Net sales**	$1,454,000 − 16,300 1,437,700
2. Compute cost of goods sold: Beginning inventory + **Net purchases** = Goods available for sale − Ending inventory = Cost of goods sold		Beginning inventory + **Net purchases** = Goods available − Ending inventory = Cost of goods sold	372,000 568,300 940,300 − 312,000 628,300
3. Compute **gross profit** (gross margin) Net sales − Cost of goods sold		Net sales − Cost of goods sold = **Gross profit**	1,437,700 − 628,300 809,400
4. Compute **operating income** Gross profit − **Operating expenses**		Gross profit − Total **operating expenses** = **Operating income**	809,400 − 279,400 530,000
5. Compute **net income before taxes** Operating income − Other expenses (if any)		Operating income − Interest expense = **Net income before taxes**	530,000 − 7,500 522,500
6. Compute **net income after taxes** Net income before taxes − Taxes		Net income before taxes − Taxes = **Net income after taxes**	522,500 − 80,900 $441,600

Section 25.2 Vertical and Horizontal Analysis of Income Statements

KEY TERMS **Vertical analysis** **Horizontal analysis**

How can a business know if net profit is sufficient in light of sales volume, or whether operating expenses are too high? **Vertical analysis** of an income statement tells how sales income was spent. Each item is expressed as percent of net sales.

Show

$$\frac{\text{item to be analyzed}}{\text{net sales}}$$

in percent form, rounded to one-tenth of a percent.

Analyze operating expenses.

Operating expenses = 106,000
Net sales = 892,000

$$\frac{106000}{892000} = 0.1188 = 11.9\%$$

Operating expenses were nearly 12% of net sales.

Example 1 Refer to the Bremmer Home Improvements income statement in figure 25.1. What percent of net sales is net income before taxes?

$$\frac{\text{net income before taxes}}{\text{net sales}} = \frac{522500}{1437700} = 0.3634 = 36.3\%$$

Net income before taxes is 36.3% of net sales.

While vertical analysis provides information about one income statement, **horizontal analysis** allows comparison between statements. A company using horizontal analysis can determine whether performance was better (or worse) in one year compared with another.

The earlier period is the base.

Rate of increase (or decrease)

$$= \frac{\text{new value} - \text{old value}}{\text{old value}},$$

in percent, rounded to a tenth of a percent. (Decreases will be negative.)

Gross profit 1990 = \$821,500
Gross profit 1991 = \$973,000

$$\frac{973000 - 821500}{821500} = \frac{151500}{821500} = 0.1844 = 18.4\% \text{ increase}$$

Example 2 Refer again to figure 25.1. Total operating expenses for the year were \$279,400. If the figure for the previous year was \$251,200, what was the rate of increase?

$$\frac{\text{new value} - \text{old value}}{\text{old value}} = \frac{279400 - 251200}{251200} = \frac{28200}{251200} = 0.1123 = 11.2\%$$

Operating expenses increased by 11.2%.

PRACTICE ASSIGNMENT

Plan: All problems in this assignment pertain to the income statement for Bremmer Home Improvements (figure 25.1). For vertical analysis of the income statement, the denominator of each fraction is net sales. The numerator is the income statement item that is being analyzed. For horizontal analysis, find the rate of increase or decrease of each item from one year to the next, where the earlier year is the base. Convert each fraction to percent form, rounded to the nearest tenth of a percent.

1. Cost of goods sold is what percent of net sales?

2. What percent of net sales is payroll?

3. If cost of goods sold the previous year was $639,100, what was the rate of decrease?

4. Last year's payroll totaled $201,400. What was the rate of increase?

Section **25.3 The Balance Sheet**

KEY TERMS **Current assets** **Long-lived assets** **Current liabilities** **Long-term liabilities** **Owner's equity**

Assets are values a company owns. Liabilities are values a company owes.

An income statement is the report of a company's activities for a specific period. A balance sheet stops the action at a particular point in time, usually the last day of the year. Businesses use balance sheets to compare assets with liabilities. A solvent company should have assets greater than liabilities.

Figure 25.2 shows an illustration of a balance sheet in which the parts are listed vertically. Another format lists assets on the left side of the page, and liabilities and owner's equity on the right.

Balance Sheet Steps

1. Sum **current assets:** (Current assets can be converted to cash or consumed within one year.)	Cash	$ 25,600
	+ Accounts receivable	21,700
	+ Supplies	4,200
	+ Inventory	312,000
	= **Current assets**	363,500
2. Sum **long-lived assets** (fixed assets): (Fixed assets are used for more than one year.)	Furniture	21,000
	+ Land	83,000
	+ Building	76,500
	= **Long-lived assets**	180,500
3. Compute total assets: Current assets + Long-lived assets	Current assets	363,500
	+ Long-lived assets	180,500
	= Total assets	$544,000
4. Sum **current liabilities:** (Current liabilities are due in one year.)	Accounts payable	28,100
	+ Taxes payable	3,900
	+ Notes payable	10,000
	= **Current liabilities**	42,000
5. Sum **long-term liabilities:** (Long-term liabilities are due in more than one year.)	Mortgage	35,000
	+ Long-term notes	15,700
	= **Long-term liabilities**	50,700
6. Compute total liabilities: Current liabilities + Long-term liabilities	Current liabilities	42,000
	+ Long-term liabilities	50,700
	= Total liabilities	$92,700
7. Compute **owner's equity** (capital or net worth) (Owner's equity is how much a business is worth.) Beginning equity + Net income (income statement) − Withdrawal (money withdrawn by owners, such as dividends)	Beginning equity	311,000
	+ Net income	441,600
	− Withdrawals	−301,300
	= **Owner's equity**	$451,300
8. Add owner's equity to total liabilities. (Equity + liabilities should equal total assets.)	Owner's equity	451,300
	+ Total liabilities	92,700
	= Total assets	$544,000

Figure 25.2
Balance Sheet

BREMMER HOME IMPROVEMENTS
Balance Sheet
December 31, 199X

Assets			
Current Assets			
Cash	$ 25,600		
Accounts Receivable	21,700		
Supplies	4,200		
Inventory	312,000		
Total Current Assets		$363,500	
Long-lived Assets			
Furniture	21,000		
Land	83,000		
Building	76,500		
Total Long-lived Assets		180,500	
Total Assets			$544,000
Liabilities			
Current Liabilities			
Accounts Payable	28,100		
Taxes Payable	3,900		
Notes Payable	10,000		
Total Current Liabilities		42,000	
Long-term Liabilities		50,700	
Total Liabilities			92,700
Owner's Equity			
Owner's Equity, January 1, 199X		311,000	
Net Income	441,600		
Less Withdrawals	301,300		
		140,300	
Owner's Equity, December 31, 199X			451,300
Total Liabilities + Owner's Equity			$544,000

Section 25.4 Vertical and Horizontal Analysis of Balance Sheets

How much of a company's assets are taken up by inventory? How significant are its long-term debts? Vertical analysis of a balance sheet provides answers to these questions. Assets are expressed as percent of the total assets; liabilities are expressed as percent of (liabilities + equity).

Show

$$\frac{\text{assets to be analyzed}}{\text{total assets}}$$

or

$$\frac{\text{liabilities to be analyzed}}{\text{liabilities + equity}}$$

in percent form, rounded to one-tenth of a percent.

Analyze inventory.

Inventory = $271,000
Total assets = $609,000

$$\frac{271000}{609000} = 0.4450 = 44.5\%$$

Almost 45% of total assets are inventory.

Example 1 Refer to the Bremmer Home Improvements balance sheet in figure 25.2. What percent of (liabilities + equity) are long-term liabilities?

$$\frac{\text{long-term liabilities}}{\text{liabilities + equity}} = \frac{50700}{544000}$$
$$= 0.0932 = 9.3\%$$

Vertical analysis provides information about one balance sheet.

How does a firm know if it is in better condition today than it was a year ago? By using horizontal analysis, a company can compare the dollar amount of items on one balance sheet with those of other balance sheets.

The earlier period is the base.

$$\text{Rate of increase (or decrease)} = \frac{\text{new value} - \text{old value}}{\text{old value}}$$

in percent, rounded to a tenth of a percent. (Decreases will be negative.)

Owner's equity 1990 = \$523,700
Owner's equity 1991 = \$586,500

$$\frac{586500 - 523700}{523700} = \frac{62800}{523700}$$
$$= 0.1199$$
$$= 12.0\% \text{ increase}$$

Example 2 Refer again to figure 25.2. Current assets were \$363,500 on December 31. The corresponding figure for the previous year was \$371,200. What was the rate of decrease?

$$\frac{\text{new value} - \text{old value}}{\text{old value}} = \frac{363500 - 371200}{371200}$$
$$= \frac{-7700}{371200}$$
$$= -0.0207 = -2.1\%$$

Current assets decreased by 2.1%.

PRACTICE ASSIGNMENT

Plan: All exercises in this assignment pertain to the balance sheet for Bremmer Home Improvements in figure 25.2. For vertical analysis, the denominator of each fraction is total assets or (total liabilities + owner's equity) (which should be the same as total assets), depending on the item being analyzed. The numerator is the balance statement item being analyzed. For horizontal analysis, find the rate of increase or decrease of each item from one year to the next, where the earlier year is the base.

Convert each fraction to percent form, rounded to the nearest tenth of a percent.

1. Accounts payable is what percent of total liabilities and owner's equity?

2. What percent of total assets is long-lived assets?

3. The balance sheet of the previous year recorded \$19,800 for accounts receivable. By what percent did accounts receivable increase?

4. At the end of the previous year, current liabilities were \$44,500. By what rate did current liabilities decrease?

Section 25.5 Financial Ratios

KEY TERMS **Current ratio** **Acid test ratio** **Equity to debt ratio** **Earnings to equity ratio**

Business owners, investment bankers, and creditors monitor the ratios obtained from financial statements. These ratios are especially useful when compared with the corresponding ratios in the industry. Typical ratios are listed in the following table. These ratios are often expressed as number:1.

Name	Formula	Description
Current Ratio (Working Capital Ratio)	$\frac{\text{current assets}}{\text{current liabilities}}$	Indication of firm's ability to pay liabilities. At least 2:1 is desirable.
Acid Test Ratio (Quick Ratio)	$\frac{\text{quick assets}}{\text{current liabilities}}$	Stronger indication of ability to pay liabilities. Quick assets are cash and other assets that can be converted quickly to cash (short-term investments, accounts receivable). At least 1:1 is desirable.
Equity to Debt Ratio	$\frac{\text{owner's equity}}{\text{total liabilities}}$	Indication of the safety of claims by creditors. A high ratio is desirable.
Earnings to Equity Ratio	$\frac{\text{net income}}{\text{owner's equity}}$	Of particular interest to investors. Usually expressed as a decimal fraction.

Example 1 Fasco, Inc. has current assets of $58,300, of which $34,900 are quick assets. Current liabilities are $23,600, and total liabilities are $178,500. The company's equity is $452,000. Last year's net income totaled $48,600.

$$\text{Current ratio} = \frac{\text{current assets}}{\text{current liabilities}} = \frac{58300}{23600} = 2.5 = 2.5{:}1$$

2.5:1 is a better ratio than 2:1.

$$\text{Acid test ratio} = \frac{\text{quick assets}}{\text{current liabilities}} = \frac{34900}{23600} = 1.5 = 1.5{:}1$$

1.5:1 is a better ratio than 1:1.

$$\text{Equity to debt ratio} = \frac{\text{owner's equity}}{\text{total liabilities}} = \frac{452000}{178500} = 2.5 = 2.5{:}1$$

For every $2.50 in capital, there are liabilities of $1.00.

$$\text{Earnings to equity ratio} = \frac{\text{net income}}{\text{owner's equity}} = \frac{48600}{452000} = 0.108$$

Each $1.00 of capital generated $0.11 of income.

PRACTICE ASSIGNMENT

Plan: All exercises in this assignment pertain to the Bremmer Home Improvements balance sheet in figure 25.2. Determine the four illustrated financial ratios. For the current, quick (acid test), and equity to debt ratios, round the resulting number to the nearest tenth and show the ratios in the form *x.x*:1. Express the earnings to equity ratio as a decimal fraction, rounded to hundredths.

1. What is the current ratio? ______ : 1

2. What is the acid test ratio? ______ : 1

3. What is the equity to debt ratio? ______ : 1

4. What is the earnings to equity ratio? _______

Challenger

Land-Mitchell, Inc. owns several subsidiary companies including Positrex Electronics. Of Land-Mitchell's $365,000,000 in total assets, $62,500,000 are Positrex assets. Of their $160,000,000 in total liabilities, $32,000,000 are Positrex liabilities. The parent corporation, Land-Mitchell, would like to improve its equity to debt ratio by selling a subsidiary. International Communications Industries, Inc. is interested in purchasing Positrex. To increase its equity to debt ratio to at least 2:1, how much must Land-Mitchell receive from the sale of Positrex?

CHAPTER 25 SUMMARY

Concept	*Example*	*Procedure*	*Formula*
SECTION 25.1 The Income Statement			
Income statement	Net sales $1,500,000 Cost of goods sold − 800,000 Gross profit on sales = 700,000 Operating expenses − 250,000 Operating income = 450,000 Other expenses − 25,000 Net income before taxes = 425,000 Income taxes − 80,000 Net income after taxes = $ 345,000	1. Show net sales. 2. Compute cost of goods sold. 3. Subtract cost of goods sold from net sales to obtain gross profit. 4. Subtract operating expenses to obtain operating income. 5. Subtract other expenses to obtain net income before taxes. 6. Subtract income taxes to obtain net income after taxes.	
SECTION 25.2 Vertical and Horizontal Analysis of Income Statements			
Vertical analysis of income statement	Operating expenses = $250,000 Net sales = $1,500,000 $\frac{250000}{1500000} = 0.1666\ldots = 16.7\%$ Operating expenses are 16.7% of net sales.	Calculate the percent of net sales represented by each item.	$\text{Rate} = \frac{\text{item}}{\text{net sales}}$
Horizontal analysis of income statement	1991 operating expenses = $250,000 1990 operating expenses = $185,000 $\frac{250000 - 185000}{185000} = 0.3513513 = 35.1\%$ Operating expenses increased 35.1%.	Calculate the rate of increase or decrease of each item over a period.	Rate of increase (decrease) $= \frac{\text{new value} - \text{old value}}{\text{old value}}$ alg: new[−]old[=][÷]old[=] arith: new[+]old[−][÷]old[=] (new−old)/old
SECTION 25.3 The Balance Sheet			
Balance sheet	Current assets $300,000 Long-lived assets + 150,000 Total assets = $450,000 Current liabilities 40,000 Long-term liabilities + 50,000 Total liabilities 90,000 Owner's equity 360,000 Total liabilities + 90,000 Liabilities + equity $450,000	1. Sum current assets. 2. Sum long-lived assets. 3. Compute total assets. 4. Sum current liabilities. 5. Sum long-term liabilities. 6. Compute total liabilities. 7. Compute owner's equity. 8. Add liabilities to equity. This sum should equal total assets.	
SECTION 25.4 Vertical and Horizontal Analysis of Balance Sheets			
Vertical analysis of balance sheet	Inventory = $185,000 Total assets = $450,000 $\frac{185000}{450000} = 0.4111\ldots = 41.1\%$ Inventory is 41.1% of total assets.	Calculate the percent of total assets each class of assets represents. Calculate the percent of (liabilities + equity) each class of liabilities represents.	$\text{Rate} = \frac{\text{asset class}}{\text{total assets}}$ $\text{Rate} = \frac{\text{liability}}{\text{total liabilities + equity}}$
Horizontal analysis of balance sheet	1991 current assets = $300,000 1990 current assets = $325,000 $\frac{300000 - 325000}{325000} = -0.076923 = -7.7\%$ Current assets decreased 7.7%.	Calculate the rate of increase or decrease of each item over a period.	Rate of increase (decrease) $= \frac{\text{new value} - \text{old value}}{\text{old value}}$ alg: new[−]old[=][÷]old[=] arith: new[+]old[−][÷]old[=] (new−old)/old

CHAPTER 25 SUMMARY—(*Continued*)

Concept	*Example*	*Procedure*	*Formula*
SECTION 25.5 Financial Ratios			
Financial ratios	Current assets = \$300,000 Current liabilities = \$40,000 $\text{Current ratio} = \frac{300000}{40000} = 7.5{:}1$	Typical Financial Ratios: Current ratio: An indication of a firm's ability to pay its debts. Acid test ratio: A stronger criterion of ability to pay debts. Equity to debt ratio: An indication of safety to creditors. Earnings to equity ratio: An indication of profitability to investors.	Current ratio $= \frac{\text{current assets}}{\text{current liabilities}}$ Acid test ratio $= \frac{\text{quick assets}}{\text{current liabilities}}$ Equity to debt ratio $= \frac{\text{owner's equity}}{\text{total liabilities}}$ Earnings to equity ratio $= \frac{\text{net income}}{\text{owner's equity}}$

Chapter 25 **Spreadsheet Exercise**

Balance Sheet

Spreadsheet software was originally developed to computerize financial reports. These computer programs replaced tediously completed paper worksheets with electronic procedures that allow automatic calculations and corrections.

You will appreciate the electronic spreadsheet when you set up a balance sheet with horizontal analysis. Entering the formulas into the spreadsheet is much easier than computing each total or ratio individually.

Your chapter 25 template will appear as follows:

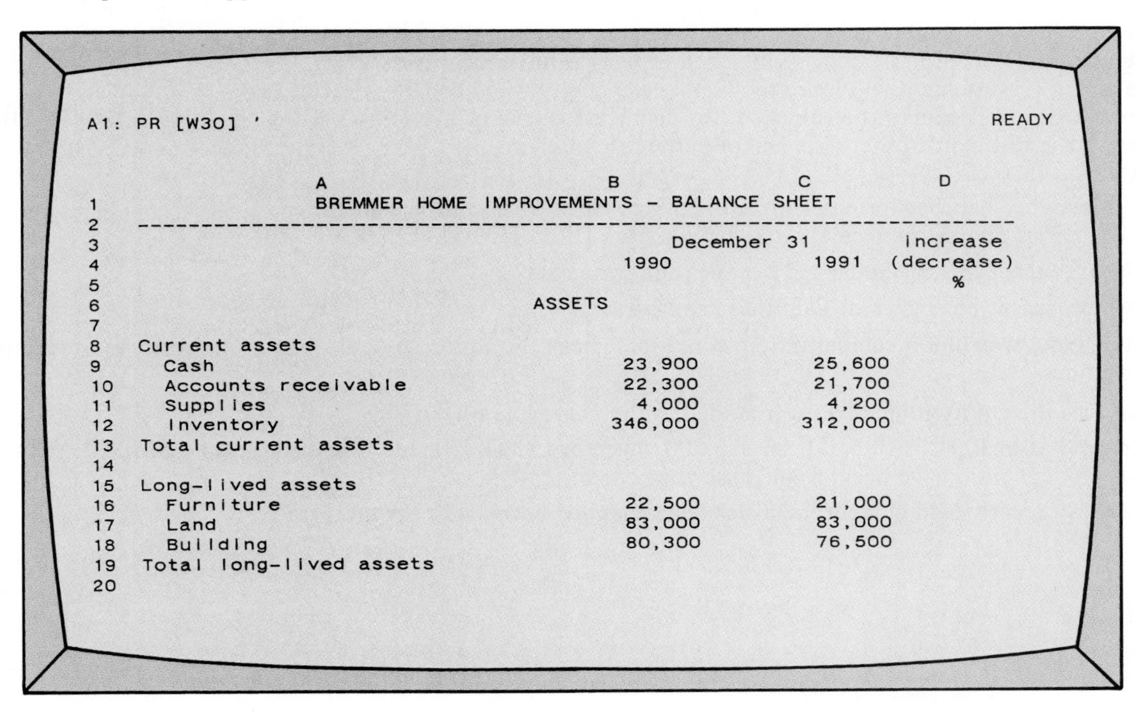

A1: PR [W30] '　　　　READY

	A	B	C	D
1	BREMMER HOME IMPROVEMENTS - BALANCE SHEET			
2	--			
3		December 31		Increase
4		1990	1991	(decrease)
5				%
6	ASSETS			
7				
8	Current assets			
9	Cash	23,900	25,600	
10	Accounts receivable	22,300	21,700	
11	Supplies	4,000	4,200	
12	Inventory	346,000	312,000	
13	Total current assets			
14				
15	Long-lived assets			
16	Furniture	22,500	21,000	
17	Land	83,000	83,000	
18	Building	80,300	76,500	
19	Total long-lived assets			
20				

Column B contains the first year's data.

Column C contains the second year's data.

Spreadsheet Exercise Steps

1. Row 13: Current Assets
   Compute the current assets for 1990.
   Complete @SUM(________ .. ________). Store the formula in B13 and copy it from B13 to C13.
2. Row 19: Long-lived Assets
   Compute the long-lived assets for 1990.
   Complete @ SUM(________ .. ________). Store the formula in B19 and copy it from B19 to C19.
3. Row 21: Total Assets
   Compute the total assets for 1990.
   Complete + ________ + ________ . Store the formula in B21 and copy it from B21 to C21.
4. Row 29: Current Liabilities
   Compute the current liabilities for 1990.
   Complete @ SUM(________ .. ________). Store the formula in B29 and copy it from B29 to C29.
5. Row 33: Total Liabilities
   Compute the total liabilities for 1990.
   Complete + ________ + ________ . Store the formula in B33 and copy it from B33 to C33.

6. Row 35: Equity
The balance sheet must always balance:

Equity + liabilities = assets.

Therefore:

Equity = assets − liabilities.

(In reality, the procedure for determining equity is more complex, but assume here that the statement is in balance.)
Complete + _______ − _______ . Store the formula in B35 and copy it from B35 to C35.

7. Row 37: Liabilities + Equity
Complete + _______ + _______ . Store the formula in B37 and copy it from B37 to C37.

8. Column D: Rates of Change between the Two Years
Compute the rate of change from the old value and new value.
New value is in C9; old value is in B9.
Because column D is not in percent format, multiply the quotient by 100.
The use of the IF statement will leave the blank cells empty when copying. Recall that the statement @IF(condition,*x*,*y*) produces the value *x* if the condition is true or the value *y* if the condition is false. (If there is no value, leave the cell blank; otherwise, compute the rate of change.)
The statement is @IF(C9=" ", " ", (_______ − _______) / _______ * _______).
Be sure to use the spacebar for entering the blank between each pair of quotes.
Complete the formula. Store it in D9 and copy it from D9 to D10..D37.

9. **a.** By what percent has cash changed between the two years?

   b. By what percent have current liabilities changed?

10. Equity is the net worth of a company. Some creditors check the equity to debt ratio, an indication of a company's creditworthiness.

    a. What was the equity to debt ratio (rounded to the nearest tenth) in 1991?

    b. Suppose that in 1991, cash had been $23,300, inventory $394,000, and accounts payable $25,300. What would the equity to debt ratio have been, rounded to the nearest tenth?

    c. By what percent would current liabilities have changed between the two years?

NAME ______________________________

CLASS SECTION ____________________ DATE ________________

SCORE ____________ MAXIMUM __250__ PERCENT ____________

CHAPTER 25 ASSIGNMENT

1. Complete the following income statement. ***(20 points)***

ROBO ELECTRONICS

Income Statement
For the Year Ended December 31, 199X

Gross Sales		992,000	
Returns and Allowances		51,500	
Net Sales			______
Inventory, January 1, 199X		154,700	
Purchases	227,100		
Returns and Allowances	36,300		
Net Purchases		______	
Goods Available for Sale		______	
Inventory, December 31, 199X		118,900	
Cost of Goods Sold			______
Gross Profit on Sales			______
Payroll		174,000	
Rent		56,000	
Utilities		10,800	
Insurance		8,700	
Supplies Expense		1,300	
Advertising		2,600	
Travel		3,400	
Depreciation		2,400	
Total Operating Expenses			______
Operating Income			______
Net Income before Income Taxes			______
Income Taxes			40,800
Net Income after Income Taxes			______

2. Complete the following balance sheet. ***(30 points)***

Bestdrive Tire Company

Balance Sheet
December 31, 199X

Assets			
Current Assets			
Cash	$ 13,200		
Accounts Receivable	9,500		
Supplies	1,100		
Inventory	84,500		
Total Current Assets		______	
Long-lived Assets			
Plant and Equipment	69,000		
Land	77,000		
Total Long-lived Assets		______	
Total Assets			______
Liabilities			
Current Liabilities			
Accounts Payable	21,700		
Taxes Payable	4,500		
Notes Payable	15,000		
Total Current Liabilities		______	
Long-term Liabilities		50,700	
Total Liabilities			______
Owner's Equity			
Owner's Equity, January 1, 199X		$138,900	
Net Income	288,500		
Less Withdrawals	265,000		
		______	
Owner's Equity, December 31, 199X		______	
Total Liabilities + Owner's Equity			______

3. Complete the following vertical analysis. ***(30 points)***

Laverne's Clothing Outlet

Income Statement
For the Year Ended December 31, 199X

			Amount	Percent
Gross Sales		837,500		
Less Returns and Allowances		10,200		
Net Sales			$827,300	100.0%
Inventory, January 1, 199X		168,500		
Purchases	405,700			
Less Returns and Allowances	12,100			
Net Purchases		393,600		
Goods Available for Sale		562,100		
Less Inventory, December 31, 199X		217,300		
Cost of Goods Sold			344,800	______
Gross Profit on Sales			482,500	______
Payroll		125,500		______
Rent		48,000		______
Utilities		9,700		______
Insurance		3,500		______
Supplies Expense		1,100		______
Depreciation		1,200		______
Total Operating Expenses			189,000	______
Operating Income			293,500	______
Interest Expense			500	______
Net Income before Income Taxes			293,000	______
Income Taxes			32,600	______
Net Income after Income Taxes			______	______

4. Complete the following horizontal analysis. Enclose decreases in parentheses. *(40 points)*

Sue-Ted Stationery

Comparative Income Statement for the Years 1990 and 1991

	1990	1991	Increase or (Decrease) Amount	Increase or (Decrease) Percent
Net Sales	$457,800	$492,300		
Cost of Goods Sold	− 289,900	− 305,800		
Gross Profit on Sales	167,900	186,500		
Operating Expenses	− 102,400	− 109,700		
Operating Income	65,500	76,800		
Interest Expense	− 900	− 800		
Income before Taxes	64,600	76,000		
Taxes	− 11,600	− 13,500		
Income after Taxes	53,000	62,500		

5. Complete the following vertical analysis. *(40 points)*

Stockport Enterprises

Balance Sheet
December 31, 199X

	Assets	Percent
Cash	15,600	
Securities	17,500	
Accounts Receivable	36,700	
Supplies	8,200	
Inventory	48,000	
Current Assets	$126,000	
Fixtures and Furniture	35,500	
Equipment	79,500	
Land	146,000	
Buildings	285,000	
Long-lived Assets	$546,000	
Total Assets	$672,000	100.0%

	Liabilities	Percent
Accounts Payable	31,700	
Taxes Payable	6,500	
Notes Payable	30,000	
Dividends Payable	12,300	
Current Liabilities	$80,500	
Long-term Liabilities	$146,500	
Total Liabilities	$227,000	
	Equity	
Capital, December 31, 199X	$445,000	
Total Liabilities and Equity	$672,000	100.0%

6. Complete the horizontal analysis for Stockport Enterprises. Enclose decreases in parentheses.

(50 points)

Stockport Enterprises

Comparative Balance Sheet

	December 31		Increase or (Decrease)	
	1990	1991	Amount	Percent
Assets				
Cash	13,900	15,600	______	______
Securities	11,500	17,500	______	______
Accounts Receivable	33,200	36,700	______	______
Supplies	7,900	8,200	______	______
Inventory	51,100	48,000	______	______
Current Assets	117,600	126,000	______	______
Fixtures and Furniture	30,700	35,500	______	______
Equipment	65,600	79,500	______	______
Land	146,000	146,000	______	______
Buildings	300,000	285,000	______	______
Long-lived Assets	542,300	546,000	______	______
Total Assets	659,900	672,000	______	______
Liabilities				
Accounts Payable	30,900	31,700	______	______
Taxes Payable	5,200	6,500	______	______
Notes Payable	27,000	30,000	______	______
Dividends Payable	12,300	12,300	______	______
Current Liabilities	75,400	80,500	______	______
Long-term Liabilities	140,700	146,500	______	______
Total Liabilities	216,100	227,000	______	______
Equity				
Capital, end of year	443,800	445,000	______	______
Total Liabilities + Equity	659,900	672,000	______	______

For problems 7 and 8, refer to the Stockport Enterprises comparative balance sheet in problem 6.

7. Complete the following ratios for 1990 and 1991: Round to the nearest tenth.

	1990	1991
a. Current ratio:	_______ : 1	_______ : 1
b. Acid test ratio:	_______ : 1	_______ : 1
c. Equity to debt ratio:	_______ : 1	_______ : 1

(30 points)

8. If net income was $39,500 in 1990 and $43,700 in 1991, what were the earnings to equity ratios for 1990 and 1991? Express results as decimal fractions, rounded to the nearest hundredth.

a. 1990: ________

b. 1991: ________ *(10 points)*

PART

7 COMPOUND INTEREST AND APPLICATIONS

Chapter 26 Compound Interest, Present and Future Values

OBJECTIVES

- Determine capital growth when interest is periodically computed on an increasing principal
- Compute the principal that will result in a given future value at compound interest

SECTIONS

INTRODUCTION

Simple interest is the basic charge paid for the use of someone else's money and is calculated once for the entire term of a loan. Compound interest, on the other hand, requires that interest be calculated each period, *after* interest for the previous period is added to the principal. Effectively, this is paying interest on interest. The difference between the initial principal and the final amount is called compound interest, which is greater than simple interest.

Section 26.1 Compound Interest

KEY TERMS **Compounding periods** **Periodic rate** **Nominal rate**

Simple interest is calculated once on the principal for the entire term of a loan. Compound interest requires that the term of the loan be divided into **compounding periods:** years, half-years, quarters, months, or days. At the end of each compounding period, the simple interest for that period is added to the amount accumulated during the previous periods. The interest rate applied at the end of each period—the **periodic rate**—is the stated annual rate, called the **nominal rate,** divided by the number of periods per year.

Monthly Compounding
Nominal rate = 12%
Periodic rate = 1%

Type of Compounding	Number of Periods per Year
Annual	1
Semiannual	2
Quarterly	4
Monthly	12
Daily	365 (or 360)

Initial principal = $1,000
Nominal rate = 8%
Quarterly compounding periodic rate = 2%

Quarter	Principal	Periodic Simple Interest	Amount
1	$1,000.00	$20.00	$1,020.00
2	1,020.00	20.40	1,040.40
3	1,040.40	20.81	1,061.21
4	1,061.21	21.22	1,082.43
5	1,082.43	21.65	1,104.08
6	1,104.08	22.08	1,126.16

Amount after $1\frac{1}{2}$ years = $1,126.16
Compound interest = $126.16

Simple interest for $1\frac{1}{2}$ years
$= 1000 \times 0.08 \times 1.5$
$= 120.00$

Example 1 Quarterly Compounding

Initial principal = $500

Nominal rate = 6.5% = 0.065

Quarterly rate = 6.5% ÷ 4 = 1.625% = 0.01625

First quarterly interest = 500 × 0.01625
= 8.125 = $8.13

Quarter	Principal	Interest	Amount
1	$500.00	$8.13	$508.13

key: 500 [×] .01625 [=]
display: 8.12565
edited: $8.13

Second quarterly interest = 508.13 × 0.01625
= 8.2571125 = $8.26

Quarter	Principal	Interest	Amount
2	$508.13	$8.26	$516.39

key: 508.13 [×] .01625 [=]
display: 8.2571125
edited: $8.26

Third quarterly interest = 516.39 × 0.01625
= 8.3913375 = $8.39

Quarter	Principal	Interest	Amount
3	$516.39	$8.39	$524.78

key: 516.39 [×] .01625 [=]
display: 8.3913375
edited: $8.39

Compound interest for the first three quarters
= last amount − initial principal
= 524.78 − 500.00
= 24.78

PRACTICE ASSIGNMENT

Plan: Number of periods in term = number of years in term × number of periods in one year
The periodic rate is the annual rate (nominal rate) divided by the number of periods per year. To derive the compound amount in problem 3, compute the periodic interest, round it to the nearest cent, and add it to the accumulated amount in each period.

1. Calculate the number of periods in each of the following terms. Use 365-day years for periods shown in days.

Period	Term	Number of Periods in the Term
a. Year	17 years	______
b. Month	4 years	______
c. Day	10 years	______
d. Half-year	$6\frac{1}{2}$ years	______
e. Quarter	$6\frac{1}{2}$ years	______
f. Month	$7\frac{3}{4}$ years	______

2. Show the periodic interest rate in percent form and in decimal form. If the percent form contains a mixed number, show its fractional part as a common fraction (for example: $2\frac{7}{8}$%). Where necessary, round the decimal form to six places after the decimal point.

Nominal Rate	Periods	Periodic Rate in % Form	Periodic Rate in Decimal Form
a. 12%	Half-years	______%	______
b. 12%	Months	______%	______
c. 8%	Quarters	______%	______
d. 8%	Months	______%	______
e. $6\frac{3}{4}$%	Quarters	______%	______
f. $6\frac{3}{4}$%	Days	______%	______

3. A six-month loan of $2,000 requires $7\frac{1}{2}$% interest compounded monthly. Compute the monthly figures. The periodic interest rate is _____% or __________.

End of Month	Monthly Interest	Amount at End of Month
1	$________	__________
2	________	__________
3	________	__________
4	________	__________
5	________	__________
6	________	__________

What is the compound interest for the term?

Section 26.2 The Compound Amount

KEY TERMS **Exponentiation** **Future value** **Amount of 1**

Consider the amount produced by $1,000 at eight percent interest compounded quarterly for one year.

Beginning with 1000 × 1.02, at the end of each quarter, the amount produced by the simple interest is

$$P(1 + RT) = P \times (1 + 0.08 \times 1/4)$$
$$= P \times 1.02,$$

where P is the previously computed principal. After four quarters, the compound amount is 1000 × 1.02 × 1.02 × 1.02 × 1.02.

This methods works for terms that have a few periods. For terms that have many compounding periods, however, more efficient methods of calculating the compound amount are available.

Successive multiplication by the same factor is accomplished by **exponentiation:**

$$1.02 \times 1.02 \times 1.02 \times 1.02 = 1.02^4$$

The base is 1.02.
The exponent is 4.

Compute the compound amount or **future value** using the exponentiation technique as follows:

$$A = P(1 + i)^n$$

where i is the periodic interest rate and n is the number of compounding periods.

This method can also produce daily compound amounts (the number of periods per year is 365 or 360); daily compounding is covered in chapter 27 (savings accounts).

Calculator method: The calculator must have the exponentiation key [y^x].*

key: 1000 [×] 1.02 [y^x] 4 [=]
display: 1082.4322
edited: $1,082.43

Table method: The **amount-of-1** table (table 26.1) shows the amount that $1 becomes at periodic rate i for n periods.

For $i = 2\%$, $n = 4$, a 6-place table shows 1.082432.
$1 grows to 1.082432.
$1,000 grows to 1000 × 1.082432 = $1082.43.

If the principal P is more complex, use your calculator to multiply P × table factor.

*Compounding by exponentiation is possible on almost any calculator by keying (in this example): 1.02 [×] [=] [=] [=] [×] 1000 [=]. The number of equal signs immediately following the first multiplication is one less than the exponent. When the exponent is large, however, this procedure is likely to result in keying errors and is not usually recommended.

Table 26.1 Amount-of-1

Periods n	$\frac{1}{2}$%	$\frac{2}{3}$%	$\frac{3}{4}$%	Periodic Interest Rate 1%	$1\frac{1}{2}$%	2%	3%
1	1.005000	1.006667	1.007500	1.010000	1.015000	1.020000	1.030000
2	1.010025	1.013378	1.015056	1.020100	1.030225	1.040400	1.060900
3	1.015075	1.020134	1.022669	1.030301	1.045678	1.061208	1.092727
4	1.020151	1.026935	1.030339	1.040604	1.061364	1.082432	1.125509
5	1.025251	1.033781	1.038067	1.051010	1.077284	1.104081	1.159274
6	1.030378	1.040673	1.045852	1.061520	1.093443	1.126162	1.194052
7	1.035529	1.047610	1.053696	1.072135	1.109845	1.148686	1.229874
8	1.040707	1.054595	1.061599	1.082857	1.126493	1.171659	1.266770
9	1.045911	1.061625	1.069561	1.093685	1.143390	1.195093	1.304773
10	1.051140	1.068703	1.077583	1.104622	1.160541	1.218994	1.343916
11	1.056396	1.075827	1.085664	1.115668	1.177949	1.243374	1.384234
12	1.061678	1.083000	1.093807	1.126825	1.195618	1.268242	1.425761
15	1.077683	1.104804	1.118603	1.160969	1.250232	1.345868	1.557967
18	1.093929	1.127048	1.143960	1.196147	1.307341	1.428246	1.702433
21	1.110420	1.149740	1.169893	1.232392	1.367058	1.515666	1.860295
24	1.127160	1.172888	1.196414	1.269735	1.429503	1.608437	2.032794
30	1.161400	1.220592	1.251272	1.347849	1.563080	1.811362	2.427262
36	1.196681	1.270237	1.308645	1.430769	1.709140	2.039887	2.898278
48	1.270489	1.375666	1.431405	1.612226	2.043478	2.587070	4.132252
60	1.348850	1.489846	1.565681	1.816697	2.443220	3.281031	5.891603
120	1.819397	2.219640	2.451357	3.300387	5.969323	10.765163	34.710987

Periods n	4%	5%	6%	Periodic Interest Rate 7%	8%	9%	10%
1	1.040000	1.050000	1.060000	1.070000	1.080000	1.090000	1.100000
2	1.081600	1.102500	1.123600	1.144900	1.166400	1.188100	1.210000
3	1.124864	1.157625	1.191016	1.225043	1.259712	1.295029	1.331000
4	1.169859	1.215506	1.262477	1.310796	1.360489	1.411582	1.464100
5	1.216653	1.276282	1.338226	1.402552	1.469328	1.538624	1.610510
6	1.265319	1.340096	1.418519	1.500730	1.586874	1.677100	1.771561
7	1.315932	1.407100	1.503630	1.605781	1.713824	1.828039	1.948717
8	1.368569	1.477455	1.593848	1.718186	1.850930	1.992563	2.143589
9	1.423312	1.551328	1.689479	1.838459	1.999005	2.171893	2.357948
10	1.480244	1.628895	1.790848	1.967151	2.158925	2.367364	2.593742
11	1.539454	1.710339	1.898299	2.104852	2.331639	2.580426	2.853117
12	1.601032	1.795856	2.012196	2.252192	2.518170	2.812665	3.138428
15	1.800944	2.078928	2.396558	2.759032	3.172169	3.642482	4.177248
18	2.025817	2.406619	2.854339	3.379932	3.996019	4.717120	5.559917
21	2.278768	2.785963	3.399564	4.140562	5.033834	6.108808	7.400250
24	2.563304	3.225100	4.048935	5.072367	6.341181	7.911083	9.849733
30	3.243398	4.321942	5.743491	7.612255	10.062657	13.267678	17.449402
36	4.103933	5.791816	8.147252	11.423942	15.968172	22.251225	30.912681
48	6.570528	10.401270	16.393872	25.728907	40.210573	62.585237	97.017234
60	10.519627	18.679186	32.987691	57.946427	101.257064	176.031292	304.481640

Example 1 What will $500 amount to in five years invested at 12 percent interest compounded monthly?

Periodic rate $i = 1\% = 0.01$

Number of periods $n = 5 \times 12 = 60$

Compound amount $A = P(1 + i)^n = 500 \times 1.01^{60}$
$= 908.34835 = \$908.35$

Calculator method:
key: 500 [×] 1.01 [y^x] 60 [=]
display: 908.34835
edited: $908.35

Table method:
For $i = 1\%$, $n = 60$, table 26.1 shows 1.816697
key: 500 [×] 1.816697 [=]
display: 908.3485
edited: $908.35

PRACTICE ASSIGNMENT

Plan: Unless the table method is specifically indicated, you may use a calculator or table 26.1 to find the compound amount (future value). When using the table, you can, of course, multiply $P \times$ table factor on your calculator.

1. Use table 26.1 to find the following.

	i	n	$(1 + i)^n$ Amount of 1
a.	3%	3	________
b.	1%	24	________
c.	10%	8	________
d.	$\frac{3}{4}$%	15	________
e.	$\frac{1}{2}$%	120	________
f.	1.5%	12	________

2. Use the factors found in problem 1 to compute the following.

	i	n	Principal	Amount	Interest
a.	3%	3	3,000.00	________	________
b.	1%	24	12,000.00	________	________
c.	10%	8	10,650.00	________	________
d.	$\frac{3}{4}$%	15	700.00	________	________
e.	$\frac{1}{2}$%	120	945.00	________	________
f.	1.5%	12	428.35	________	________

3. How much will $2,500 at 6% compounded monthly amount to in 3 years?

4. Twelve years ago Marjorie Ross deposited $3,436.75 into an account paying 6% interest compounded quarterly. If there were no other deposits or withdrawals, how much is in the account now?

5. The rapid growth produced by compounding applies not only to money but also to other fields, such as population studies. In 1990, 10,000 people lived in the town of Northbank. To increase employment opportunities, the town officials implemented a plan that attracted more business and industry to Northbank. As a result, population grew 5% annually. Project Northbank's population by the year 2020, rounded to the nearest thousand.

Section 26.3 Present and Future Values at Compound Interest

KEY TERM **Present value**

If the maturity value of a two-year note at eight percent interest compounded quarterly is $3,135.23, what is its face value? The value of a loan or investment at the beginning of a term (the principal or face value) is called the **present value.**

Future value or maturity value is the value at the end of a term. It is the amount due.

In some cases, you may need to find future value (amount) given present value (principal).

Example 1 What is the maturity value of \$4,000 invested for 2 years at 6% interest compounded quarterly?

$$A = P(1 + i)^n = 4000 \times 1.015^8$$

$$\text{Future value} = 4000 \times 1.015^8$$

Calculator with exponential key:

key: 4000 [×] 1.015 [y^x] 8 [=]
display: 4505.9703
edited: \$4,505.97

Table method: (table 26.1)
$i = 1\frac{1}{2}\%$, $n = 8$, Table displays 1.126493

key: 4000 [×] 1.126493 [=]
display: 4505.972
edited: \$4,505.97

In other cases, you may need to find present value (principal), given future value (amount).

Example 2 What principal is needed to obtain \$5,000 in 2 years if money earns 6% interest compounded quarterly?

$$P = \frac{A}{(1 + i)^n} = \frac{5000}{1.015^8}$$

$$\text{Present value} = 5000 \div 1.015^8$$

Calculator with exponential key:

key: 5000 [÷] 1.015 [y^x] 8 [=]
display: 4438.5556
edited: \$4,438.56

Table method (table 26.1)*
$i = 1\frac{1}{2}\%$, $n = 8$, Table displays 1.126493

key: 5000 [÷] 1.126493 [=]
display: 4438.554
edited: \$4,438.55

Future value is greater than present value. The value of an investment grows over time because it earns interest, but it grows faster with compound interest than with simple interest.

Present and Future Value at Compound Interest

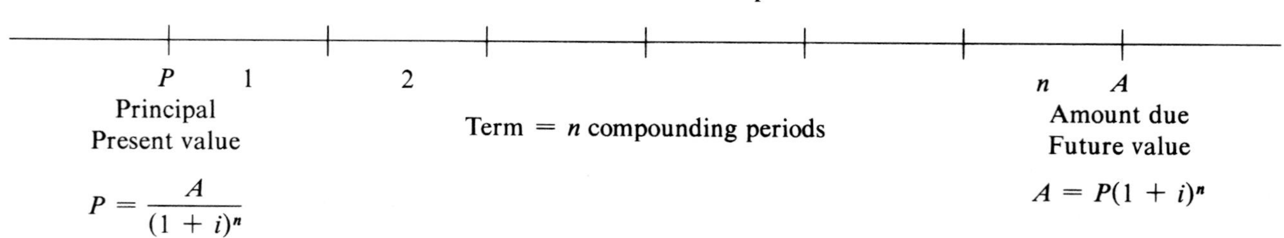

*Present-value-of-1 tables (also called present-worth-of-1 tables) display the value needed to produce \$1 at the rate i in n periods. These entries are the reciprocals of the amount-of-1 entries. The reciprocal of x is $1/x$: the reciprocal of 1.126493 = 1/1.126493 = 0.887711.

Example 3 Find the principal that produces $5,000 when invested for eight months at nine percent compounded monthly.

Periodic rate $i = \frac{9}{12}\% = \frac{3}{4}\% = 0.0075$
Number of periods $n = 8$
Present value $= A \div (1 + i)^n = 5000 \div 1.0075^8$
Calculator method:

key: 5000 [÷] 1.0075 [y^x] 8 [=]
display: 4709.877
edited: $4,709.88

Table method: (table 26.1)
$i = \frac{3}{4}\%$, $n = 8$, Table displays 1.061599.

key: 5000 [÷] 1.061599 [=]
display: 4709.8763
edited: $4,709.88

PRACTICE ASSIGNMENT

Plan: You may use a calculator with an exponentiation key to find compound present values. For the table method, divide the future value by the amount-of-1 factor shown for the given i and n in table 26.1.

1. Use table 26.1 to find the following present values of 1 (reciprocals of amount of 1), accurate to six decimal places.

Present value of 1

i	n	$\frac{1}{(1+i)^n}$
a. 2%	3	________
b. 1%	24	________
c. 4%	15	________
d. $\frac{1}{2}\%$	18	________
e. 7%	48	________
f. 1.5%	12	________

2. The maturity value of a two-year loan at 8% interest compounded quarterly was $5,632.46. What was the principal?

3. To enlarge and modernize its office, a small firm plans to spend $7,500 after five years. It would prefer not to borrow for this purpose. Money can be invested now at 8% interest compounded semiannually. How much money should the firm invest now?

4. Betty Jo Carson received $10,000 from a 15-year-old trust fund, an investment that had accrued interest at 12% compounded semiannually. What was the original investment?

5. If you can invest money at 8% interest compounded quarterly, should you spend $4,000 cash to make home improvements now, or should you invest your money for a year and spend $4,350 then? Check one:
$4,000 cash: _____ $4,350 in one year: _____

Challenger

(Attempt this problem only if you have a calculator with an exponentiation key.)

The maturity value of a ten-year note is $10,000. The note specifies interest at 8.13 percent compounded semiannually. If the note is discounted at any time during the term, the price is established by computing the present value.

a. How much did the maker of the note borrow?

b. If the note is discounted eight years prior to maturity at 7.86 percent compounded semiannually, what is the price of the note?

c. Suppose the note is discounted eight years and 94 days prior to maturity at 7.86 percent. Since the 94 days are a fraction of a term, calculate with a fractional exponent. What is the price of the note? (Use a 365-day year.)

CHAPTER 26 SUMMARY

Concept	*Example*	*Procedure*	*Formula*
SECTION 26.1 Compound Interest			
Nominal interest rate Periodic interest rate	Nominal rate = 8% Quarterly periods Periodic rate = 2%	The term is divided into compounding periods (years, half-years, quarters, months, days).	Periodic rate $= \dfrac{\text{nominal rate}}{\text{number of periods per year}}$ nom.rate[÷]periods[=] nomrate/periods
SECTION 26.2 The Compound Amount			
Compound amount	$1,000 at 8% compounded quarterly for 1.5 years amounts to: $1000 \times 1.02^6 = 1126.16$	At the end of each period the simple interest is computed at the periodic rate and added to the previous amount.	Compound amount = principal $x(1 + i)^n$, where i is the periodic rate and n is the number of periods in the term P[×]1 + i[y^x]n[=] P*(1+i)^n
Compound interest	Compound interest = 1126.16 − 1000 = 126.16	The compound interest is the difference between the compound amount and the initial principal.	Interest = amount − principal
SECTION 26.3 Present and Future Values at Compound Interest			
Future value	Amount = $1,126.16 Future value = amount = $1,126.16	The future value is the value at the end of the term (amount due or maturity value).	Same as compound amount formula in section 26.2.
Present value	Future value = $5,000 Interest is 8% compounded quarterly for 1.5 years Present value $= \dfrac{5000}{1.02^6} = 4439.86$	The present value is the value at the beginning of the term (principal or face value).	Present value $= \dfrac{\text{future value}}{(1 + i)^n}$ futval[÷]1 + i[y^x]n[=] futval/(1+i)^n

Chapter 26 **Spreadsheet Exercise**

Compounding Effects

Compounding interest means paying interest on interest. The more frequently interest is compounded, the faster the original principal grows.

Daily compounding is not difficult with computers. How much greater is the yield with daily rather than monthly compounding? Is the compounding effect appreciable on a large principal?

This spreadsheet displays the effects of compounding on a principal of $10,000 invested for five years at the nominal rates of six percent, seven percent, and eight percent. Your chapter 26 template will appear as follows:

```
A1: PR [W11] '                                                        READY

            A         B        C          D          E        F         G         H
1       THE EFFECT OF COMPOUNDING ON A PRINCIPAL OF $10,000 FOR 5 YEARS
2   ---------------------------------------------------------------------------
3                              6%                    7%                  8%
4   ---------------------------------------------------------------------------
5    type of    times   effect.               effect.              effect.
6    interest  p.year    rate     interest     rate     interest    rate    interest
7   ---------------------------------------------------------------------------
8   simple         0.2
9   annual         1.0
10  semi annual    2.0
11  quarterly      4.0
12  monthly       12.0
13  daily        365.0
14  continuous
15  ---------------------------------------------------------------------------
16
17
18
19
20
```

Column A displays the type of compounding. Column B displays the number of times per year interest is compounded. Seven types of interest are in rows 8–14.

Columns C–D are reserved for six percent, columns E–F for seven percent, and columns G–H for eight percent. Row 3 contains the rates: six percent in C3, seven percent in E3, and eight percent in G3.

Columns C, E, and G show the effective rate, which is the annually compounded rate that produces the same gain as the given rate. The two extreme cases are: no compounding (simple interest) and continuous compounding. The effective rate can be extended to simple interest, in which case it is lower than the nominal rate. Higher mathematics prove that if we spent all our available time compounding (continuous compounding), the yield would not exceed a certain limit.

Columns D, F, and H display the compound interest.

Spreadsheet Exercise Steps

1. Column C: Effective Rates Other Than for Continuous Compounding
   The effective rate r equals $(1 + i)^m - 1$ where i is the periodic rate and m is the number of times compounded per year.
   The nominal rate is in C3 (absolute row address). Divide by the number of times compounded per year in B8 to obtain the periodic rate (absolute column address).
   Exponentiation is indicated by the ^ symbol.
   The spreadsheet formula is (_____ + _____ $ _____ / $ _____) ^ $ _____ − _____ .
   Complete the formula and store it in C8.

2. Column D: Interest Other Than for Continuous Compounding
   Compound interest $= P(1 + i)^n - P$ where P is the principal, i is the periodic rate, and n is the number of compounding periods in the term.
   $P = 10000$; the nominal rate is in C3 (coded as in step 2 with an absolute row address); the number of times compounded per year is in B8 (coded as in step 2 with an absolute column address). In five years, the number of periods is 5 × the number in B8.
   The spreadsheet formula is:
   10000 * (____ + ____ $ ____ / $ ____) ^ (____ * $ ____) − ________ .
   Complete the formula and store it in D8.
3. Copy the formulas from C8..D8 to C9..D13.
4. Column C, Row 14: Effective Rate for Continuous Compounding
   The justification for the formulas in steps 4 and 5 is shown in higher mathematics.
   The effective rate r for continuous compounding equals $e^j - 1$ where e is the non-repeating decimal 2.7182818 . . . (an irrational number), and j is the nominal rate.
   The nominal rate is in C3, as above, and e^j is coded with the exponential function EXP(j).
   The spreadsheet formula is @EXP(C$3)−1. Store it in C14.
5. Column D Row 14: Interest with Continuous Compounding
   Continuously compounded interest $= P(e^{jt}) - P$ where P is the principal, e is the same number as in step 4, j is the nominal rate, and t is the term.
   The spreadsheet formula is 10000*@ EXP(C$3*5)−10000.
   Store it in D14.
6. Copy all of the cells in columns C–D to columns E–F and columns G–H: copy from C8..D14 to E8..F14 and from C8..D14 to G8..H14.
7. **a.** What is the increase in interest income between quarterly and monthly compounding at each of the three nominal rates?
   6% ________ 7% ________ 8% ________

   b. What is the increase in interest income between annual and daily compounding at each of the three nominal rates?
   6% ________ 7% ________ 8% ________

   c. What is the upper limit of the effective rate that corresponds to each of the three nominal rates?
   6% ________ 7% ________ 8% ________

8. Change the nominal interest rates to 9%, 10%, and 11% (enter in decimal form).
   What is the upper limit of the effective rate that corresponds to each of the three nominal rates?
   9% ________ 10% ________ 11% ________

NAME ______________________________

CLASS SECTION ________________ DATE ________________

SCORE ____________ MAXIMUM 100 PERCENT ____________

CHAPTER 26 ASSIGNMENT

1. A principal of $2,500 is deposited at 8% for 3 years. Find the compound amount and the compound interest in the following cases:

Compounding Period	Amount	Interest
a. Annual	________	________
b. Semiannual	________	________
c. Quarterly	________	________
d. Monthly	________	________

(20 points)

2. To receive $3,000 in 3 years at 6%, how much must you invest in the following situations:

Compounding Period	Investment
a. Annual	________
b. Semiannual	________
c. Quarterly	________
d. Monthly	________

(10 points)

3. A business borrowed $7,500 for four years at 10% interest compounded semiannually. What was the total amount repaid? *(5 points)*

4. On July 1, 1990, Tom Karolinski deposited $4,000 into a bank paying 8% interest compounded quarterly at the end of each March, June, September, and December.

a. What will the compound amount be on January 1, 1995?

b. What will the compound interest be?

(10 points)

5. Bruce Metzger repaid $784.39 on a six-month loan at 9% compounded monthly. What was the amount of the loan? *(5 points)*

6. R & S Notions owed $3,000 on June 1. The loan was compounded quarterly at 12%. If it was settled on March 1 with no penalty for early payment, how much was paid? *(10 points)*

7. In the year 2001, a new father invests $1,000 for 30 years at 8% compounded quarterly for his daughter. How much money will she receive in 2031, rounded to the nearest hundred dollars? *(10 points)*

8. The Pembrokes invested money at 6% compounded monthly. By September 1, 1994, they expected to have $6,000. An emergency arose, however, and they were forced to liquidate their investment on October 1, 1993. How much did they receive at that time? *(10 points)*

9. Mindmax Educational Software, Inc. borrowed $35,000 for three years at 7% compounded annually. Upon maturity, the company refinanced the loan at 8% compounded monthly for one additional year. What was the complete settlement after the four years (to the nearest dollar)? ***(10 points)***

10. The Corsos can purchase a vacant lot for $10,500 now or wait a year and purchase a similar one for $11,000. They can afford to buy now, but are considering an investment opportunity at 8% compounded monthly. Which is the better financial plan? Should the Corsos buy a lot now or should they wait a year? ***(10 points)***

Chapter 27 Savings Accounts

OBJECTIVES

- Distinguish between various types of savings accounts
- Determine interest on savings accounts

SECTIONS

INTRODUCTION

Banks and credit unions offer a variety of plans to encourage their customers to save money. The basic day-to-day account pays a fixed rate of interest and has no transaction limitations. Money market accounts, NOW accounts, and certificates of deposit (CDs) offer higher interest rates than day-to-day accounts, but also carry minimum balance requirements, transaction limitations, or time period stipulations. Customers may take advantage of more than one type of account to meet their specific banking needs.

Section 27.1 Day-to-Day Accounts

KEY TERMS **Day-to-day savings accounts** **Daily simple interest** **Daily compounded interest**

Traditionally, savings accounts paid quarterly compounded interest that was recorded in savers' passbooks. Usually, no interest was paid on money withdrawn between compounding dates.

Today, **day-to-day savings accounts** may still be called passbook accounts, although savers may no longer receive passbooks; the interest from these accounts is relatively low.

Day-to-day accounts are convenience accounts. You may deposit or withdraw any amount at any time. Interest rates are low when compared to other types of savings accounts. Because the balance may change daily, interest is calculated as **daily simple interest** or **daily compounded interest.**

Brownsville First National Bank
Day-to-day Savings Account

- 5% interest compounded daily
- Unlimited transactions
- Minimum opening deposit of $50
- No minimum balance
- Interest rate increases to $5\frac{1}{4}$% if withdrawals are limited to 6 every 3 months with $1 charge for each extra withdrawal

Example 1 Compute simple interest on a savings account balance of $313.25 for 10 days at $5\frac{1}{4}$ percent.

$$I = PRT = 313.25 \times 0.0525 \times 10/365$$
$$= 0.450565 = 0.45$$

key: 313.25 [×] .0525 [×] 10 [÷] 365 [=]
display: 0.450565
edited: $0.45

Example 2 Daily Simple Interest at 5 percent

Date	Deposit	Withdrawal	Balance	Days	Interest
1/01			500.00		
1/14	105.00		605.00	13	500.00 × .05 × 13/365 = 0.89
1/18		92.50	512.50	4	605.00 × .05 × 4/365 = 0.33
1/25		72.15	440.35	7	512.50 × .05 × 7/365 = 0.49
1/31	300.00		740.35	6	440.35 × .05 × 6/365 = 0.36
					2.07

The accumulated interest may not be added to the balance until the end of the quarter.

Table 27.1 Daily Amount-of-1

Day	$4\frac{1}{2}$%	$4\frac{3}{4}$%	5%	*Nominal Rate* $5\frac{1}{4}$%	$5\frac{1}{2}$%	$5\frac{3}{4}$%	6%
1	1.000123	1.000130	1.000137	1.000144	1.000151	1.000158	1.000164
2	1.000247	1.000260	1.000274	1.000288	1.000301	1.000315	1.000329
3	1.000370	1.000390	1.000411	1.000432	1.000452	1.000473	1.000493
4	1.000493	1.000521	1.000548	1.000575	1.000603	1.000630	1.000658
5	1.000617	1.000651	1.000685	1.000719	1.000754	1.000788	1.000822
6	1.000740	1.000781	1.000822	1.000863	1.000904	1.000946	1.000987
7	1.000863	1.000911	1.000959	1.001007	1.001055	1.001103	1.001151
8	1.000987	1.001042	1.001096	1.001151	1.001206	1.001261	1.001316
9	1.001110	1.001172	1.001234	1.001295	1.001357	1.001419	1.001480
10	1.001234	1.001302	1.001371	1.001439	1.001508	1.001576	1.001645
11	1.001357	1.001432	1.001508	1.001583	1.001659	1.001734	1.001810
12	1.001480	1.001563	1.001645	1.001727	1.001810	1.001892	1.001974
13	1.001604	1.001693	1.001782	1.001871	1.001961	1.002050	1.002139
14	1.001727	1.001823	1.001920	1.002016	1.002112	1.002208	1.002304
15	1.001851	1.001954	1.002057	1.002160	1.002263	1.002366	1.002469
16	1.001974	1.002084	1.002194	1.002304	1.002414	1.002524	1.002633
17	1.002098	1.002215	1.002331	1.002448	1.002565	1.002681	1.002798
18	1.002222	1.002345	1.002469	1.002592	1.002716	1.002839	1.002963
19	1.002345	1.002476	1.002606	1.002736	1.002867	1.002997	1.003128
20	1.002469	1.002606	1.002743	1.002881	1.003018	1.003155	1.003293
21	1.002592	1.002736	1.002881	1.003025	1.003169	1.003313	1.003458
22	1.002716	1.002867	1.003018	1.003169	1.003320	1.003471	1.003623
23	1.002839	1.002997	1.003155	1.003313	1.003472	1.003630	1.003788
24	1.002963	1.003128	1.003293	1.003458	1.003623	1.003788	1.003953
25	1.003087	1.003259	1.003430	1.003602	1.003774	1.003946	1.004118
26	1.003210	1.003389	1.003568	1.003746	1.003925	1.004104	1.004283
27	1.003334	1.003520	1.003705	1.003891	1.004076	1.004262	1.004448
28	1.003458	1.003650	1.003843	1.004035	1.004228	1.004420	1.004613
29	1.003582	1.003781	1.003980	1.004180	1.004379	1.004579	1.004778
30	1.003705	1.003911	1.004118	1.004324	1.004530	1.004737	1.004943
31	1.003829	1.004042	1.004255	1.004469	1.004682	1.004895	1.005108

Example 3 Compute the next compound amount on an account balance of $313.25 for 10 days at $5\frac{1}{4}$ percent.

$$A = P(1 + i)^d$$
$$= 313.25 \times (1 + .0525/365)^{10}$$
$$= 313.70086 = 313.70$$

key w/paren: 313.25 [×] [(] 1 [+] .0525 [÷] 365 [)] [y^x] 10 [=]
key w/o paren: .0525 [÷] 365 [+] 1 [=] [y^x] 10 [×] 313.25 [=]
display: 313.70086
edited: $313.70

Daily amount-of-1-table method (table 27.1):
$5\frac{1}{4}$% column displays 1.001439 for 10 days

key: 313.25 [×] 1.001439 [=]
display: 313.70077
edited: $313.70

Example 4 Daily Compounded Interest at 5 percent

Date	Deposit	Withdrawal	Days	Compound Amount	Balance
1/01					500.00
1/14	105.00		13	$500.00 \times (1 + .05/365)^{13} = 500.89$	605.89
1/18		92.50	4	$605.89 \times (1 + .05/365)^{4} = 606.22$	513.72
1/25		72.15	7	$513.72 \times (1 + .05/365)^{7} = 514.21$	442.06
1/31	300.00		6	$442.06 \times (1 + .05/365)^{6} = 442.42$	742.42

PRACTICE ASSIGNMENT

Plan: Daily simple interest is computed with the simple interest formula

$$I = PRT = P \times R \times d/365.$$

P is previous balance
R is interest rate
d is number of days during which the balance did not change (the number of days between two transactions)

Next balance = previous balance + deposit − withdrawal

The interest is computed daily but is not added to the balance until the end of the quarter. Round all interest figures to the nearest cent.

Daily compounded interest is added to the previous balance. Therefore, the next amount is computed with the compound amount formula

$$A = P(1 + i)^d.$$

P is previous balance
i is the periodic rate = nominal rate/365
d is the number of days during which the balance did not change (the number of days between two transactions)

Next balance = new amount + deposit − withdrawal

If possible, use a calculator with an exponentiation key. Otherwise, find the amount of 1 in table 27.1 under the nominal rate in the row that corresponds to the number of days. Then use your calculator to multiply the amount of 1 by the previous balance. Round all amounts to the nearest cent.

1. Compute the simple interest on the following daily balances:

	Balance	Interest Rate	Number of Days	Interest
a.	300.00	6%	20	______
b.	513.25	5%	10	______
c.	87.30	$5\frac{3}{4}\%$	25	______
d.	104.76	$4\frac{3}{4}\%$	31	______
e.	1,245.00	$5\frac{1}{4}\%$	8	______
f.	851.50	$5\frac{1}{2}\%$	27	______

2. Complete the following passbook account record assuming daily simple interest at $5\frac{1}{2}\%$.

Date	Deposit	Withdrawal	Balance	Days	Interest
8/01			450.00		
8/06	275.00		______	____	______ × ______ × ______ = ______
8/11		22.23	______	____	______ × ______ × ______ = ______
8/18		36.35	______	____	______ × ______ × ______ = ______
8/27		32.85	______	____	______ × ______ × ______ = ______
8/31	150.00		______	____	______ × ______ × ______ = ______
					Accumulated interest = ______

3. Compute the next amount at daily compounded interest for the following:

	Balance	Interest Rate	Number of Days	Amount
a.	300.00	6%	20	________
b.	513.25	5%	10	________
c.	87.30	$5\frac{3}{4}\%$	25	________
d.	104.76	$4\frac{3}{4}\%$	31	________
e.	1,245.00	$5\frac{1}{4}\%$	8	________
f.	851.50	$5\frac{1}{2}\%$	27	________

4. Complete the following passbook account record assuming daily compounded interest at $5\frac{1}{2}\%$.

Date	Deposit	Withdrawal	Days	Compound Amount			Balance
6/01							1,150.00
6/14		413.00	____	________________	=	________	________
6/15		192.50	____	________________	=	________	________
6/17		83.45	____	________________	=	________	________
6/25	500.00		____	________________	=	________	________
6/30		150.00	____	________________	=	________	________

Section 27.2 Other Types of Savings Accounts

KEY TERMS **Money market account** **NOW account** **Certificate of deposit**

Other types of savings accounts carry higher interest rates and pay simple, daily compounded, quarterly compounded, or annually compounded interest. The type of interest and rate of interest depend on the competitiveness of the local banking market. NOW accounts provide checking privileges.

Compare the following descriptions of savings account plans on the left with an example of one bank's options in the boxes on the right.

Money Market Accounts

The number of deposits or withdrawals may be limited. A minimum balance is required. Interest rates are high and may vary weekly or daily. Usually, simple interest is computed on balances.

Money Market Account
· $1,000 minimum balance · Current interest rate · $1 per month service charge · 6 withdrawals per month · $5 for each withdrawal over 6

NOW Accounts

Deposits or withdrawals may be made at any time. NOW accounts provide checking privileges. (A check is called a **n**egotiable **o**rder of **w**ithdrawal—NOW.)

A minimum balance is required. Interest rates are high and may vary. Usually, simple interest is computed on balances.

NOW Account
· $1,000 minimum balance · Current interest rate · $0.25 per check charge · $5 per month service charge
Super NOW Account
· $1,500 minimum balance · $5 charge if balance falls below $1,500 · Current interest rate · Unlimited checks

Time Deposit Accounts

A time deposit account may be called a **certificate of deposit** (CD). Money must be left in the account for a fixed period of time. A substantial penalty is levied for early withdrawals. CDs do not offer checking. Interest rates vary. When interest rates are rising nationally, they tend to increase as the terms increase, as shown here; when interest rates are dropping nationally, they tend to decrease as the terms increase. Interest may be simple or compounded in various ways.

Certificates of Deposit

Period	**Rate**
0– 30 days	6.0%
31– 60 days	6.5%
61– 90 days	6.8%
91–120 days	7.2%
121–150 days	7.5%
151–180 days	8.0%
181–270 days	8.3%
271–365 days	8.5%
2 years	8.8%
3–4 years	9.0%
5 years	9.25%

· $500 or multiples.
· If over $50,000, consult manager for rates.

Example 1 A $2,000 certificate of deposit (CD) is purchased for 3 months (90 days) at 6.8 percent interest. What is the maturity value? What is the gain? Compute simple exact interest for this CD.

$$\begin{aligned} A &= P(1 + RT) \\ &= 2000 \times (1 + 0.068 \times 90/365) \\ &= 2033.5342 = 2033.53 \end{aligned}$$

key w/paren: 2000 [×] [(] 1 [+] .068 [×] 90 [÷] 365 [)] [=]
key w/o paren: .068 [×] 90 [÷] 365 [+] 1 [=] [×] 2000 [=]
key arith: .068 [×] 90 [÷] 365 [=] [+] 1 [+] [×] 2000 [=]
display: 2033.5342
edited: $2,033.53 maturity value
$33.53 interest (gain)

Example 2 A $2,000 certificate of deposit is purchased for one year at 8.5 percent interest. What is the maturity value? What is the gain?

Compute quarterly compounded interest for this CD.

$$\begin{aligned} A &= P(1 + i)^n \\ &= 2000 \times (1 + 0.085/4)^4 \\ &= 2175.4959 = 2175.50 \end{aligned}$$

key w/paren: 2000 [×] [(] 1 [+] .085 [÷] 4 [)] [y^x] 4 [=]
key w/o paren: .085 [÷] 4 [+] 1 [=] [y^x] 4 [×] 2000 [=]
display: 2175.4959
edited: $2,175.50 maturity value
$175.50 interest (gain)

Suppose your calculator does not feature the exponential key and you cannot find the rate in an amount-of-1 table. Since the exponent is small, key the following sequence:

key alg: .085 [÷] 4 [+] 1 [=] [×] [=] [=] [=] [×] 2000 [=]
key arith: .085 [÷] 4 [=] [+] 1 [+] [×] [=] [=] [=] [×] 2000 [=]
display: 2175.4959
edited: $2,175.50

The number of equal signs immediately following the first multiplication is one less than the exponent. The display before the first multiplication is repeatedly multiplied by itself.

Example 3 Consider the following NOW account:

$1,000 minimum balance
Current interest rate
$0.25 per check
$5 per month service charge

Some banks compute simple interest for the number of days during which a balance was maintained with the rate that was in effect on the day prior to the transaction or rate change.

Service charges are subtracted from the balances.

Balance = previous balance + deposits − withdrawals − charges

Interest rates are as follows:

1/01–1/04	7.2%
1/05–1/11	7.1%
1/12–1/18	7.0%
1/19–1/25	6.9%
1/26–1/31	7.0%

Interest must be computed on 1/5, 1/12, 1/19, and 1/26 as well as on transaction days.

Date	Deposit	Withdrawal	Service Charge	Balance	Interest
1/01				1800.00	
1/05				1800.00	1800.00 × .072 × 4/365 = 1.42
1/10		250.00		1550.00	1800.00 × .071 × 5/365 = 1.75
1/12				1550.00	1550.00 × .071 × 2/365 = 0.60
1/16		192.50	0.25	1357.25	1550.00 × .070 × 4/365 = 1.19
1/19				1357.25	1357.25 × .070 × 3/365 = 0.78
1/26		272.25	0.25	1084.75	1357.25 × .069 × 7/365 = 1.80
1/31	800.00		5.00	1879.75	1084.75 × .070 × 5/365 = 1.04
					Accumulated interest = 8.58

PRACTICE ASSIGNMENT

Plan: The gain on money market or NOW accounts is often simple interest on the balances. The rates on money market or NOW accounts may vary each week. The gain on certificates of deposit is simple interest, quarterly compounded interest, or annually compounded interest. For certificates of deposit, use interest rates listed on page 375. To compute compound amounts, use a calculator with exponentiation or follow the alternative sequence for small exponents shown in example 2.

1. Compute the maturity value of a 60-day certificate of deposit with face value of $1,500. Assume simple interest.

2. Karpas Cleaners purchased a one-year certificate of deposit with face value of $2,500. What will the store receive at maturity if quarterly compounded interest is paid?

3. Compute the account record for a NOW account that pays 6.5% simple interest on the balances, and charges $0.25 per check and a monthly service charge of $5. The second and third withdrawals are checks.

Date	Deposit	Withdrawal	Service Charge	Balance	Interest
3/01				1,800.75	
3/10		250.00		______	______ × ______ × ______ = ______
3/18		221.50CK	0.25	______	______ × ______ × ______ = ______
3/28		83.35CK	0.25	______	______ × ______ × ______ = ______
3/31	450.00		5.00	______	______ × ______ × ______ = ______
					Accumulated interest = ______

4. A money market account charges $4 for each withdrawal above one per month and pays simple interest on the balances. Complete the account record for the month. Interest rates are as follows:

9/01–9/06	7.35%
9/07–9/13	7.37%
9/14–9/20	7.37%
9/21–9/27	7.39%
9/28–9/30	7.38%

Interest must be computed on 9/7, 9/14, 9/21, and 9/28 as well as on transaction days.

Date	Deposit	Withdrawal	Service Charge	Balance	Interest
9/05	1800.00			1800.00	
9/07				1800.00	1800.00 × _____ × _____ = _____
9/12		322.50		1477.50	1800.00 × _____ × _____ = _____
9/14				1477.50	1477.50 × _____ × _____ = _____
9/21				1477.50	1477.50 × _____ × _____ = _____
9/25		149.39	4.00	1324.11	1477.50 × _____ × _____ = _____
9/28				1324.11	1324.11 × _____ × _____ = _____
9/30	465.00			1789.11	1324.11 × _____ × _____ = _____
					Accumulated interest = _____

Challenger

(Attempt this problem **only** if you have a calculator with an exponentiation key.)

Consider the rules for early withdrawal of certificates of deposit. Once you agree to purchase a certificate of deposit, you cannot withdraw the money prior to maturity without paying a substantial penalty. If you were faced with an emergency, you might be forced to withdraw early. What would the cash value of your certificate be?

Suppose that on March 1, 1990, you purchased a 4-year certificate of deposit with face value of $2,000 at 9 percent interest. Your bank stipulates that if you withdraw early, no interest will be paid for the first 90 days, and the passbook interest rate of 5 percent compounded daily will be paid from the purchase date plus 90 up to the withdrawal date.

a. What is the maturity date?

b. Assuming quarterly compounding, what is the maturity value?

c. What amount will you receive if you withdraw your money on November 20, 1992?

d. What is the 9 percent quarterly compounded maturity value from the purchase date to the early withdrawal date? (Calculate with a fractional exponent.)

e. What is the effective penalty?

CHAPTER 27 SUMMARY

Concept	*Example*	*Procedure*	*Formula*
SECTION 27.1 Day-to-Day Accounts			
Passbook account	Unlimited number of transactions allowed; 5% simple interest computed daily. Interest on balance of \$500 for 10 days = 500 × 0.05 × 10/365 = \$0.68	Passbook accounts are day-to-day convenience accounts. The low-rate interest is computed daily and either added to the balance each day (daily compounding) or accumulated and added to the balance at the end of each quarter (simple interest).	Daily simple interest = PRT = daily rate × balance × $d/365$ bal[×]rate[×]d[÷]365[=] bal*rate*d/365 Daily compounded amount = $P(1 + \text{nom.rate}/365)^d$ = daily balance × $(1 + \text{nom.rate}/365)^d$ nomrate[÷]365[+]1[=][y^x]d[×]bal[=] or bal[×][(]1[+]nomrate[÷]365[)][y^x]d[=] bal*(1+nomrate/365)^d
SECTION 27.2 Other Types of Savings Accounts			
Money market accounts	\$1,000 minimum balance 6 withdrawals per month \$1 per month service charge Current interest rate	Money market accounts pay high interest rates that vary weekly or daily. The number of transactions is limited, and a minimum balance is often required. Usually, simple interest is computed on the balances.	
NOW accounts	\$1,500 minimum balance Unlimited checks Current interest rate	NOW accounts permit the writing of *n*egotiable *o*rders of *w*ithdrawal (NOW) which, in effect, are checks. Interest rates (applied to the balance) vary. Usually, a minimum balance is required.	
Certificates of deposit	2-year \$1,000 certificate of deposit pays 8.8% compounded quarterly. Maturity value = $1000 \times (1 + 0.088/4)^8$ = 1190.17	Money must be left in the account for a fixed period of time. There are penalties for early withdrawal. Interest rates depend on the length of the term. Interest is usually simple or compounded quarterly.	Simple interest for d days = face value × rate × $d/365$ faceval[×]rate[×]d[÷]365[=] faceval*rate*d/365 Compound amount for t years = face value × $(1 + \text{nom.rate}/m)^{mt}$ where m is the number of periods in the year nomrate[÷]m[+]1[=][y^x]mt[×]faceval[=] or faceval[×][(]1[+]nomrate[÷]m[)][y^x]mt[=] faceval*(1+nomrate/m)^(m*t)

Chapter 27 **Spreadsheet Exercise**

Variable Rate Savings Account

Most banks use computers to compute interest on the daily balances of savings accounts with interest rates that change weekly. The chapter 27 template enables you to set up the necessary computations for a variable rate savings account. It appears as follows:

```
A1: PR [W13]                                                          READY

           A           B         C         D      E        F          G     H       I
1                             VARIABLE RATE SAVINGS ACCOUNT
2    ---------------------------------------------------------------------------------------
3      comments      date   deposits   withdr.  fee    balance      rate  days  interest
4    ---------------------------------------------------------------------------------------
5    opening bal.   5/01                                2500.00
6                   5/05               645.00                       6.98%
7    rate change    5/05                                            6.98%
8    rate change    5/12                                            6.99%
9                   5/15               207.60                       7.01%
10                  5/18               476.95                       7.01%
11   rate change    5/19                                            7.01%
12                  5/20     250.00                                 7.03%
13                  5/21               500.00                       7.03%
14   rate change    5/26                                            7.03%
15                  5/28    1000.00                                 7.02%
16   closing bal.   5/31                                            7.02%
17   ---------------------------------------------------------------------------------------
18   + interest
19   = new balance
20
```

The spreadsheet contains the transactions of a variable rate savings account in rows 6–15. Column B displays the transaction dates, column C the deposits, and column D the withdrawals.

Column G: Interest Rates

The following interest rates were in effect:

5/01–5/04	6.98%
5/05–5/11	6.99%
5/12–5/18	7.01%
5/19–5/25	7.03%
5/26–5/31	7.02%

Interest is computed for the number of days during which a balance was maintained, with the rate that was in effect on the day prior to the transaction or rate change. These interest rates have been entered.

Spreadsheet Exercise Steps

1. A fee of $2.00 is charged for each withdrawal exceeding two per month. Enter these fees into column E.
2. Column F: Balance
   New balance = previous balance + deposit − withdrawal − fee
   The previous balance is in F5, a possible deposit in C6, possible withdrawal in D6, and possible fee in E6.
   The spreadsheet formula begins with a cell address and must be preceded by +.
   The spreadsheet formula is
   + ______ + ______ − ______ − ______ .
   Complete the formula, store it in F6, and copy it from F6 to F7..F16.

3. Column H: Days
The number of days during which a balance is maintained is the difference between the transaction or rate change date and the previous transaction or rate change date. The first transaction date is in B6; the opening date is in B5.
The spreadsheet formula is
@DATEVALUE(______) −
@ DATEVALUE(______).
Complete the formula and store it in H6.

4. Column I: Interest
Compute the simple exact interest on the previous balance with the rate in effect for the number of days. The previous balance is in F5, the rate in G6, and the days in H6.
The spreadsheet formula must be preceded by +.
The spreadsheet formula is
+ ______ * ______ * ______ / ______ .
The daily interest will display with four decimal places for greater accuracy. Complete the formula and store it in I6.

5. Copy the formulas from H6..I6 to H7..I16.

6. Total interest = sum of interest earned.
The interest amounts are in cells I6..I16.
Complete
@ SUM(______ .. ______) and store it in I18.

7. Enter a monthly fee of $3.00 into E16.
New balance = closing balance + interest
The closing balance is in F16 and the interest in I18.
The spreadsheet formula must be preceded by +.
The formula is + ______ + _____ .
Complete the formula and store it in F19.

8. What is the total interest for the month? ________

9. What is the opening balance for the next month?
____

CHAPTER 27 ASSIGNMENT

NAME ____________________

CLASS SECTION ____________ DATE ____________

SCORE ________ MAXIMUM 125 PERCENT ________

---

1. If the simple interest rate is $5\frac{3}{4}$%, how much interest will a deposit of \$72.55 accumulate in 25 days?
(5 points)

2. How much will \$904.60 be worth in 27 days if the account pays $5\frac{1}{2}$% interest compounded daily?
(5 points)

3. Compute the next balance and the interest, assuming daily simple interest at $5\frac{1}{4}$%.

Date	Deposit	Withdrawal	Balance	Days	Interest
5/23			265.25		
5/28	322.00		______	____	______ × ______ × ______ = ______

(10 points)

4. Compute the daily compounded amount and the next balance at $5\frac{1}{4}$%.

Date	Deposit	Withdrawal	Days	Compound Amount	Balance
9/12					416.00
9/25		215.00	____	____________ = ______	______

(10 points)

5. On May 14, Ellen Malone deposited \$478 into a new passbook account that paid $5\frac{1}{4}$% daily simple interest. On May 20, she deposited \$50.75. What was the accumulated interest on May 31? *(10 points)*

6. Broad Street Photography has a passbook account that pays $4\frac{3}{4}$% interest compounded daily. If the bookkeeper deposited \$787.50 on October 1 and withdrew \$400 on October 28, how much was left in the account on October 28? *(10 points)*

7. Compute the next balance and the interest for a money market account that pays 7.15% simple interest.

Date	Deposit	Withdrawal	Balance	Days	Interest
5/23			1465.25		
5/28		122.75	______	____	______ × ______ × ______ = ______

(10 points)

Use the following certificate of deposit interest rates for the next two exercises.

Period	Rate
2 years	8.8%
3–4 years	9.0%
5 years	9.25%

8. Robard Trophies purchased a three-year certificate of deposit with face value of $2,500. If interest is compounded annually, how much will the CD be worth at maturity? *(10 points)*

9. Ortega and Sons purchased a $4,000 two-year certificate of deposit. If interest is compounded semiannually, how much interest will be collected at maturity? *(15 points)*

10. Complete the following record for a NOW account that requires $1,000 minimum balance, pays 6.5% simple interest on the balances, allows free checking, and charges $10 per month.

Date	Deposit	Withdrawal	Charge	Balance	Interest
3/01				2300.75	
3/10		350.00		______	______ × ______ × ______ = ______
3/18		312.50CK		______	______ × ______ × ______ = ______
3/28		112.25CK		______	______ × ______ × ______ = ______
3/31	550.00		10.00	______	______ × ______ × ______ = ______
					Accumulated interest = ______

(40 points)

Chapter 28 Annuities, Present and Future Values

OBJECTIVES

- Determine the value of a series of equal payments at a future date
- Find the lump sum that would produce a series of equal payments

SECTIONS

INTRODUCTION

So far, you have observed only lump sum investments. An annuity consists of a series of equal payments at regular time intervals. While annuities are often associated with retirement plans, a series of regular payments to a bank for the purpose of owning a home (a mortgage) is also an annuity.

Section 28.1 The Future Value of an Ordinary Annuity

KEY TERMS **Simple annuity** **Annuity certain** **Ordinary annuity** **Future value of an annuity**

This chapter covers the **simple annuity** in which payment periods coincide with interest compounding periods.

Annuity with Six Periods

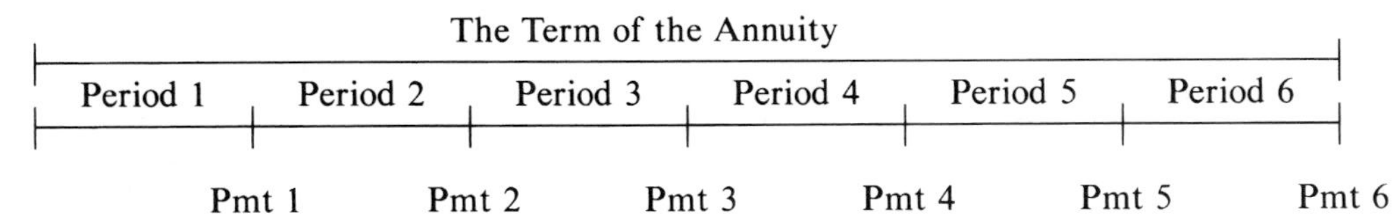

An annuity that begins and ends on known dates is an **annuity certain.** The interval between payments is the period. In an **ordinary annuity,** each payment is rendered at the end of the period. Although the technical term is simple, ordinary annuity certain, in this chapter, the term *annuity* suffices.

You may want to know the value of an annuity at the end of its term—the amount or the **future value of an annuity.** The amount of an annuity equals the sum of the payments plus accumulated interest.

Ordinary Annuity

\$250 was deposited at the end of each of 6 quarters.
Nominal interest rate = 8%.

Payment	Number of Periods Deposited	Compound Interest	Amounts
1	5	$250 \times 1.02^5 =$	276.02
2	4	$250 \times 1.02^4 =$	270.61
3	3	$250 \times 1.02^3 =$	265.30
4	2	$250 \times 1.02^2 =$	260.10
5	1	$250 \times 1.02 =$	255.00
6	0		250.00
		Amount or future value =	1577.03

The preceding procedure for determining future value is too cumbersome for a large number of periods. A simpler method follows. In this text the amount of an annuity with n periods is designated as $A(n)$.

$$A(n) = \text{payment} \times \frac{(1+i)^n - 1}{i}$$

Calculator Method

If you are not using a specialized financial calculator, you will require an exponentiation key. (Do not key parentheses with these formulas.)

Table Method

A table called the amount-of-1-per-period table shows the amount that $1 per period will grow to over n periods at periodic rate i. (See table 28.1.)

Apply this formula to the ordinary annuity on page 383.

$$A(6) = 250 \times \frac{1.02^6 - 1}{0.02}$$

key: 1.02 [y^x] 6 [−] 1 [=] [÷] .02 [×] 250 [=]
display: 1577.0302
edited: $1,577.03

Table method: for $i = 2\%$, $n = 6$, a 6-place amount-of-1-per-period table (table 28.1) lists 6.308121.

key: 250 [×] 6.308121 [=]
display: 1577.0302
edited: $1,577.03

Table 28.1 Amount-of-1-per-period

Periods *n*	$\frac{1}{2}$%	$\frac{2}{3}$%	$\frac{3}{4}$%	*Periodic Interest Rate* 1%	$1\frac{1}{2}$%	2%	3%
1	1.000000	1.000000	1.000000	1.000000	1.000000	1.000000	1.000000
2	2.005000	2.006667	2.007500	2.010000	2.015000	2.020000	2.030000
3	3.015025	3.020044	3.022556	3.030100	3.045225	3.060400	3.090900
4	4.030100	4.040178	4.045225	4.060401	4.090903	4.121608	4.183627
5	5.050251	5.067113	5.075565	5.101005	5.152267	5.204040	5.309136
6	6.075502	6.100893	6.113631	6.152015	6.229551	6.308121	6.468410
7	7.105879	7.141566	7.159484	7.213535	7.322994	7.434283	7.662462
8	8.141409	8.189176	8.213180	8.285671	8.432839	8.582969	8.892336
9	9.182116	9.243771	9.274779	9.368527	9.559332	9.754628	10.159106
10	10.228026	10.305396	10.344339	10.462213	10.702722	10.949721	11.463879
11	11.279167	11.374099	11.421922	11.566835	11.863262	12.168715	12.807796
12	12.335562	12.449926	12.507586	12.682503	13.041211	13.412090	14.192030
16	16.614230	16.825437	16.932282	17.257864	17.932370	18.639285	20.156881
18	18.785788	19.057191	19.194718	19.614748	20.489376	21.412312	23.414435
20	20.979115	21.318800	21.491219	22.019004	23.123667	24.297370	26.870374
24	25.431955	25.933190	26.188471	26.973465	28.633521	30.421862	34.426470
36	39.336105	40.535558	41.152716	43.076878	47.275969	51.994367	63.275944
48	54.097832	56.349915	57.520711	61.222608	69.565219	79.353519	104.408396
60	69.770031	73.476856	75.424137	81.669670	96.214652	114.051539	163.053437
120	163.879347	182.946035	193.514277	230.038689	331.288191	488.258152	1123.699571
240	462.040895	589.020416	667.886870	989.255365	2308.854370	5744.436758	40128.420931

Table 28.1 *(Continued)*

Periods *n*	*4%*	*5%*	*6%*	*Periodic Interest Rate* *7%*	*8%*	*9%*	*10%*
1	1.000000	1.000000	1.000000	1.000000	1.000000	1.000000	1.000000
2	2.040000	2.050000	2.060000	2.070000	2.080000	2.090000	2.100000
3	3.121600	3.152500	3.183600	3.214900	3.246400	3.278100	3.310000
4	4.246464	4.310125	4.374616	4.439943	4.506112	4.573129	4.641000
5	5.416323	5.525631	5.637093	5.750739	5.866601	5.984711	6.105100
6	6.632975	6.801913	6.975319	7.153291	7.335929	7.523335	7.715610
7	7.898294	8.142008	8.393838	8.654021	8.922803	9.200435	9.487171
8	9.214226	9.549109	9.897468	10.259803	10.636628	11.028474	11.435888
9	10.582795	11.026564	11.491316	11.977989	12.487558	13.021036	13.579477
10	12.006107	12.577893	13.180795	13.816448	14.486562	15.192930	15.937425
11	13.486351	14.206787	14.971643	15.783599	16.645487	17.560293	18.531167
12	15.025805	15.917127	16.869941	17.888451	18.977126	20.140720	21.384284
16	21.824531	23.657492	25.672528	27.888054	30.324283	33.003399	35.949730
18	25.645413	28.132385	30.905653	33.999033	37.450244	41.301338	45.599173
20	29.778079	33.065954	36.785591	40.995492	45.761964	51.160120	57.274999
24	39.082604	44.501999	50.815577	58.176671	66.764759	76.789813	88.497327
36	77.598314	95.836323	119.120867	148.913460	187.102148	236.124723	299.126805
48	139.263206	188.025393	256.564529	353.270093	490.132164	684.280411	960.172338
60	237.990685	353.583718	533.128181	813.520383	1253.213296	1944.792133	3034.816395

Example 1 For four years, a family deposited $500 at the end of each three-month period into an account that paid six percent quarterly compounded interest. How much was in the account at the end of the four years (measuring time from the beginning of the first period)?

Periodic rate $i = 6\% \div 4 = 1\frac{1}{2}\% = 0.015$

Number of periods $n = 4 \times 4 = 16$

$$A(16) = 500 \times \frac{1.015^{16} - 1}{0.015} = 8966.18$$

key: 1.015 [y^x] 16 [−] 1 [=] [÷] .015 [×] 500 [=]
display: 8966.1848
edited: $8,966.18

Table method: for $i = 1.5\%$, $n = 16$, table 28.1 lists 17.932370.

key: 500 [×] 17.932370 [=]
display: 8966.185
edited: $8,966.19

PRACTICE ASSIGNMENT

Plan: Unless the table method is specifically indicated, you may use either a calculator or table 28.1 to find the future value. When using the table, multiply payment × factor. The total interest is the difference between the amount of the annuity and the sum of the payments.

1. Use table 28.1 to find the following:

	Intervals	Term	Periodic Interest Rate	Amount-of-1 per-period
a.	Years	9 years	9%	________
b.	Months	8 months	1%	________
c.	Quarters	5 years	$1\frac{1}{2}\%$	________
d.	Half-years	10 years	4%	________
e.	Quarters	6 years	2%	________
f.	Months	3 years	$\frac{2}{3}\%$	________

2. Use the factors found in problem 1 to complete the following:

	Intervals	Term	Periodic Interest Rate	Payments	Future Value
a.	Years	9 years	9%	5,000.00	________
b.	Months	8 months	1%	500.00	________
c.	Quarters	5 years	$1\frac{1}{2}$%	1,500.00	________
d.	Half-years	10 years	4%	775.00	________
e.	Quarters	6 years	2%	832.25	________
f.	Months	3 years	$\frac{2}{3}$%	156.83	________

3. Compute the future value and the total interest for each of the following annuities:

	Payments	Intervals	Term	Periodic Interest Rate	Future Value	Total Interest
a.	125.00	Months	10 months	$\frac{1}{2}$%	________	________
b.	4,000.00	Years	3 years	10%	________	________
c.	835.50	Half-years	5 years	3%	________	________
d.	645.25	Quarters	9 years	2%	________	________
e.	917.50	Quarters	21 months	3%	________	________
f.	261.17	Months	$1\frac{1}{2}$ years	$\frac{3}{4}$%	________	________

4. A payment of $1,000 was made at the end of each 6-month period for a 10-year term, measuring time from the beginning of the first period. What single payment at the end of the 10 years would be equally acceptable if money earns 6% compounded semiannually?

5. Cindi Maxwell is a singer whose records are frequently at the top of the charts. To prepare for retirement, she puts $50,000 at the end of every six months into a fund that is expected to yield an interest rate of 8% compounded semiannually. How much (to the nearest thousand) will the fund contain if she contributes for nine years, measuring time from the beginning of the first six months?

Section **28.2** **The Present Value of an Ordinary Annuity**

The present value of an annuity represents the value of the annuity at the beginning of the term, the lump sum that, if deposited in the beginning of the term, would produce the annuity payments. It is the sum of the present values of its payments.

Present and Future Value of an Ordinary Annuity with Six Periods

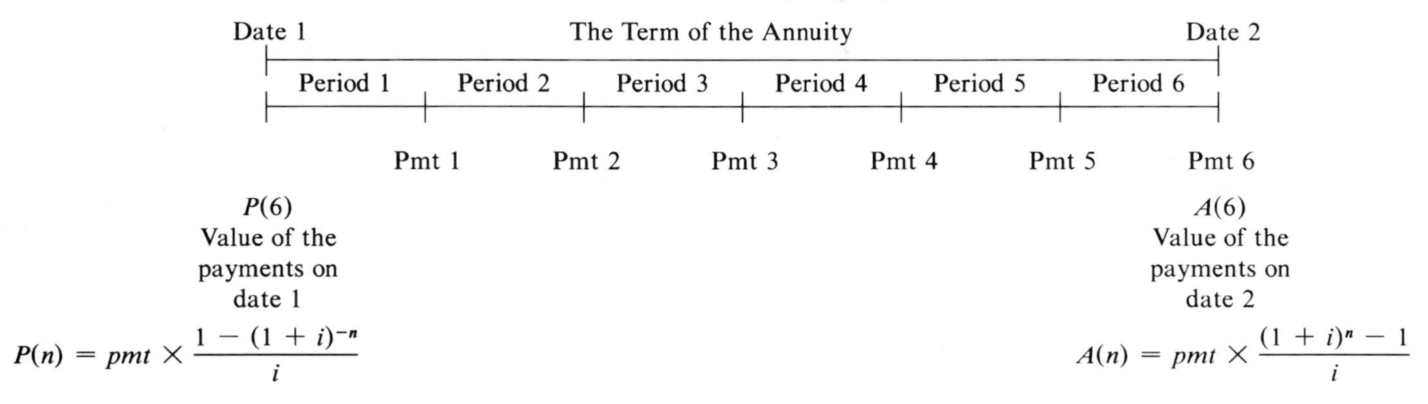

$$P(n) = pmt \times \frac{1 - (1 + i)^{-n}}{i} \qquad A(n) = pmt \times \frac{(1 + i)^{n} - 1}{i}$$

To find the present value, multiply the payment by the present-value-of-1-per-period. This factor is not the reciprocal of the amount-of-1-per-period. Rather, the negative exponent is a short way of designating a reciprocal.

$$(1 + i)^{-n} \text{ is the reciprocal of } (1 + i)^{n}.$$

That is,

$$(1 + i)^{-n} = \frac{1}{(1 + i)^{n}}$$

To compute the present value on a general calculator, it is advisable to use the change-of-sign key [+/−] as well as the exponential key. If these keys are not available, use a present-worth-of-1-per-period table (table 28.2).

Table 28.2 Present-worth-of-1-per-period

Periods	*Periodic Interest Rate*						
n	$\frac{1}{2}$%	$\frac{2}{3}$%	$\frac{3}{4}$%	1%	$1\frac{1}{2}$%	2%	3%
1	0.995025	0.993377	0.992556	0.990099	0.985222	0.980392	0.970874
2	1.985099	1.980176	1.977723	1.970395	1.955883	1.941561	1.913470
3	2.970248	2.960440	2.955556	2.940985	2.912200	2.883883	2.828611
4	3.950496	3.934212	3.926110	3.901966	3.854385	3.807729	3.717098
5	4.925866	4.901535	4.889440	4.853431	4.782645	4.713460	4.579707
6	5.896384	5.862452	5.845598	5.795476	5.697187	5.601431	5.417191
7	6.862074	6.817005	6.794638	6.728195	6.598214	6.471991	6.230283
8	7.822959	7.765237	7.736613	7.651678	7.485925	7.325481	7.019692
9	8.779064	8.707189	8.671576	8.566018	8.360517	8.162237	7.786109
10	9.730412	9.642903	9.599580	9.471305	9.222185	8.982585	8.530203
11	10.677027	10.572420	10.520675	10.367628	10.071118	9.786848	9.252624
12	11.618932	11.495782	11.434913	11.255077	10.907505	10.575341	9.954004
16	15.339925	15.128481	15.024313	14.717874	14.131264	13.577709	12.561102
18	17.172768	16.908944	16.779181	16.398269	15.672561	14.992031	13.753513
20	18.987419	18.665902	18.508020	18.045553	17.168639	16.351433	14.877475
24	22.562866	22.110544	21.889146	21.243387	20.030405	18.913926	16.935542
36	32.871016	31.911806	31.446805	30.107505	27.660684	25.488842	21.832252
48	42.580318	40.961913	40.184782	37.973959	34.042554	30.673120	25.266707
60	51.725561	49.318433	48.173374	44.955038	39.380269	34.760887	27.675564
120	90.073453	82.421481	78.941693	69.700522	55.498454	45.355389	32.373023
240	139.580772	119.554292	111.144954	90.819416	64.795732	49.568552	33.305667
300	155.206864	129.564523	119.161622	94.946551	65.900901	49.868502	33.328638
360	166.791614	136.283494	124.281866	97.218331	66.353242	49.959922	33.332536

Table 28.2 *(Continued)*

Periods *n*	*Periodic Interest Rate* 4%	5%	6%	7%	8%	9%	10%
1	0.961538	0.952381	0.943396	0.934579	0.925926	0.917431	0.909091
2	1.886095	1.859410	1.833393	1.808018	1.783265	1.759111	1.735537
3	2.775091	2.723248	2.673012	2.624316	2.577097	2.531295	2.486852
4	3.629895	3.545951	3.465106	3.387211	3.312127	3.239720	3.169865
5	4.451822	4.329477	4.212364	4.100197	3.992710	3.889651	3.790787
6	5.242137	5.075692	4.917324	4.766540	4.622880	4.485919	4.355261
7	6.002055	5.786373	5.582381	5.389289	5.206370	5.032953	4.868419
8	6.732745	6.463213	6.209794	5.971299	5.746639	5.534819	5.334926
9	7.435332	7.107822	6.801692	6.515232	6.246888	5.995247	5.759024
10	8.110896	7.721735	7.360087	7.023582	6.710081	6.417658	6.144567
11	8.760477	8.306414	7.886875	7.498674	7.138964	6.805191	6.495061
12	9.385074	8.863252	8.383844	7.942686	7.536078	7.160725	6.813692
16	11.652296	10.837770	10.105895	9.446649	8.851369	8.312558	7.823709
18	12.659297	11.689587	10.827603	10.059087	9.371887	8.755625	8.201412
20	13.590326	12.462210	11.469921	10.594014	9.818147	9.128546	8.513564
24	15.246963	13.798642	12.550358	11.469334	10.528758	9.706612	8.984744
36	18.908282	16.546852	14.620987	13.035208	11.717193	10.611763	9.676508
48	21.195131	18.077158	15.650027	13.730474	12.189136	10.933575	9.896926
60	22.623490	18.929290	16.161428	14.039181	12.376552	11.047991	9.967157

Example 1 Recall the family in example 1 of section 28.1. For four years they deposited $500 at the end of each 3-month period into an account that paid interest at 6 percent compounded quarterly. At the end of the four years, the value of the account (future value) was $8,966.18. What lump sum, deposited at the beginning of the four years (present value), would have resulted in the same amount (future value)?

Periodic rate $i = 6\% \div 4 = 1\frac{1}{2}\% = 0.015$
Number of periods $n = 4 \times 4 = 16$

$$P(16) = 500 \times \frac{1 - 1.015^{-16}}{0.015} = 7065.63$$

key: 1 [−] 1.015 [y^x] 16 [+/−] [=] [÷] .015 [×] 500 [=]
display: 7065.632
edited: $7,065.63

Table method: for $i = 1.5\%$, $n = 16$, a 6-place present-worth-of-1-per-period table (table 28.2) lists 14.131264.

key: 500 [×] 14.131264 [=]
display: 7065.632
edited: $7,065.63

PRACTICE ASSIGNMENT

Plan: Unless the table method is specifically indicated, you may use either the calculator or table 28.2 to find the present value. When using the table, multiply payment × table factor.

1. Use table 28.2 to find the following:

Intervals	**Term**	**Periodic Interest Rate**	**Present-value-of 1-per-period**
a. Years	7 years	9%	________
b. Months	8 months	1%	________
c. Half-years	6 years	3%	________
d. Years	20 years	6%	________
e. Quarters	18 months	2%	________
f. Months	$1\frac{1}{2}$ years	$\frac{1}{2}\%$	________

2. Use the factors found in problem 1 to complete the following:

	Intervals	Term	Periodic Interest Rate	Payments	Annuity Present Value
a.	Years	7 years	9%	5,600.00	______
b.	Months	8 months	1%	500.00	______
c.	Half-years	6 years	3%	3,825.00	______
d.	Years	20 years	6%	4,750.00	______
e.	Quarters	18 months	2%	1,055.00	______
f.	Months	$1\frac{1}{2}$ years	$\frac{1}{2}$%	262.25	______

3. Compute the present value for each of the following cases:

	Payments	Intervals	Term	Periodic Interest Rate	Present Value
a.	125.00	Months	10 months	$\frac{1}{2}$%	______
b.	1,515.00	Years	12 years	6%	______
c.	550.00	Quarters	5 years	$1\frac{1}{2}$%	______
d.	613.83	Months	10 years	$\frac{2}{3}$%	______
e.	2,040.00	Half-years	$5\frac{1}{2}$ years	5%	______
f.	953.50	Quarters	27 months	3%	______

4. Audrey Miller wants to buy a boat. If she can invest money at 6% interest compounded monthly, should she pay $15,000 cash or pay a $3,000 down payment plus $1,000 per month for 1 year?
Check one: $15,000 cash: _____
$3,000 down plus $1,000 monthly: _____

Section 28.3 The Annuity Due

An annuity that requires each payment at the beginning of the period rather than at the end is called an annuity due. The annuity due is not very different from the ordinary annuity.

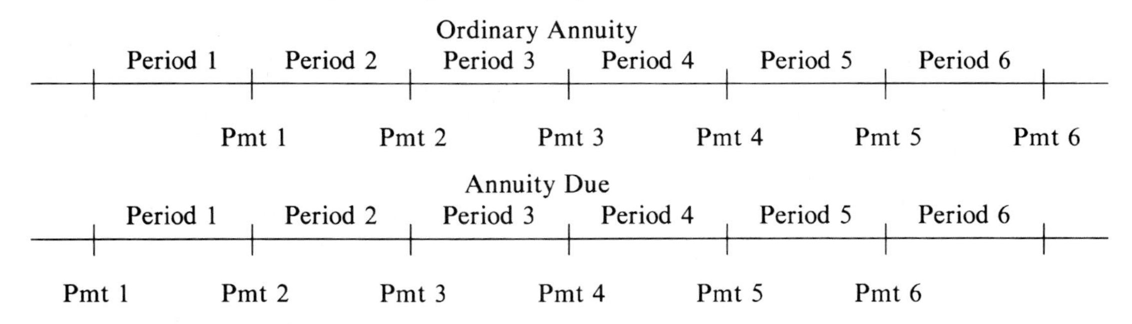

In an annuity due, each payment is rendered one period earlier than in an ordinary annuity. Therefore, each payment in an annuity due receives interest for one more period. Because the amount of each payment contains one more $(1 + i)$ factor, the future value of the annuity due is the future value of the ordinary annuity multiplied by $(1 + i)$.

$$A(n,\text{due}) = A(n,\text{ord}) \times (1 + i) = pmt \times \frac{(1 + i)^n - 1}{i} \times (1 + i)$$

When computing present value, each payment's present value contains one less $(1 + i)$ reciprocal. The present value of the annuity due is the present value of the ordinary annuity multiplied by $(1 + i)$. Other methods are more efficient for hand computation, but the preceding formula is easy to remember, and the extra multiplication can be performed conveniently with a calculator.

$$P(n,\text{due}) = P(n,\text{ord}) \times (1 + i)$$
$$= pmt \times \frac{1 - (1 + i)^{-n}}{i} \times (1 + i)$$

Deciding whether you need to find the present value or the future value can be difficult. The present value or the future value is a lump sum financially equivalent to the annuity payments. An equivalent value *before* the payments is the present value. An equivalent value *after* the payments is the future value.

Example 1 A consumer paid $65 at the beginning of each month for two years to retire a debt. If 18 percent interest compounded monthly was charged, what was the original debt? In this case you need to find the value *before* the payments, the present value.

$i = 0.18 \div 12 = 0.015$

$1 + i = 1.015$

$n = 12 \times 2 = 24$

$$P(24,\text{due}) = P(24,\text{ord}) \times 1.015$$
$$= 65 \times \frac{1 - 1.015^{-24}}{0.015} \times 1.015$$
$$= 1321.51$$

key: 1 [−] 1.015 [y^x] 24 [+/−] [=] [÷] .015 [×] 65 [×] 1.015 [=]
display: 1321.506
edited: $1,321.51

The present-worth-of-1-per-period table (table 28.2) for $i = 0.015$, $n = 24$ lists 20.030405.

key: 20.030405 [×] 65 [×] 1.015 [=]
display: 1321.506
edited: $1,321.51

Example 2 The proprietor of a small store deposited $1,200 at the beginning of each quarter into an investment plan paying 12 percent interest compounded quarterly. How much was the investment worth at the end of five years? Here you need to find the value *after* the payments, the future value.

$i = 0.12 \div 4 = 0.03$

$1 + i = 1.03$

$n = 4 \times 5 = 20$

$$A(20,\text{due}) = A(20,\text{ord}) \times 1.03$$
$$= 1200 \times \frac{1.03^{20} - 1}{0.03} \times 1.03$$
$$= 33211.78$$

key: 1.03 [y^x] 20 [−] 1 [=] [÷] .03 [×] 1200 [×] 1.03 [=]
display: 33211.783
edited: $33,211.78

The amount-of-1-per-period table (table 28.1) for $i = 0.03$, $n = 20$ lists 26.870374.

key: 26.870374 [×] 1200 [×] 1.03 [=]
display: 33211.782
edited: $33,211.78

PRACTICE ASSIGNMENT

Plan: In the following problems you must find either the present value or the future value of an annuity due.

To find the present value of an annuity due, compute the present value of the ordinary annuity and multiply it by $(1 + i)$.
To find the future value of an annuity due, compute the future value of the ordinary annuity and multiply it by $(1 + i)$.

If your calculator does not feature the exponentiation and change-of-sign keys, use the amount-of-1-per-period table (table 28.1) or the present-worth-of-1-per-period table (table 28.2) to find the appropriate factor, and multiply it by the payment and by $(1 + i)$.

1. The premium for a life insurance policy is $165 per quarter, payable at the beginning of each quarter. The annual premium, paid in advance, should be less than $4 \times 165 = 660$ because it represents the present value of a four-payment annuity due. What is the annual premium if the insurance company credits the policyholder with 6% interest compounded quarterly?

2. Jason Baumann invested $500 at the beginning of each year for 10 years. If the annually compounded interest rate was 8%, how much was his investment worth at the end of 10 years?

3. The owner of Skippy's Soft-Serve pays a premium of $1,535.62 per year at the beginning of each year for a comprehensive insurance policy. How much is the single premium payment for a five-year policy if the interest rate is 7%?

4. The McKennas are getting out of debt by paying $175 at the beginning of each month for the next three years, which includes 12% interest compounded monthly. How much would they save if they paid the debt now in one lump sum?

Challenger

In some circumstances, you may want to know how long it would take to accumulate a certain amount of money with an annuity. If you pay $100 at the end of each month into a fund that earns nine percent interest compounded monthly, how long will it take to obtain $10,000?

Without interest, it would take 100 months or $8\frac{1}{3}$ years, but the nine percent interest will reduce the time considerably. Use the following formula to find n, the number of periods (months in this case):

$$n = \frac{\log\left(1 + \dfrac{FV \times i}{pmt}\right)}{\log(1 + i)}$$

FV is future value.

Log(x) stands for logarithm base 10 (also called common logarithm).

The logarithm of a number is the exponent that raises the base to that number.

For example, $\log 1000 = 3$, because $10^3 = 1000$.

If your calculator has the [log] key, you can obtain the logarithm of x base 10 by keying x [log].

Assume payments of $100 at the end of each month into a fund earning nine percent interest compounded monthly. How long will it take to accumulate $10,000?

a. Write a proper keying sequence for the formula, and compute the number of periods.

b. The result is not a whole number. How much time in years and months will it take to accumulate $10,000?

CHAPTER 28 SUMMARY

Concept	*Example*	*Procedure*	*Formula**
SECTION 28.1 The Future Value of an Ordinary Annuity			
Ordinary annuity	Six monthly payments of $100, each rendered at the end of the month; each payment draws interest at 6% compounded monthly.	Any series of constant payments, each rendered at the end of the payment period, is an ordinary annuity. An annuity with payment periods that coincide with interest periods is a simple annuity.	
Future value of an ordinary annuity	In the preceding example, the value of the annuity at the end of the 6-month term is $100 \times \frac{1.005^6 - 1}{0.005} = 607.55$	The annuity's value at the end of its term is called the amount or future value. It is the sum of the payment amounts.	$A(n) = pmt \times \frac{(1+i)^n - 1}{i}$ 1+i[y^x]n[−]1[=][÷]i[×]pmt[=]* pmt∗((1+i)^n−1)/i
SECTION 28.2 The Present Value of an Ordinary Annuity			
Present value of an ordinary annuity	In the preceding example, the value of the annuity at the beginning of the 6-month term is $100 \times \frac{1 - 1.005^{-6}}{0.005} = 589.64$	The annuity's value at the beginning of its term is called the present value. It is the sum of the present values of the payments.	$P(n) = pmt \times \frac{1 - (1+i)^{-n}}{i}$ 1[−]1+i[y^x]n[+/−][=] [÷]i[×]pmt[=]* pmt∗(1−(1+i)^(−n))/i
SECTION 28.3 The Annuity Due			
Annuity due	Six monthly payments of $100, each rendered at the beginning of the month; each payment draws interest at 6% compounded monthly.	If each payment is rendered at the beginning of the period, the series is called an annuity due. Each payment draws interest for one additional period.	
Future value of an annuity due	In the preceding example, the value of the annuity at the end of the 6-month term is $100 \times \frac{1.005^6 - 1}{0.005} \times 1.005$ $= 610.59$	Because each payment draws interest for one additional period, the future value of the annuity due equals future value of ordinary annuity × (1 + i).	$A(n,\text{due}) = A(n,\text{ord}) \times (1+i)$ $= pmt \times \frac{(1+i)^n - 1}{i} \times (1+i)$ 1+i[y^x]n[−]1[=][÷] i[×]pmt[×]1+i[=]* pmt∗((1+i)^n−1)/i∗(1+i)
Present value of an annuity due	In the preceding example, the value of the annuity at the beginning of the 6-month term is $100 \times \frac{1 - 1.005^{-6}}{0.005} \times 1.005$ $= 592.59$	Because each payment draws interest for one additional period, the present value of the annuity due equals present value of ordinary annuity × (1 + i).	$P(n,\text{due}) = P(n,\text{ord}) \times (1+i)$ $= pmt \times \frac{1 - (1+i)^{-n}}{i} \times (1+i)$ 1[−]1+i[y^x]n[+/−][=][÷] i[×]pmt[×]1+i[=]* pmt∗(1−(1+i)^(−n))/i∗(1+i)

*In calculator sequences, 1+i without brackets or parentheses means that 1+i is computed mentally and entered; for example, 1+.06 is entered as 1.06.

Chapter 28 **Spreadsheet Exercise**

How Many Years Will it Take to Accumulate $100,000?

Should you save money? Economists encourage the habit, but the gradual decline of the dollar's purchasing power discourages saving. Most people consider $100,000 a significant amount of money. Can an individual accumulate $100,000 within a reasonable period of time?

This exercise introduces two more financial functions: CTERM and TERM. You can use these functions to help plan your future. The chapter 28 template appears as follows:

```
A1: PR [W10]                                                              READY

          A          B          C          D          E          F          G          H
 1                HOW MANY YEARS NEEDED TO ACCUMULATE              $ 100,000
 2   ----------------------------------------------------------------------------------
 3        MONTHLY COMPOUNDED INTEREST              ANNUITY OF MONTHLY PAYMENTS
 4   ----------------------------------------------------------------------------------
 5                     interest rates                                interest rates
 6   ----------------------------------------------------------------------------------
 7                   8.00%      9.00%     10.00%              8.00%      9.00%    10.00%
 8   ----------------------------------------------------------------------------------
 9       initial                                   monthly
10     investment                                  payments
11
12          100                                        100
13          250                                        150
14          500                                        200
15         1000                                        250
16         2000                                        300
17   ----------------------------------------------------------------------------------
18
19
20
```

Cell G1 contains the future value. Row 7 contains interest rates.

Column A contains present values to be invested at monthly compounded interest. Column E contains the monthly annuity payments that accumulate the future value at monthly compounded interest. All of these figures can be varied.

Spreadsheet Exercise Steps

1. Column B: The Number of Years Needed to Accumulate the Future Value with an Initial Investment and Monthly Compounded Interest

   @CTERM computes the number of compounding periods required for a present value to grow to a future value at compound interest. The general form is:

   @ CTERM(periodic rate,future value,present value)

   The annual rate is in B7 (absolute row address). Divide by 12 to obtain the monthly rate.
   The future value is in G1 (absolute row and column address).
   The present value is in A12 (absolute column address).
   The CTERM function computes the number of periods. Divide by 12 to obtain the number of years.
   The spreadsheet formula is
   @ CTERM(_____ $ _____ / _____ , $ _____ $ _____ , $ _______)/ _____ .

   Complete the formula, store it in B12, and copy it from B12 to B12..D16.

2. Column F: The Number of Years Needed to Accumulate the Future Value with Monthly Payments and Monthly Compounded Interest

@TERM computes the number of payment periods required for an ordinary annuity to accumulate a future value.
The general form is:

@TERM(payment,periodic rate,future value)

The payment is in E12 (absolute column address).
The annual rate is in F7 (absolute row address).
The future value is in G1 (absolute row and column address).
The TERM function computes the number of periods.
Divide by 12 to obtain the number of years.
The spreadsheet formula is
@TERM($ _______ , _____ $ _____ /
______ , $ _____ $ _____)/ ______ .
Complete the formula, store it in F12, and copy it from F12 to F12..H16.

3. Compound interest is not enough to accumulate $100,000 within a moderate time period. If you can contribute monthly payments at a high interest rate, the number of years required to accumulate $100,000 is reasonably small.

a. How many years will it take for $1,000 to grow to $100,000 at 10% interest compounded monthly?

b. If you start saving $150 per month at age 25 and earn 9% interest compounded monthly, how old will you be when you have accumulated $100,000?

c. Change the annuity interest rate from 10% to 12% (decimal form) and the future value to $50,000. If you start saving $300 per month at age 25 and earn 12% interest compounded monthly, how old will you be when you have accumulated $50,000?

NAME ______________________________

CLASS SECTION ____________ DATE ____________

SCORE ________ MAXIMUM __100__ PERCENT ________

CHAPTER 28 ASSIGNMENT

1. Suppose you invest $250 at the end of each month. How much will your investment be worth after 5 years, calculating from the beginning of the first month at 8% interest compounded monthly? ***(5 points)***

2. Leslie Perez invests $125 per month at 6% interest compounded monthly. How much interest will be accumulated after 10 years, measuring time from the beginning of the first month? ***(10 points)***

3. John Arroya received a student loan of $1,250 every six months for four years. What was the cash value of the loan if money earned 6% interest compounded semiannually? ***(5 points)***

4. If Susan Giordano deposits $55 at the beginning of each month into an account paying 6% interest compounded monthly, how much will she have after five years? ***(5 points)***

5. A stereo set may be purchased with a $100 cash down payment and $34.58 per month for 12 months. What is the equivalent cash price if money earns 18% interest compounded monthly? ***(10 points)***

6. The beneficiary of an inheritance is entitled to $635 per month for ten years payable at the beginning of each month or an equivalent lump sum payment immediately. What is the lump sum payment (to the nearest dollar) if money earns 8% interest compounded monthly? ***(10 points)***

7. The Swensens purchased a home with a down payment of $10,000 and a 30-year fixed rate mortgage with monthly payments of $632.74. What was the cash price of the home (to the nearest $100) if the lending institution receives 12% interest compounded monthly? ***(10 points)***

8. The monthly premium on an automobile insurance policy is $103.15 payable at the beginning of each month. The policyholder may choose to pay an annual premium at the beginning of the year. What is the annual premium based on 6% interest compounded monthly? ***(5 points)***

9. When Danny Kovacs was 10 years old, he received an inheritance of $1,500 per year with the stipulation that the money be deposited into an annuity until he reached the age of 21. The first deposit was made on his 10th birthday, and the last one on his 21st birthday. How much was in the fund immediately after his 21st birthday if the annually compounded interest rate was 7%? ***(10 points)***

10. A store in a new suburban mall can be rented for $925 per month payable at the beginning of each month. The developers will credit 12% interest compounded monthly if a two-year lease is signed and the entire rent is paid in a lump sum in advance. How much should be paid in advance? ***(5 points)***

11. TCM, Inc. bought out the National Furniture chain and agreed to pay National Furniture's president $30,000 at the end of each of the next six years if he agreed to retire immediately. What is the equivalent lump sum cash payment (to the nearest $1,000), assuming that money earns 7% interest compounded annually? ***(10 points)***

12. Morgan Strand receives income of $225 every three months from bonds. He deposits this income into an annuity at 6% interest compounded quarterly. What will the value of this fund be on March 31, 1995, if the first deposit was made on December 31, 1990, the last deposit was made on March 31, 1995, and the interest was credited on March 31, June 30, September 30, and December 31? ***(15 points)***

Chapter 29 Amortization and Sinking Funds

OBJECTIVES

- Determine annuity payments used to retire a loan
- Become acquainted with the real estate mortgage
- Determine annuity payments for a sinking fund

SECTIONS

INTRODUCTION

In this chapter you will become familiar with amortization and sinking funds. Mortgages are a common form of long-term borrowing that allow consumers to acquire and use relatively expensive items, such as homes and cars, while paying for them over an extended period of time in prearranged amounts at specified intervals. Amortization is the process of repaying long-term debt with equal, regular payments.

The sinking fund is an alternative to borrowing, whereby a business or consumer anticipating a major purchase determines how much money will be needed by a certain date for that purpose. Money is *sunk* into the fund in prearranged amounts at specified intervals. At the end of the time period, the fund is complete and the money is available.

Section 29.1 Finding the Payment Given the Present Value

KEY TERM **Amortization**

What Regular Equal Payments Are Necessary to Retire a Long-Term Debt?

Suppose you have a five-year auto loan of $25,000 at 12 percent interest compounded monthly. How much are your monthly payments? Remember from chapter 28 that a loan repaid in equal payments over a period of time is an ordinary annuity, so the loan is the present value of the annuity. The annuity concept provides many mathematical and computational shortcuts. To determine the size of the payments, use the inverted present or future value formula. In this case, you know that the present value is $25,000.

$$\text{Payment} = P(n) \times \frac{i}{1 - (1 + i)^{-n}}$$

If a calculator with the required keys is not available, you can obtain the size of the payments by dividing the loan (present value) by the present-worth-of-1-per-period factor. (If tabulated, the factors are merely the reciprocals of the present-worth-of-1-per-period factors.)

Example 1 The price of a new car is $18,700. The purchaser pays $1,700 in cash and secures a three-year loan at six percent interest compounded monthly for the remaining $17,000. What are the monthly payments?

$$\text{Loan} = \text{price} - \text{down payment} = 18700 - 1700 = 17000$$

Periodic rate $i = \frac{1}{2}\% = 0.005$

Number of periods $= 3 \times 12 = 36$

$$\text{Payment} = P(n) \times \frac{i}{1 - (1 + i)^{-n}}$$

$$= 17000 \times \frac{0.005}{1 - 1.005^{-36}} = 517.17$$

key: 1 − 1.005 [y^x] 36 [+/−] [=] [M+]
17000 [×] .005 [÷] [MR] [=]
display: 517.17294
edited: $517.17

The present-value-of-1-per-period in table 28.2 under $i = \frac{1}{2}\%$, $n = 36$ is 32.871016.

key: 17000 [÷] 32.871016 [=]
display: 517.17294
edited: $517.17

PRACTICE ASSIGNMENT

Plan: In the following problems, the loan is the present value of an annuity. Find the size of the payments that corresponds to the given present value, interest rate, term, and periods. In the case of mortgages, the periods are usually months, and the interest is assumed to be compounded monthly.

Unless otherwise indicated, compute with a calculator or use table 28.2. If your calculator does not feature the exponential and change-of-sign keys, use table 28.2. Divide the present value by the appropriate factor.

1. Use table 28.2 to find the present-value-of-1-per-period.

Period	Term	Periodic Interest Rate	Present-Value-of-1-per-period
a. Months	8 months	1%	______
b. Quarters	5 years	$1\frac{1}{2}\%$	______
c. Half-years	9 years	3%	______
d. Years	20 years	6%	______
e. Quarters	6 years	2%	______
f. Months	$1\frac{1}{2}$ years	$\frac{1}{2}\%$	______

2. Use the factors found in problem 1 to complete the following.

Period	Term	Periodic Interest Rate	Loan	Payment Rounded to Cent
a. Months	8 months	1%	6,500	______
b. Quarters	5 years	$1\frac{1}{2}\%$	14,825	______
c. Half-years	9 years	3%	42,000	______
d. Years	20 years	6%	55,000	______
e. Quarters	6 years	2%	29,500	______
f. Months	$1\frac{1}{2}$ years	$\frac{1}{2}\%$	7,250	______

3. Compute the periodic payments for each of the following debts.

	Contracted Debt	Down Payment	Payment Period and Compounding Period	Term	Nominal Interest Rate	Periodic Payment Rounded to Cent
a.	5,400	400	Monthly	1 year	6%	______
b.	5,400	400	Quarterly	1 year	6%	______
c.	22,800	800	Monthly	3 years	12%	______
d.	22,800	800	Monthly	3 years	9%	______
e.	17,250	1,000	Quarterly	5 years	12%	______
f.	6,750	750	Quarterly	$4\frac{1}{2}$ years	8%	______

4. A couple purchased a mobile home at a cost of \$25,000. They put \$2,500 down and financed the remainder over 10 years with a 12% mortgage. What are the monthly payments?

Section 29.2 The Fixed-Rate Mortgage

KEY TERMS **Mortgage** **Closing costs**

How Is Amortization Different from Paying Off an Installment Plan?

By now you may have noticed that amortizing a loan seems similar to paying off an installment plan.

Recall that one approach to paying off an installment plan is to pay an equal portion of the principal plus interest on the unpaid balance each month. This method, however, generates unequal payments.

Amortization equalizes the payments. A portion of each payment pays interest, and the remaining portion pays off the principal, reducing the principal balance. With each payment period, the interest portion of the payment decreases while the principal portion of the payment increases. Although these portions change, the payments remain the same through the term of an amortized loan.

Compare the installment plan with amortization. A combination washer-drier costing \$840 is purchased with \$40 down. The remaining \$800 will be paid in eight monthly payments at eight percent interest.

Interest on the Unpaid Balance

Monthly payment on principal = 800 ÷ 8 = 100
Interest = balance × 0.08 ÷ 12
Payment = payment on principal + interest

End of	Balance	Interest	Payment	New Balance
Month 1	800	5.33	105.33	700
Month 2	700	4.67	104.67	600
Month 3	600	4.00	104.00	500
Month 4	500	3.33	103.33	400
Month 5	400	2.67	102.67	300
Month 6	300	2.00	102.00	200
Month 7	200	1.33	101.33	100
Month 8	100	0.67	100.67	0
Total interest		24.00		

Amortization

$$\text{Payment} = 800 \times \frac{0.08/12}{1-(1+0.08/12)^{-8}} = 103.02$$

Interest = balance × 0.08 ÷ 12

Balance	Payment	Interest	Payment on Principal	New Balance
800.00	103.02	5.33	97.69	702.31
702.31	103.02	4.68	98.34	603.97
603.97	103.02	4.03	98.99	504.98
504.98	103.02	3.37	99.65	405.33
405.33	103.02	2.70	100.32	305.01
305.01	103.02	2.03	100.99	204.02
204.02	103.02	1.36	101.66	102.36
102.36	103.04	0.68	102.36	0.00
Total interest		24.18		

Notice that in the amortization schedule on the right, the last payment was increased by $0.02 to compensate for rounding.

Although the simple interest formula is applied each month to calculate how much of the payment is for interest, the payments include compound interest appropriate for long-term debt. The total finance charge for the amortization is higher than that for the installment plan, but the nominal interest rate is still eight percent. Equal payments are convenient, but balance reduction on the principal is small during the early years of the loan (an advantage to the lender).

Example 1 A loan of $5,000 is amortized in 10 months at nine percent.

$$I = 3/4\% = 0.0075$$

$$n = 10$$

$$\text{Payment} = 5000 \times \frac{0.0075}{1 - 1.0075^{-10}}$$

$$= 520.86$$

key: 1 − 1.0075 [y^x] 10 [+/−] [=] [M+]
5000 [×] .0075 [÷] [MR] [=]
display: 520.85614
edited: $520.86

The present-value-of-1-per-period in table 28.2 under $i = 3/4\%$, $n = 10$ is 9.599580.

key: 5000 [÷] 9.59958 [=]
display: 520.85612
edited: $520.86

The amortization schedule follows.

Payment Number	Balance	Payment	Interest	Payment on Principal	New Balance
0					5,000.00
1	5,000.00	520.86	37.50	483.36	4,516.64
2	4,516.64	520.86	33.87	486.99	4,029.65
3	4,029.65	520.86	30.22	490.64	3,539.01
4	3,539.01	520.86	26.54	494.32	3,044.69
5	3,044.69	520.86	22.84	498.02	2,546.67
6	2,546.67	520.86	19.10	501.76	2,044.91
7	2,044.91	520.86	15.34	505.52	1,539.39
8	1,539.39	520.86	11.55	509.31	1,030.08
9	1,030.08	520.86	7.73	513.13	516.95
10	516.95	520.83	3.88	516.95	0.00

Most people who buy a home do so with a mortgage loan. A **mortgage** is an amortized loan that a bank or trust company makes on a house, condominium, or other real estate. Besides equal payments, other features distinguish the mortgage from the installment loan:

1. The mortgage is a long-term debt, which is often, but not exclusively, incurred through the purchase of real estate (commercial or private).
2. The mortgage is secured by the property for which the mortgage was written. A mortgaged home, for example, is pledged to the lending institution as security and remains pledged until the loan is terminated. If the borrower fails to meet the terms of the mortgage, the lender has the right to take possession of the home (foreclosure).
3. A mortgage requires initiation expenses called **closing costs.** These costs are paid at initiation of the loan and are separate from the mortgage payments. Closing costs for a real estate mortgage may be flat fees or percent of the loan.

Examples of Closing Costs

- Property appraisal
- Property inspection
- Property surveyance
- Title search and title insurance (title is legal proof of ownership)
- Attorney's fees
- Lender's fees
- Origination fee
- "Points" are percent on the loan and are charged before the loan is obtained. For example, 2 points = 2%.

With increasing urbanization, condominiums have become popular dwellings in cities and suburbs. A condo is an apartment unit purchased by the resident or an investor. The condominium combines the advantages of home ownership with the convenience of apartment dwelling. The buyer obtains a mortgage, takes advantage of federal and state tax interest deductions, and pays local real estate taxes. The owner pays a separate association fee that covers exterior maintenance, lawn and pool care, and so forth. A condominium owner has the right to sell the unit or remodel it (with limitations). Many condominium communities have shared facilities that may include swimming pools, tennis courts, and club houses.

PRACTICE ASSIGNMENT

Plan: Mortgage payments are the payments of an ordinary annuity; the mortgage loan constitutes the present value. The computation is the same as in section 29.1. The amortization schedule shows the interest portion of the payment, the principal portion of the payment, and the resulting new balance. Assume monthly payments. The interest portion of the payment is computed with the simple interest formula:

$$I = PRT = \text{balance} \times \text{interest rate} \div 12$$

The principal portion of the payment is payment − interest.

$$\text{New balance} = \text{previous balance} - \text{payment on principal}$$

If necessary, adjust the last payment to compensate for rounding. Closing costs are flat fees or percent of the loan.

1. Compute the payments on a mortgage loan of $70,000 under the following conditions:

Term	Interest Rate	Payment
a. 20 years	9%	$________
b. 25 years	9%	________
c. 30 years	9%	________
d. 20 years	12%	________
e. 25 years	12%	________
f. 30 years	12%	________

2. Compute the payment and show the amortization schedule for a 1-year loan of $10,000 at 8% interest.
Payment: ________

Payment Number	Balance	Interest	Payment	Payment on Principal	New Balance
0					10,000.00
1	10,000.00				
2					
3					
4					
5					
6					
7					
8					
9					
10					
11					
12					0.00

Problems 3 through 5 pertain to the purchase of a home for $103,500.

3. The down payment was 15% of the purchase price, rounded upward to the next hundred. The remainder was financed with a 25-year mortgage at 9% interest.

   a. How much was the down payment?

   b. How much was the mortgage?

4. Closing costs included two points to the mortgage company. How much did the buyer pay on points?

5. The home buyer incurred the following additional closing costs:

Title search and insurance	$145.00
Termite inspection	112.00
Surveyor's fee	35.00
Attorney's fee	55.00
Title stamps	5.00
Prorated property insurance	436.35

   What were the total closing costs, including points?

Section **29.3 The Variable Rate Mortgage**

The traditional real estate mortgage features a fixed interest rate and fixed payments. In times of financial stability, this arrangement is sensible to both lender and borrower.

When interest rates rose precipitously in the late 1970s, banks were caught with low-rate long-term mortgages while compelled to pay high rates to depositors. As a response, banks began to offer mortgages with rates subject to change, the variable rate mortgage.

The rates are sometimes pegged to a bench mark figure, such as a Federal Reserve rate or the rate quoted for certain U.S. Treasury securities. Sometimes, the lender guarantees that the rate will not rise beyond a certain figure. A rate ceiling protects the borrower from steep rate increases. A variable rate mortgage without a rate ceiling could cause financial difficulties for a borrower in the face of rapidly rising rates.

Lenders notify borrowers of rate changes and the consequent changes in the monthly payments. Suppose the rate on a 20-year variable rate mortgage changes after two years from 11 percent to 11.25 percent. The borrower can check the accuracy of the new payments by computing the payments of a new annuity.

Step 1
Compute the outstanding principal prior to the rate change.

Balance = value of loan − value of rendered payments

Step 2
Find the annuity payments on the balance computed in step 1 with the new rate for the remaining periods.

Example 1 The rate on a 20-year $50,000 variable rate mortgage changed from 11 percent to 11.25 percent after two years. The payments had been $516.09.

Step 1: Compute the outstanding principal.

Value of loan − value of payments
= compound interest amount − annuity amount

(at old rate for elapsed periods).

Old $i = 0.11 \div 12 = 0.0091666$

Elapsed $n = 2 \times 12 = 24$

$$\text{Loan} \times (1+i)^n = 50000 \times 1.0091666^{24} = 62241.33$$

$$\text{Payment} \times \frac{(1+i)^n - 1}{i} = 516.09 \times \frac{1.0091666^{24} - 1}{0.0091666} = 13784.01$$

$$\text{Outstanding principal} = 48457.32$$

key: 50000 [×] 1.0091666 [y^x] 24 [=] [M+]
1.0091666 [y^x] 24 [−] 1 [=]
[÷] .0091666 [×] 516.09 [=] [M−] [MR]
display: 48457.314
rounded: 48457.31
Clear M-memory.

Step 2: The balance computed in step 1 is the new mortgage loan. Compute the new payments at the new rate for the remaining periods.

New $i = 0.1125 \div 12 = 0.009375$

Remaining $n = 240 - 24 = 216$

$$\text{Payment} = P(n) \times \frac{i}{1 - (1 + i)^{-n}}$$

$$= 48457.31 \times \frac{0.009375}{1 - 1.009375^{-216}}$$

key: 1 [−] 1.009375 [y^x] 216 [+/−] [=] [M+]
48457.31 [×] .009375 [÷] [MR] [=]
display: 524.12413
edited: $524.12

Conclusion: If the interest rate of a 20-year $50,000 mortgage rises from 11 percent to 11.25 percent after two years, the payments increase from $516.09 to $524.12 for the next 18 years (assuming the rate will not change again).

PRACTICE ASSIGNMENT

Plan: This assignment requires a calculator that features both exponential and change-of-sign keys. Compute the new payments for a variable rate mortgage after the rate has changed.

Step 1: Compute the outstanding principal:

Value of loan − value of payments = compound interest amount − annuity amount

(at the old rate for the number of elapsed periods).

Step 2: Compute the new payments for the new loan (the outstanding principal) at the new rate for the number of remaining periods.

1. Compute the old payments by the method of sections 29.1 and 29.2.

	Mortgage	Term in Years	Rate	Old Payments
a.	40,000	20	8.5%	________
b.	45,000	20	8.5%	________
c.	50,000	25	8.6%	________
d.	50,000	25	8.7%	________
e.	85,000	30	8.8%	________

2. Use the information from problem 1 to compute the outstanding principals after three years have elapsed.

	Mortgage	Term in Years	Rate	Old Payments	Compound Amount	Annuity Amount	Outstanding Principal
a.	40,000	20	8.5%	________	________	________	________
b.	45,000	20	8.5%	________	________	________	________
c.	50,000	25	8.6%	________	________	________	________
d.	50,000	25	8.7%	________	________	________	________
e.	85,000	30	8.8%	________	________	________	________

3. Use the information from problem 2 to compute the new payments after three years have elapsed.

Mortgage	Term in Years	Remaining Months	Old Payments	Outstanding Principal	New Rate	New Payments
a. 40,000	20	______	______	______	8.65%	______
b. 45,000	20	______	______	______	8.65%	______
c. 50,000	25	______	______	______	8.80%	______
d. 50,000	25	______	______	______	9.10%	______
e. 85,000	30	______	______	______	9.10%	______

4. A variable rate 25-year mortgage for $55,000 at 9.35% interest changed to 9.54% after two years. The old payments were $474.81.

a. What was the outstanding principal?

b. What are the new payments?

Section 29.4 Sinking Funds

What Equal Regular Payments Are Necessary to Acquire a Specified Fund?

Instead of borrowing, an organization or individual may accumulate a sum of money needed at some future date. Money amassed in this way is called a sinking fund. Businesses may use sinking funds to replace worn out equipment or to purchase new machinery. In some circumstances, it is advantageous to retire a loan with a sinking fund rather than prior to maturity.

The term *sinking fund* can be misleading. It does not mean sinking in the sense of decreasing (in fact, the fund is increasing); it means that a business or individual is sinking money into the fund.

A sinking fund schedule displays what is in the fund. It shows the growth of the fund by listing the payments, the credited interest, and the accumulated fund at the end of each period. Each periodic simple interest is computed on the previous balance.

As the money in the fund increases, the credited interest increases. The fund grows faster each period than during the previous period until the objective has been reached.

A sinking fund of $10,000 is accumulated in five years and earns 8% interest compounded semiannually.
Semiannual payment = $832.91

Payment Number	Payment	Interest	Accumulated Fund
0			0.00
1	832.91	0.00	832.91
2	832.91	33.32	1,699.14
3	832.91	67.97	2,600.02
4	832.91	104.00	3,536.93
5	832.91	141.48	4,511.32
6	832.91	180.45	5,524.68
7	832.91	220.99	6,578.58
8	832.91	263.14	7,674.63
9	832.91	306.99	8,814.53
10	832.89	352.58	10,000.00

To find the regular equal payments needed to accumulate a certain fund (the objective) the annuity concept again provides a convenient computational solution. It is similar to the amortization payment solution, except that in this case the future value is known rather than the present value.

$$\text{Payment} = A(n) \times \frac{i}{(1+i)^n - 1}$$

Example 1 A small business plans to purchase new computers for $18,700. The sinking fund pays 8% interest compounded quarterly. How much must be deposited into the fund each quarter to reach the purchase price in two years?

$i = 8\% \div 4 = 2\% = 0.02$

$n = 2 \times 4 = 8$

$$\begin{aligned}\text{Payment} &= A(n) \times \frac{i}{(1+i)^n - 1}\\ &= 18700 \times \frac{0.02}{1.02^8 - 1}\\ &= 2178.73\end{aligned}$$

key: 1.02 [y^x] 8 [−] 1 [=] [M+]
.02 [÷] [MR] [×] 18700 [=]
display: 2178.7333
edited: $2,178.73

If a calculator with an exponentiation key is not available, divide the future value by the amount-of-1-per-period factor in table 28.1. Entry in table 28.1 at $i = 2\%$, $n = 8$, is 8.582969.

key: 18700 [÷] 8.582969 [=]
display: 2178.7333
edited: $2,178.73

Sinking Fund Schedule

Payment Number	Payment	Interest	Accumulated Fund
0			0.00
1	2,178.73	0.00	2,178.73
2	2,178.73	43.57	4,401.03
3	2,178.73	88.02	6,667.78
4	2,178.73	133.36	8,979.87
5	2,178.73	179.60	11,338.20
6	2,178.73	226.76	13,743.69
7	2,178.73	274.87	16,197.29
8	2,178.76	323.95	18,700.00

PRACTICE ASSIGNMENT

Plan: Calculate the sinking fund payments and draw up the sinking fund schedule. If your calculator does not feature the required keys, use table 28.1. The schedule shows the payments, interest, and accumulated fund. The fund begins with $0. There is no interest at the end of the first period because the payment is not rendered until the end of the period. After the first period, each interest is computed with the simple interest formula on the previous balance and is added to the fund. After the last payment, the fund should contain the objective. If this figure is not exact due to rounding, adjust the last payment to reach the objective.

1. Compute the periodic payments for each of the following sinking funds.

	Sinking Fund Objective	Period	Term	Nominal Interest Rate	Payments
a.	12,500	Monthly	1 year	6%	________
b.	12,500	Quarterly	1 year	6%	________
c.	35,000	Yearly	4 years	8%	________
d.	20,000	Monthly	$1\frac{1}{2}$ years	9%	________
e.	27,500	Quarterly	2 years	8%	________
f.	25,000	Quarterly	$2\frac{1}{2}$ years	6%	________

2. Langley County plans to retire its bonded indebtedness of $100,000 by depositing a sufficient amount into a sinking fund at the end of each year for the next 10 years. How much is the annual sinking fund payment if the money draws interest at 10% compounded annually?

3. Draw up the sinking fund schedule for problem 2.

Payment Number	Payment	Interest	Accumulated Fund
0			0.00
1	______	0.00	______
2	______	______	______
3	______	______	______
4	______	______	______
5	______	______	______
6	______	______	______
7	______	______	______
8	______	______	______
9	______	______	______
10	______	______	______

4. Best-Foot Carpet Service plans to purchase new equipment in three years with a sinking fund that earns interest at 6% compounded quarterly. The equipment costs $11,250 but is expected to increase in price at about 4% per year. How much are the quarterly sinking fund payments?

Challenger

Robby Cranston is a major league baseball player. His financial counselor has located a reliable investment fund for professional athletes that guarantees a minimum average annual yield of seven percent barring severe economic depression or war. Cranston is in his twenties and expects to retire from professional baseball at age 35.

At age 25, he will begin depositing a certain amount each year for ten years. At age 35, the accumulated amount will remain in the fund (earning at least the guaranteed return) until he reaches 65. If he does not survive to age 65, the money will be paid to a beneficiary.

What should Cranston's ten annual payments be (to the nearest dollar) in order for him to receive $5,000,000 at age 65?

CHAPTER 29 SUMMARY

Concept	*Example*	*Procedure*	*Formula**

SECTION 29.1 Finding the Payment Given the Present Value

Concept: Amortization

Example: A $25,000 debt is to be paid off in 5 years with 12% interest.
Monthly amortization payment

$$= 25000 \times \frac{0.01}{1 - 1.01^{-60}}$$

$= 556.11$

Procedure: The buyer borrows money from a financial institution, takes possession of the purchase, and repays the loan with a series of regular equal payments. The loan is the present value of an ordinary annuity.

Formula:

$$\text{Payment} = \text{loan} \times \frac{i}{1 - (1 + i)^{-n}}$$

1[−]1+i[y^x]n[+/−][=][M+] loan[×]i[÷][MR][=]*

loan*i/(1−(1+i)^(−n))

SECTION 29.2 The Fixed-Rate Mortgage

Concept: Fixed-rate mortgage

Example: 10-month amortization schedule for debt of $5,000 at 12%.

Pmt.	Int.	Payment on Principal	Balance
			5000.00
527.91	50.00	477.91	4522.09
527.91	45.22	482.69	4039.40
527.91	40.39	487.52	3551.88
527.91	35.52	492.39	3059.49
527.91	30.59	497.32	2562.17
527.91	25.62	502.29	2059.88
527.91	20.60	507.31	1552.57
527.91	15.53	512.38	1040.19
527.91	10.40	517.51	522.68
527.91	5.23	522.68	0.00

Procedure: A mortgage is a long-term debt usually arranged to finance the purchase of real estate. The interest rate and the payments remain the same throughout the term. The portion of the monthly payment for interest is computed as simple interest; the remainder of the payment reduces the principal balance.

Formula: Portion of payment for interest = previous balance × i
Principal balance reduction = payment − interest
New balance = previous balance − principal balance reduction
(The formula for payment is the same as that in 29.1.)

SECTION 29.3 The Variable Rate Mortgage

Concept: Variable rate mortgage

Example: The rate for a 20-year $50,000 mortgage changes from 11% to 11.25% after 2 years.
Monthly payments change from $516.09 to $524.12.

Procedure: The interest rate is pegged to a bench mark figure and may change every year.
Payments are computed with the new rate on the outstanding balance for the remainder of the term.

Formula: Outstanding balance = compound amount − annuity amount
(The formula for payment is the same as that in 29.1.)

SECTION 29.4 Sinking Funds

Concept: Sinking fund

Example: A sinking fund of $10,000 paid during 5 years earns 8% compounded semiannually.
Semiannual payment

$$= 10000 \times \frac{0.04}{1.04^{10} - 1}$$

$= 832.91$

Payment	Interest	Fund
		0.00
832.91	0.00	832.91
832.91	33.32	1699.14
832.91	67.97	2600.02
832.91	104.00	3536.93
832.91	141.48	4511.32
832.91	180.45	5524.68
832.91	220.99	6578.58
832.91	263.14	7674.63
832.91	306.99	8814.53
832.89	352.58	10000.00

Procedure: An organization or an individual may systematically accumulate a sum of money needed at some future date. The fund earns interest on the balances.
The amount to be accumulated, the objective, is the future value of an annuity.

Formula:

$$\text{Payment} = \text{objective} \times \frac{i}{(1 + i)^n - 1}$$

1+i[y^x]n[−]1[=][M+] objective[×]i[÷][MR][=]*

obj*i/((1+i)^n−1)

Interest = previous balance × i
Accumulation = previous balance + payment + interest

*In calculator sequences, 1+i without brackets or parentheses means that 1+i is computed mentally and entered; for example, 1+.06 is entered as 1.06.

Chapter 29 **Spreadsheet Exercise**

Annual Payment Adjustments for Variable Rate Mortgages

To cope with interest rate fluctuations, lending institutions began to issue variable rate mortgages. Consequently, the interest rates and the payments on these mortgages may change every year.

Manually recomputing the payments on a bank's variable rate mortgages each year would be an extensive task. In this exercise, you will program annual payment adjustments. The chapter 29 template appears as follows:

```
A1: PR [W5]                                                          READY

      A     B          C           D        E       F      G       H          I
1   -------------------------------------------------------------------------
2       ANNUAL PAYMENT ADJUSTMENTS FOR VARIABLE RATE MORTGAGES
3   -------------------------------------------------------------------------
4           LOAN:   $20,000.00                    YEARS:    20
5   -------------------------------------------------------------------------
6                         outstanding    elapsed      remaining
7   year    rate          balance        periods      periods         payments
8   -------------------------------------------------------------------------
9   1986 11.500%     20,000.00           12            240
10  1987 10.500%                         12            228
11  1988  9.000%                         12            216
12  1989  9.375%                         12            204
13  1990  9.750%                         12            192
14  1991                                 12            180
15  1992                                 12            168
16  1993                                 12            156
17
18
19
20
```

The loan is in cell C4, and the term (number of years) is in cell G4. Column A displays the years, column B the rates, column D the number of elapsed periods (12 if the rates change every year), and column F the number of remaining periods (the previous number of periods less 12).

The adjusted payments are the payments of a new annuity after the outstanding balance has been established.

Spreadsheet Exercise Steps

1. Column C: Outstanding Balance
   The original outstanding balance in C9 is the loan from C4 (C9 contains +C4).
   For the following years, the outstanding balance equals value of loan at compound interest − value of paid-up annuity.

   Outstanding balance = old balance
   $\times (1 + i)^n$ − value of n payments at rate i

   where i is the old periodic rate and n the number of elapsed periods.
   The @FV(payment,i,n) function produces the value of n payments at rate i.

   The old balance is in C9, the previous nominal rate in B9, the number of elapsed periods in D9, and the old payments in H9. The spreadsheet formula begins with the + sign and is:
   + ______ * (_____ + ______ / ______) ^
   ______ − @FV(_______ , ______ / ______ ,
   ______).
   Complete the formula, store it in C10, and copy it from C10 to C11..C13. (Column C will not display correctly until column H is complete.)

2. Column H: Payments
The @ PMT(loan,*i*,*n*) function produces the payments necessary to pay off a loan at rate *i* in *n* periods, where *n* is the number of remaining periods.
The loan is in C9, the nominal rate is in B9, and the remaining periods are in F9.
The spreadsheet formula is
@ PMT(______ , ______ / ______ , ______).
Complete the formula, store it in H9, and copy it from H9 to H10..H13.
Observe that the rate began with 11.5%, decreased by 1%, decreased further by 1.5%, and began creeping up again. Consequently, the payments dropped, declined even more, and then began a slight increase.
3. By how much did the monthly payments increase when the nominal interest rate increased from 9.375% to 9.75%? ________
4. Recompute for a 30-year loan of $60,000.
   a. Change the loan in C4 to 60000 and the term of the loan in G4 to 30.
   b. Enter the following interest rates for 1991–1993. (Enter the rates in decimal form.)
      1991 10.25%
      1992 10.00%
      1993 9.50%
   c. Copy the balance formula from C13 to C14..C16 and the payment formula from H13 to H14..H16.
   d. Record the payments on the following chart.

Year	Rate	Payments
1986	11.5 %	________
1987	10.5 %	________
1988	9.0 %	________
1989	9.375%	________
1990	9.75 %	________
1991	10.25 %	________
1992	10.0 %	________
1993	9.5 %	________

NAME ____________________

CLASS SECTION ____________ DATE ____________

SCORE ________ MAXIMUM 175 PERCENT ________

CHAPTER 29 ASSIGNMENT

1. A loan of $8,000 is to be amortized with 6 equal semiannual payments. Find the size of the payments if the nominal interest rate is 8%. ***(10 points)***

2. The Spinellis purchased a home costing $85,000. They paid $5,000 down and financed the remainder with a 30-year mortgage at 9% interest. What are their monthly payments? ***(10 points)***

3. Compute the payment and show the amortization schedule for a 1-year loan of $10,000 at 9% interest.
Payment: ________

Payment Number	Balance	Interest	Payment	Payment on Principal	New Balance
0					10,000.00
1	10,000.00	____	____	____	____
2	____	____	____	____	____
3	____	____	____	____	____
4	____	____	____	____	____
5	____	____	____	____	____
6	____	____	____	____	____
7	____	____	____	____	____
8	____	____	____	____	____
9	____	____	____	____	____
10	____	____	____	____	____
11	____	____	____	____	____
12	____	____	____	____	0.00

(60 points)

4. What payment must the city of Dorchester deposit quarterly into a sinking fund that earns 8% interest compounded quarterly in order to purchase new fire fighting equipment costing $125,000 two years from now? ***(5 points)***

5. Draw up the sinking fund schedule for problem 4.

Payment Number	Payment	Interest	Accumulated Fund
0			0.00
1	________	0.00	________
2	________	________	________
3	________	________	________
4	________	________	________
5	________	________	________
6	________	________	________
7	________	________	________
8	________	________	________

(20 points)

6. A small airplane costing $100,000 is financed with a $20,000 down payment and a 6-year loan amortized with equal quarterly payments. If the nominal interest rate is 8%, how much is each payment?

(10 points)

7. Linex Printers established a sinking fund to pay off a promissory note that is due in two years. The note bears interest at 6% compounded quarterly, and its face value is $23,750. What are the quarterly sinking fund payments if the fund earns interest at 8%?

(10 points)

8. On March 1, 1991, Sally's Gift Shoppe borrowed $45,000 at 10% interest compounded semiannually. The owner intends to repay the debt in equal installments over 5 years. The first payment is due on September 1, 1991.

a. What will the semiannual payments be?

b. How much interest will have been paid by the end of the term?

(20 points)

9. A $114,750 home was purchased with a 20% down payment and a 25-year 12% mortgage. Compute the amount of the monthly mortgage payments.
Home insurance premiums are estimated at $525 per year.
Real estate taxes are estimated at $3,100 per year.
What are the effective monthly expenses including mortgage payment, insurance, and taxes, rounded to the nearest hundred dollars?

(20 points)

10. Three business partners intend to acquire a fast-food subsandwich franchise. The franchisor requires start-up capital of $25,000. The partners plan to generate the required capital in two years by means of a joint sinking fund earning 8% interest compounded quarterly. What quarterly payment must each of the partners deposit during the two years?

(10 points)

PART

8 SUPPLEMENTARY TOPICS

Chapter 30 Common Statistical Measures and Presentation Graphics

OBJECTIVES

- Use and interpret measures of central tendency: mean, median, and mode
- Design a frequency distribution
- Interpret presentation graphics

SECTIONS

INTRODUCTION

In the course of doing business, companies generate, collect, and store information. In raw form, this data is not very useful. But when raw data is interpreted and transformed into charts, graphs, summaries, or reports, the resulting statistics become an important tool that businesses use to set short- and long-term goals, to analyze staffing and physical plant needs, or to evaluate productivity or sales. Computers make it possible to store large quantities of raw data and to transform that raw data into useful charts, graphs, or reports.

Section 30.1 Common Statistical Measures

KEY TERMS **Mean** **Median** **Ranked data** **Range** **Mode** **Frequency distribution** **Grouped data**

Statistics is the practice of collecting, organizing, analyzing, interpreting, and presenting numerical data.

Summarizing is one way to organize raw data into useful form. This is done by computing totals and percent.

Example 1 An analysis of sales data from eight department store branches follows.

Sales in Rounded Millions

Stores	Quarter 1	Quarter 2	Quarter 3	Quarter 4	Year
Chicago	40	52	41	36	169
New York	49	48	47	48	192
Boston	13	15	16	18	62
Dallas	41	43	37	41	162
Atlanta	41	43	38	38	160
Los Angeles	53	63	51	51	218
San Francisco	34	40	41	47	162
Washington, D.C.	41	41	27	32	141
Total	312	345	298	311	1266

Total annual sales for all stores were $1,266 million.

Sales were greatest during the second quarter and were fairly uniform during the other quarters.

The Los Angeles store sold the most, the Boston store the least.

Some statistical measures of general trends are called measures of central tendency. Mean, median, and mode are measures of central tendency and are covered in this section.*

Mean

The **mean** is the arithmetic average and is the most commonly used measure of central tendency. Refer to the sales summary in the preceding example.

Mean = sum of values ÷ number of values
= total sales ÷ number of quarterly sales observations
= 1266 ÷ 32 = 40

The sales mean of a store for a quarter is 40. This means that the average sales of any store during any quarter is $40 million.

Median and Range

When numerical data is ranked by size, the **median** is the midpoint of that set of **ranked data;** the number of values below the median (midpoint) is the same as the number of values above it. For example, the populations (in thousands) of seven large cities in a particular state are 56, 83, 110, 143, 168, 210, 315. The median population for these cities is 143. This statistic tells you that of the seven cities, three have populations below 143,000 and three have populations above 143,000.

Ranking the department store sales from the preceding example produces the following sequence: 13 15 16 18 27 32 34 36 37 38 38 40 40 41 41 41 41 41 41 41 43 43 47 47 48 48 49 51 51 52 53 63. Because the values in this set are an even number (32), no number falls at midpoint. In this case, the average of the middle two values is the median. The middle two values are both 41; 15 values are below, and 15 values are above. The median is 41. You can conclude from this statistic that half the stores had sales below $41 million and half had sales above $41 million. The **range** is the difference between the highest and the lowest values. In example 1, the lowest quarterly sales were 13, the highest 63. The range is 63 − 13 = 50. The difference in quarterly sales between some stores was considerable.

Mode

The fact that one (or more) value occurs more frequently than any other may be useful in setting a marketing strategy. The **mode** is a value that occurs more often than any other. The mode of the preceding ranked data is 41. (There can be more than one mode.) It is significant that all three measures of the sales summary example—mean, median, and mode—are close to 40; those three measures are not always so close together.

Frequency Distribution

A computer can calculate the mean of thousands of values in very little time. To simplify hand calculations, data may be grouped into classes with the same interval. The interval is the difference between consecutive lower (or upper) limits. The frequency is the number of values in a class. **Frequency distribution** is the organization of classes and frequencies. A close estimate of the mean can usually be obtained from the frequency distribution and the frequency midpoints (where midpoint = average of two consecutive lower limits).

The **grouped data** in the table below illustrates the frequency distribution of quarterly sales from the example. The interval is somewhat arbitrary and is set at 10. The frequency of stores with quarterly sales in the interval of $10 million to $19 million is 4. The interval midpoint is $15 million.

Interval	Frequency (f)	Interval Midpoint (m)	$f \times m$
10–19	4	15	60
20–29	1	25	25
30–39	6	35	210
40–49	16	45	720
50–59	4	55	220
60–69	1	65	65
	32		1,300

$$\text{Mean of grouped data} = \frac{\text{sum of } (f \times m)}{\text{sum of } f} = \frac{1300}{32} = 41$$

*Statistics must be carefully interpreted to prevent incorrect conclusions. For example, if a survey found that the residents of area *A* consume more bottled water than the residents of area *B*, and that more artistically gifted people live in area A than in area B, a headline proclaiming that drinking bottled water improves artistic ability would be misinformation based on false conclusions drawn from statistical data.

PRACTICE ASSIGNMENT

1. Find the mean for each of the following sets of data: (Round to the nearest integer.)

a.	**b.**	**c.**	**d.**	**e.**
34	85	74	243	522
61	118	60	308	379
42	104	60	296	429
56	98	89	336	613
40	120	81	290	324
56	113	74		595
49	104	62		356
	111	77		682
	110	74		
		83		

Mean:

a. ______ **b.** ______ **c.** ______
d. ______ **e.** ______

2. Find the median for each of the sets of data in problem 1. Rank the data first.

a.		**b.**		**c.**	
34	______	85	______	74	______
61	______	118	______	60	______
42	______	104	______	60	______
56	______	98	______	89	______
40	______	120	______	81	______
56	______	113	______	74	______
49	______	104	______	62	______
		111	______	77	______
		110	______	74	______
				83	______

d.		**e.**	
243	______	522	______
308	______	379	______
296	______	429	______
336	______	613	______
290	______	324	______
		595	______
		356	______
		682	______

Median:

a. ______ **b.** ______ **c.** ______
d. ______ **e.** ______

3. Find the mode for each of the sets of data in problem 1. If there is no mode, write *none*.

a.	**b.**	**c.**	**d.**	**e.**
34	85	74	243	522
61	118	60	308	379
42	104	60	296	429
56	98	89	336	613
40	120	81	290	324
56	113	74		595
49	104	62		356
	111	77		682
	110	74		
		83		

Mode:

a. ______ **b.** ______ **c.** ______
d. ______ **e.** ______

4. The following figures show the average temperature (in degrees Fahrenheit) of selected cities for the four seasons:

City	Spring	Summer	Winter	Autumn
Almeeda	45.3	76.5	30.4	48.2
Bankley	51.6	69.7	40.1	51.8
Carlyle	42.0	70.4	29.8	45.3
Darson	55.3	79.7	48.7	56.6
Elwood	46.1	78.9	31.6	45.1

Round all temperatures to the nearest tenth of a degree.

a. What is the mean temperature of Bankley for the year?

b. What is the mean temperature of Elwood for the year?

c. What is the mean summer temperature of the five cities?

d. What is the median winter temperature of the five cities?

e. What is the mean annual temperature of the five cities?

f. What is the median annual temperature of the five cities? (Take the median of the five computed annual city means.)

g. Which city's mean annual temperature is closest to the mean annual temperature computed in step *e?*

5. Metro-Shop employs 30 telemarketers. During one week, the company recorded the following numbers of telephone orders:

Aaron	79	Gitano	75	Meany	74
Baird	97	Harvey	96	Morris	76
Bjornsen	88	Huang	85	Park	108
Butcher	78	Jacobs	106	Roszkowski	91
Carillo	104	Johnson	60	Salazar	90
Dubois	85	Joseph	99	Salley	68
Ehrhardt	90	Kelly	75	Schaffner	79
Fields	109	Kimura	76	Stevenson	83
Fogel	92	Ladd	104	Tomas	84
Garcia	81	McNamara	82	Wright	77

Group the data using intervals of ten. Use the frequency distribution to find the mean of the grouped data.

Interval	**Frequency** (f)	**Interval Midpoint** (m)	$f \times m$
60– 69	_____	_____	_____
70– 79	_____	_____	_____
80– 89	_____	_____	_____
90– 99	_____	_____	_____
100–109	_____	_____	_____
	30		_____

Mean of grouped data = _____

Section **30.2 Presentation Graphics**

KEY TERMS **Circle graph** **Bar graph** **Line graph**

Graphs are used to illustrate numerical relationships. Desktop computer software has greatly facilitated the creation of graphics. With several keystrokes, raw data can be transformed into graphics. Computer graphics can be roughly divided into two major categories:

1. Free-form graphics
2. Presentation graphics (sometimes called business graphics)

Free-form graphics allow computer users to design and produce their own graphs.

Presentation graphics software offers computer users a choice of standard graphic forms that may be constructed with very little effort. Standard presentation graphics include circle graphs (also called pie charts), bar graphs, and line graphs.

Circle Graph

The **circle graph** (pie chart) displays the relationship between a whole and its parts. The whole or base (100 percent) is the circle. The parts that make up the whole are the wedge-shaped sectors of the circle. Each sector represents a percent of the whole. Because a circle contains 360 degrees, the size of the angle in a sector that displays 25 percent of the whole is 90 degrees. Generally, the number of degrees in a sector is part/whole × 360 degrees.

The following circle graph illustrates the sales of the eight stores from the example in section 30.1. The whole circle represents the sales of all stores.

Percent of Sales by Store

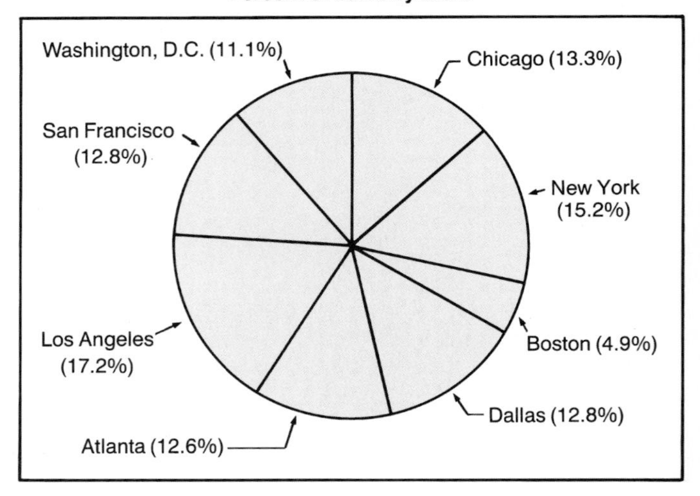

PRACTICE ASSIGNMENT 1

1. A sample of 12,000 people was surveyed for preference of toothpaste brand. The following circle graph illustrates the results of the survey:

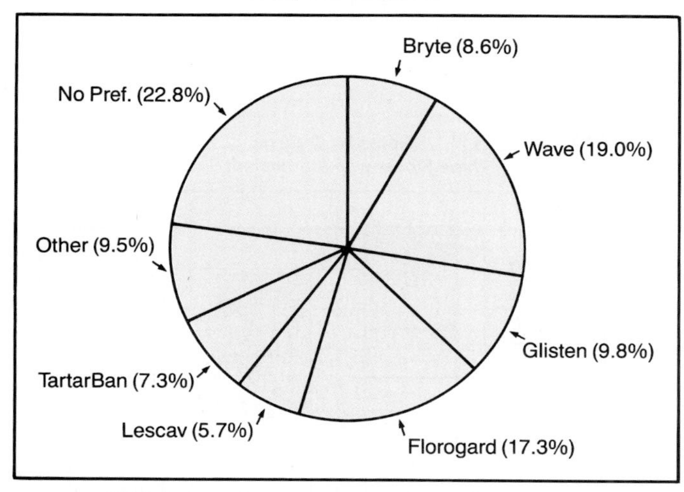

   a. Which brand was most preferred?
   b. How many people in the survey named the most preferred brand?
   c. What percent preferred the next two highest-rated brands?
   d. What percent did not prefer any of the six named brands?
   e. How many people surveyed preferred one of the six named brands?

2. Glenmont, Inc. reported annual revenue of $28,600,000. The following pie chart (circle graph) illustrates how the money was spent.

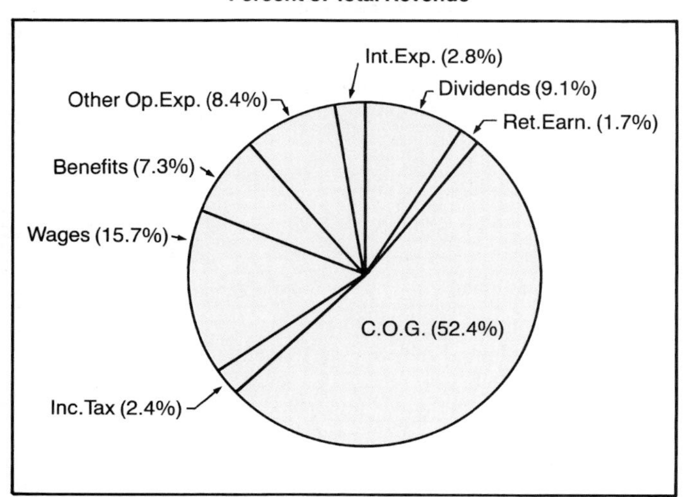

Abbreviations:

Int.Exp.	Interest expense
Op.Exp.	Operating expenses
Inc.Tax	Income taxes
Ret.Earn.	Retained earnings
C.O.G.	Cost of goods sold

   a. What used up the largest portion of revenue after cost of goods sold?
   b. What percent of total revenue was used for employee expenses?
   c. How much did the company pay in income taxes?
   d. What percent of revenue was net profit after taxes?
   e. How much were total operating expenses?

Bar Graph

The **bar graph** illustrates increases or decreases in magnitude. The horizontal axis, called the *x-axis,* represents time or some other variable. The vertical axis, called the *y-axis,* represents the magnitude of what is being measured (money, weight, number of people, and so forth). Each bar (or column) represents a magnitude produced for a particular grouping. The top of each bar indicates the magnitude on the vertical axis. The following bar graph illustrates the quarterly sales of the eight stores from the example in section 30.1.

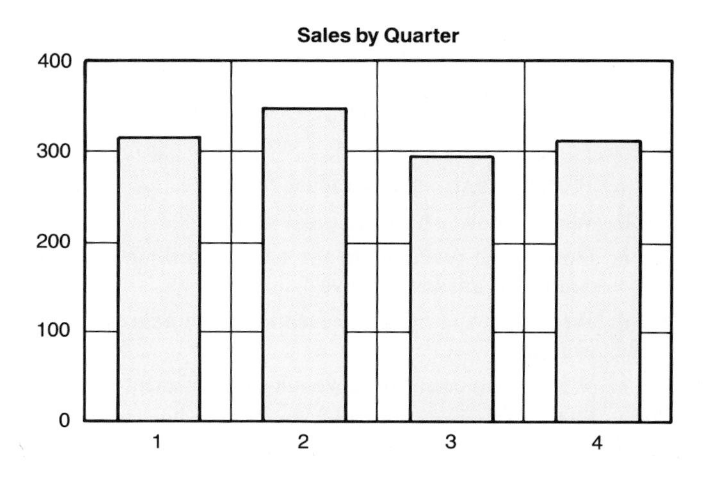

A variation of the simple bar graph is the side-by-side multiple bar graph. Several related bars are shown for each grouping. A side-by-side bar graph might display the number of rejected items among four products manufactured in a plant over a number of years.

The quarterly sales analysis from example 1 in section 30.1 may be illustrated with side-by-side bars. The following side-by-side bar graph illustrates the sales pattern of the three most successful stores, those with highest sales.

Sales by Quarter
Three Stores with the Highest Sales

70 60 50 40 30 20 10 0

1 2 3 4

Los Angeles New York Chicago

The stacked bar graph combines features of the simple and the side-by-side bar graph. Each bar is subdivided into segments. Each segment illustrates the contribution of the part to the total. The height of the complete bar corresponds to the total.

The quarterly sales analysis of the three stores with the highest sales from the example in section 30.1 is illustrated in the following stacked bar graph.

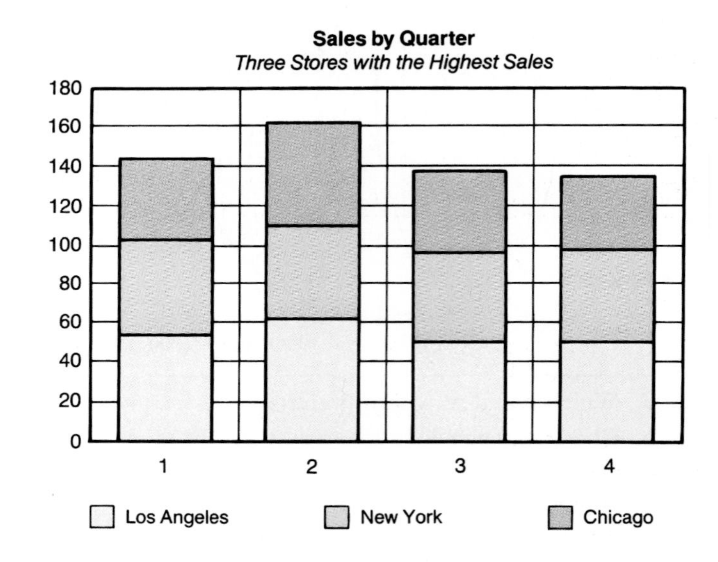

PRACTICE ASSIGNMENT 2

1. The following simple bar graph illustrates the annual revenues received by six adjacent counties.

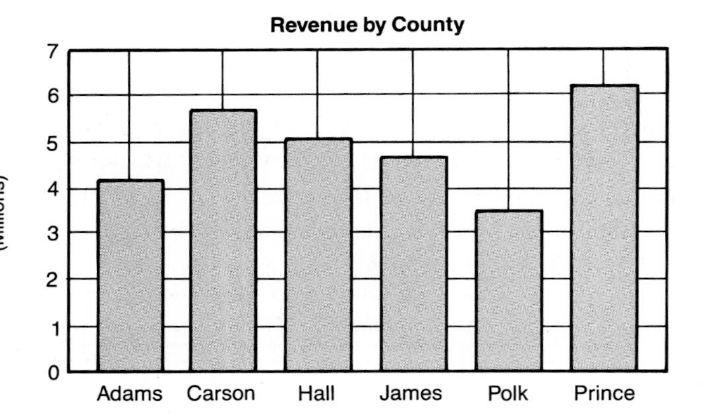

   a. Which county had the least revenue?
   b. Which county had the most revenue?
   c. How much revenue (to the nearest million) did the county in answer *b.* have?
   d. What was the median revenue to the nearest million?
   e. Which two counties' revenues were closest to the median?

2. The following side-by-side bar graph illustrates the price/earnings ratios of three corporations (Denex, Lintel, and Petroco) for a five-year period.

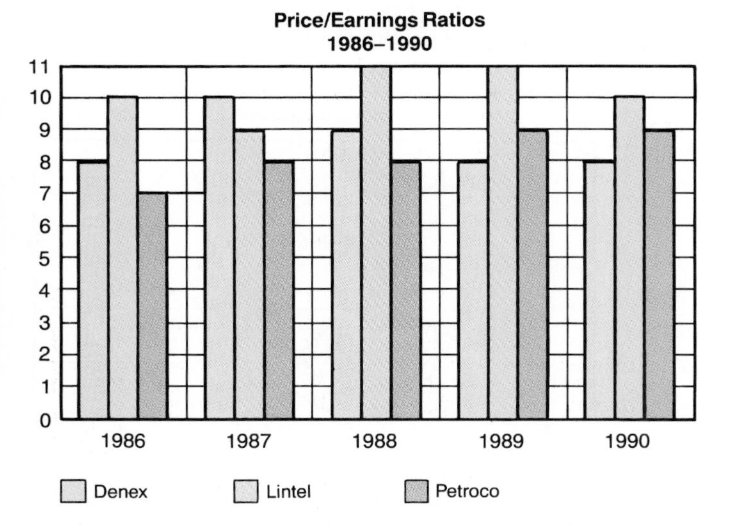

a. Which corporation had the lowest price/earnings ratio in 1987?

b. Which corporation had the lowest price/earnings ratio in 1990?

c. Which corporation had the highest ratio in 1986?

d. What was Lintel's price/earnings ratio in 1988?

e. In which year did Petroco have its lowest ratio?

3. The following stacked bar graph compares the SAT scores of the freshman classes of four colleges. The number of students receiving scores within each of four ranges is illustrated.

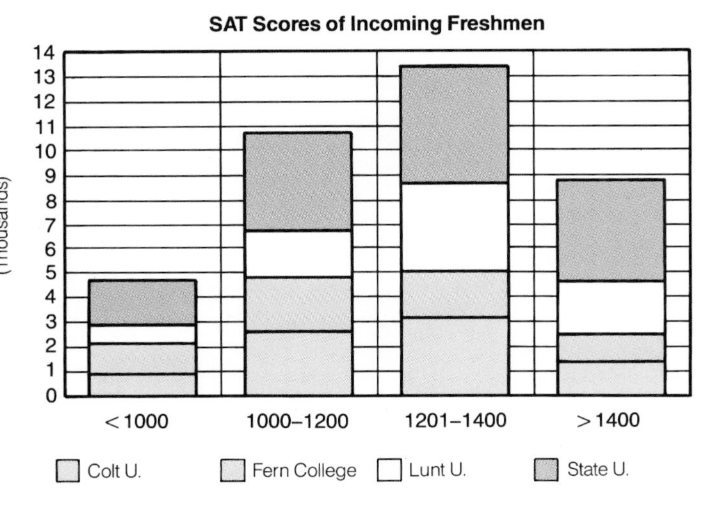

a. Which college admitted the most students in the lowest scoring range?

b. Which college admitted the fewest students in the highest scoring range?

c. How many students (to the nearest thousand) scoring between 1,000 and 1,200 did Fern College admit?

d. How many students (to the nearest thousand) in the two highest scoring ranges did State University admit?

e. Which range had the greatest number of students for all colleges combined?

Line Graph

The **line graph** is another way to illustrate increases or decreases in magnitude. The line graph is a continuous image, unlike the bar graph, which is a discrete (disconnected) image. A line is drawn through points determined by distances from the *x-axis* (horizontal) and *y-axis* (vertical).

The horizontal axis may represent time or another variable. A simple line graph has just one line. A multiple line graph has two or more lines. In chapter 24 you estimated depreciation with the straight-line, sum-of-the-years, and declining-balance methods.

Straight-line depreciations are equal each year; sum-of-the-years depreciations decline steadily; double-declining depreciations start high, decline steeply, and then flatten out. The following multiple line graph illustrates these classical depreciation patterns.

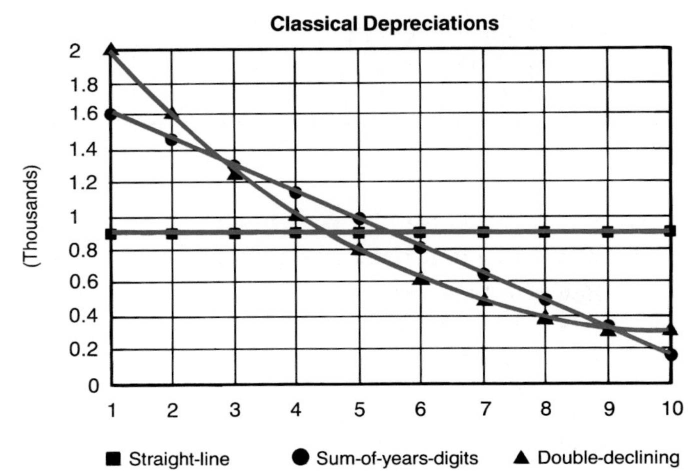

PRACTICE ASSIGNMENT 3

1. The toxicity of varying doses of chemicals (AL-60, BPP, L-CH, and Kappa) was tested on several groups of laboratory mice. The control groups were given food and water only. The survival of all the mice was charted on the following line graph.

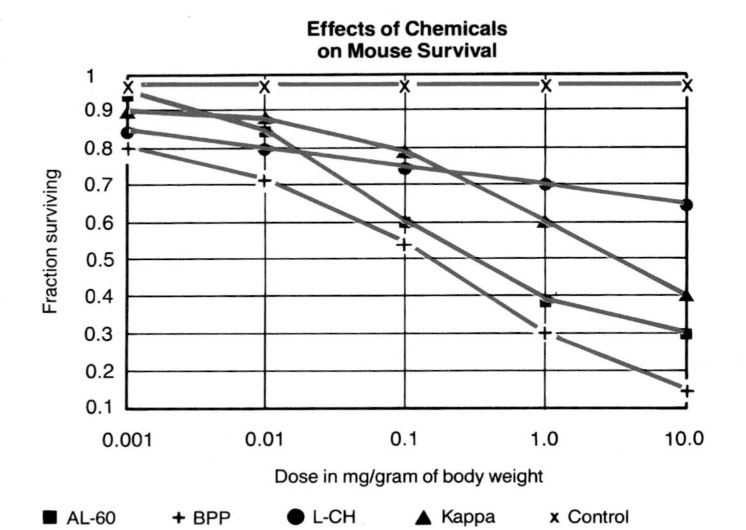

   a. On which chemical did the mice have the poorest survival record?
   b. On which of the four chemicals did the mice show the least difference in survival when dosage was increased?
   c. On which chemical did the mice have the best survival figures at a dose of 0.1?
   d. What percent of mice treated with AL-60 at a dose of 1.0 survived?
   e. Of the Kappa-treated mice, what was the difference in percent survival between the lowest and highest doses?

2. A company charted its monthly advertising, utility, telephone, and travel expenses during the year on the following graph. (The names of months are abbreviated to save space.)

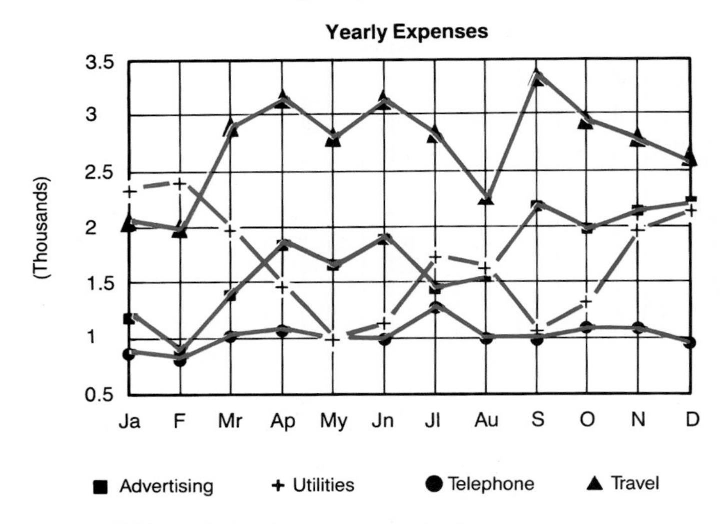

   a. Which of the four expenses is the lowest?
   b. Which month shows the lowest figure for advertising?
   c. Which month shows the highest figure for utilities?
   d. How much was spent for advertising in April (to the nearest five hundred dollars)?
   e. Between which two months did travel expenses rise most steeply?

Chapter 30 **Spreadsheet**

Graphics

Spreadsheet software packages such as Lotus 1-2-3 incorporate graphics capabilities. Retrieve file CHAP30, which contains sales analysis data.

With the spreadsheet on the screen, you can display the pie chart (circle graph), the simple bar graph, and the side-by-side bar graphs shown in this chapter by keying as follows:

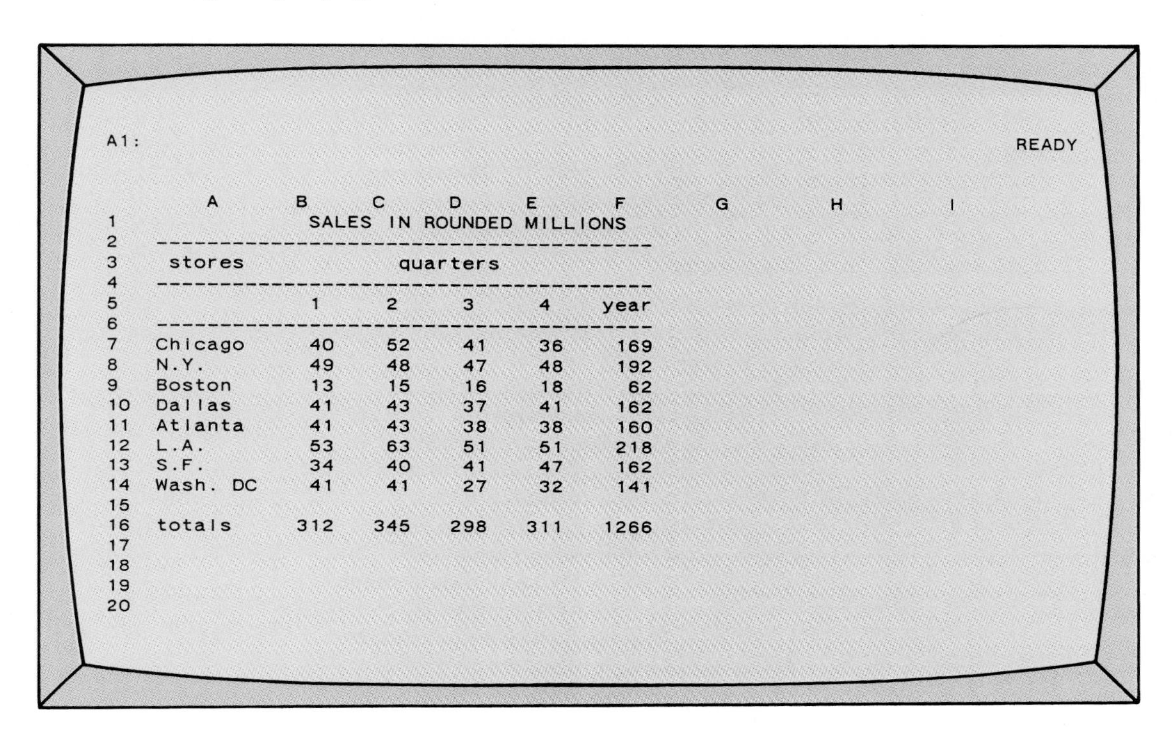

A1: READY

	A	B	C	D	E	F	G	H	I
1		SALES IN ROUNDED MILLIONS							
2	-------	-------	-------	-------	-------	-------			
3	stores		quarters						
4	-------	-------	-------	-------	-------	-------			
5		1	2	3	4	year			
6	-------	-------	-------	-------	-------	-------			
7	Chicago	40	52	41	36	169			
8	N.Y.	49	48	47	48	192			
9	Boston	13	15	16	18	62			
10	Dallas	41	43	37	41	162			
11	Atlanta	41	43	38	38	160			
12	L.A.	53	63	51	51	218			
13	S.F.	34	40	41	47	162			
14	Wash. DC	41	41	27	32	141			
15									
16	totals	312	345	298	311	1266			
17									
18									
19									
20									

Pie Chart (Circle Graph) Commands

/	
Graph[enter]	Graph menu appears
Type[enter]	Type menu appears
Pie[enter]	Graph menu reappears
X[enter]	Inquiry for *x-axis* range
A7..A14[enter]	Graph menu reappears
A[enter]	Inquiry for first data range
F7..F14[enter]	Graph menu reappears
Options[enter]	Option menu appears
Titles[enter]	Title menu appears
First[enter]	Inquiry for first title
Percent of Sales by Store[enter]	Option menu reappears
Quit[enter]	Graph menu reappears
View[enter]	Pie chart appears

When finished viewing:

[enter]	Graph menu reappears
Reset[enter]	Cancel menu appears
Graph[enter]	Graph menu reappears
Quit[enter]	

Simple Bar Graph Commands

/	
Graph[enter]	Graph menu appears
Type[enter]	Type menu appears
Bar[enter]	Graph menu reappears
X[enter]	Inquiry for *x-axis* range
B5..E5[enter]	Graph menu reappears
A[enter]	Inquiry for first data range
B16..E16[enter]	Graph menu reappears
Options[enter]	Option menu appears
Titles[enter]	Title menu appears
First[enter]	Inquiry for first title
Sales by Quarter[enter]	Option menu reappears
Quit[enter]	Graph menu reappears
View[enter]	Simple bar graph appears

When finished viewing:

[enter]	Graph menu reappears
Reset[enter]	Cancel menu appears
Graph[enter]	Graph menu reappears
Quit[enter]	

Side-by-Side Bar Graph

/	
Graph[enter]	Graph menu appears
Type[enter]	Type menu appears
Bar[enter]	Graph menu reappears
X[enter]	Inquiry for *x-axis* range
B5..E5[enter]	Graph menu reappears
A[enter]	Inquiry for first data range
B12..E12[enter]	Graph menu reappears
B[enter]	Inquiry for second data range
B8..E8[enter]	Graph menu reappears
C[enter]	Inquiry for third data range
B7..E7[enter]	Graph menu reappears
Options[enter]	Option menu appears
Titles[enter]	Title menu appears
First[enter]	Inquiry for first title
Sales by Quarter[enter]	Option menu reappears
Titles[enter]	Title menu reappears
Second[enter]	Inquiry for second title
Three Stores with Highest Sales[enter]	Option menu reappears
Legend[enter]	Legend menu appears
A[enter]	Inquiry for first legend
L.A.[enter]	Option menu reappears
Legend[enter]	Legend menu reappears
B[enter]	Inquiry for second legend
N.Y.[enter]	Option menu reappears
Legend[enter]	Legend menu reappears
C[enter]	Inquiry for third legend
Chicago[enter]	Option menu reappears
Quit[enter]	Graph menu reappears
View[enter]	Side-by-side bar graph appears

To display the stacked-bar graph following the side-by-side graph:

[enter]	Graph menu reappears
Type[enter]	Type menu reappears
Stacked-Bar[enter]	Graph menu reappears
View[enter]	Stacked-bar graph appears

When finished viewing:

[enter]	Graph menu reappears
Quit[enter]	

To display the depreciation multiple line graph, retrieve your completed Methods of Depreciation Spreadsheet for Chapter 24. Place the numbers 1, 2, 3, 4, 5, 6, 7, 8, 9, 10 into cells K11..K20.

Keystroke	Result
/	
Graph[enter]	Graph menu appears
Type[enter]	Type menu appears
Line[enter]	Graph menu reappears
X[enter]	Inquiry for *x-axis* range
K11..K20[enter]	Graph menu reappears
A[enter]	Inquiry for first data range
A11..A20[enter]	Graph menu reappears
B[enter]	Inquiry for second data range
D11..D20[enter]	Graph menu reappears
C[enter]	Inquiry for third data range
G11..G20[enter]	Graph menu reappears
Options[enter]	Option menu appears
Titles[enter]	Title menu appears
First[enter]	Inquiry for first title
Depreciation—Three Classical Methods[enter]	Option menu reappears
Legend[enter]	Legend menu appears
A[enter]	Inquiry for first legend
Str.line[enter]	Option menu reappears
Legend[enter]	Legend menu reappears
B[enter]	Inquiry for second legend
Sm-of-Yrs-Digits[enter]	Option menu reappears
Legend[enter]	Legend menu reappears
C[enter]	Inquiry for third legend
Dbl.decl.[enter]	Option menu reappears
Quit[enter]	Graph menu reappears
View[enter]	Multiple line graph appears

When finished viewing:

Keystroke	Result
[enter]	Graph menu reappears
Quit[enter]	

Chapter 31 Elements of Algebra

OBJECTIVES

- Evaluate algebraic expressions with paper and pencil
- Solve linear equations

SECTIONS

INTRODUCTION

As the chapter on employment tests (chapter 32) will demonstrate, a knowledge of simple algebra enables you to solve many business-related problems—time-work, rate-time-distance, interest, and others—that would be troublesome by any other method. Furthermore, familiarity with algebraic techniques enhances your ability to use calculators and computers.

Section 31.1 Evaluating Expressions

KEY TERMS **Operations** **Variables** **Constants** **Operator** **Evaluation** **Substitution** **Parentheses** **Expressions** **Order of operations**

Algebra is a symbolic representation of arithmetic. In algebra, addition, subtraction, multiplication, and division are the basic operations, as in arithmetic. **Operations** are indicated by the use of **variables, constants,** and **operator** symbols.

The eventual **evaluation** step requires **substitution** of numbers for the variables; then the operations are performed as usual.

A variable stands for a quantity that may change; variable h may stand for the number of hours worked.

Indicated Operations with the Variables a and b	
$+ \quad - \quad \times \quad \div$	are operators
Addition	$a + b$
Subtraction	$a - b$
Multiplication	$a \times b$ or ab
Division	$a \div b$ or $\frac{a}{b}$

A constant represents one number and does not change; 1.25 is a constant. There are also symbolic constants; K can stand for a fixed value. Variables and constants combined with operators, sometimes including **parentheses,** make up **expressions.**

Algebraic Expressions

$$3xy$$

$$a + ab$$

$$3(m + n)$$

$$\frac{2}{a} - \frac{b}{3}$$

$$\frac{3}{a + b} + \frac{4}{c + d}$$

An expression is evaluated by substituting values for the variables and performing the operations.

$$\frac{2}{a} + \frac{b}{3}$$

For $a = 5$ and $b = 3$

$$\frac{2}{5} + \frac{3}{3} = \frac{2}{5} + 1$$

$$= 1\frac{2}{5} = 1.4$$

The operations within parentheses are performed first. Then the evaluation of the expression is continued.

$3(m + n)$ is not the same expression as $3m + n$

For $m = 2$ and $n = 4$

$$3(m + n) = 3 \times (2 + 4)$$
$$= 3 \times 6 = 18$$
$$3m + n = 3 \times 2 + 4$$
$$= 6 + 4 = 10$$

Exponentiation and root extraction are extended operations. For example, a^2 (a square) implies that a is multiplied by itself; that is: $a^2 = aa = a \times a$. The **order of operations** for evaluating expressions manually, with the better calculators, and with most computer programming languages is as follows:

1. Parenthesized groupings or groups implied by numerator/denominator
2. Exponentiation or roots
3. Multiplication or division
4. Addition or subtraction
5. Left-to-right

Example 1

	$(a^2 + 4b^2) \times 5$
Substitution: $a = 2$ and $b = 3$	$(2^2 + 4 \times 3^2) \times 5$
Evaluation:	
Inside parentheses: left square	$(4 + 4 \times 3^2) \times 5$
Inside parentheses: right square	$(4 + 4 \times 9) \times 5$
Inside parentheses: multiplication	$(4 + 36) \times 5$
Inside parentheses: addition	40×5
Multiplication	200

Example 2

	$(a + 4b)^2 \times 5$
Substitution: $a = 2$ and $b = 3$	$(2 + 4 \times 3)^2 \times 5$
Evaluation:	
Inside parentheses: multiplication	$(2 + 12)^2 \times 5$
Inside parentheses: addition	$14^2 \times 5$
Square	196×5
Multiplication	980

Example 3

	$\dfrac{2y^2}{4y - 2x}$
Substitution: $x = 2$ and $y = 3$	$\dfrac{2 \times 3^2}{4 \times 3 - 2 \times 2}$
	$= 2 \times 3^2 \div (4 \times 3 - 2 \times 2)$
Evaluation:	
Inside parentheses: left multiplication	$2 \times 3^2 \div (12 - 2 \times 2)$
Inside parentheses: right multiplication	$2 \times 3^2 \div (12 - 4)$
Inside parentheses: subtraction	$2 \times 3^2 \div 8$
Square	$2 \times 9 \div 8$
Multiplication	$18 \div 8$
Division	$18/8 = 9/4 = 2\frac{1}{4}$

The numerator or denominator of a fractional expression may form a group, as if it were parenthesized. When such an expression is changed to division form (often for the purpose of creating a calculator sequence or computer code), care must be taken to insert the implied parentheses.

Fraction Form	Division Form with Parentheses
$\dfrac{a}{b}$	$a \div b$
$\dfrac{a + b}{c}$	$(a + b) \div c$
$\dfrac{a}{b + c}$	$a \div (b + c)$
$\dfrac{a + b}{c + d}$	$(a + b) \div (c + d)$
$\dfrac{ab}{c}$	$a \times b \div c$
$\dfrac{a}{bc}$	$a \div (b \times c)$
$\dfrac{ab}{cd}$	$a \times b \div (c \times d)$

PRACTICE ASSIGNMENT

1. Evaluate the following expressions for $a = 5$ and $b = 2$.

a. $a + 5a - b =$

b. $6a - 2a + 3b =$

c. $ab - 2 + a + 3 =$

d. $ab - 2b + 5 =$

e. $2ab + 3a - 10b =$

f. $5ab - a + 2ab - 3b =$

2. Evaluate the following expressions for $a = 2$, $b = 3$ and for $x = 3$, $y = 4$ in the precise order of operations. Write the fractional expressions in division form with parentheses. Substitute and show each step of the evaluation.

Examples:	$a^2 + 3(b - 1)^2$	$\dfrac{x^2}{4x + 3y}$
Substitution:	$2^2 + 3 \times (3 - 1)^2$	$3^2 \div (4 \times 3 + 3 \times 4)$
Step 1:	$2^2 + 3 \times 2^2$	$3^2 \div (12 + 3 \times 4)$
Step 2:	$4 + 3 \times 2^2$	$3^2 \div (12 + 12)$
Step 3:	$4 + 3 \times 4$	$3^2 \div 24$
Step 4:	$4 + 12$	$9 \div 24$
Step 5:	16	$\frac{9}{24} = \frac{3}{8}$

a. $5a(3b - 2)$
Substitution:
Step 1:
Step 2:
Step 3:
Step 4:

b. $(a + 5)(b + 5)$
Substitution:
Step 1:
Step 2:
Step 3:

c. $b(2b - a)(3a - 4)$
Substitution:
Step 1:
Step 2:
Step 3:
Step 4:
Step 5:
Step 6:

d. $a(b - 1)^2 - b(a - 1)^2$
Substitution:
Step 1:
Step 2:
Step 3:
Step 4:
Step 5:
Step 6:
Step 7:

e. $3(a^2 - 1)^2 + ab$
Substitution:
Step 1:
Step 2:
Step 3:
Step 4:
Step 5:
Step 6:

f. $\dfrac{y}{2x + 6}$
Substitution:
Step 1:
Step 2:
Step 3:

g. $\dfrac{8x - y}{y}$
Substitution:
Step 1:
Step 2:
Step 3:

h. $\dfrac{x + 2y}{2x - y}$
Substitution:
Step 1:
Step 2:
Step 3:
Step 4:
Step 5:

i. $\dfrac{3y^2}{3y - 2x}$
Substitution:
Step 1:
Step 2:
Step 3:
Step 4:
Step 5:
Step 6:

j. $\dfrac{x^2 + 7}{2xy}$
Substitution:
Step 1:
Step 2:
Step 3:
Step 4:
Step 5:

Section 31.2 Solving Equations

KEY TERMS **Equation** **Inverse operations** **Terms** **Like terms** **Factor**

In addition to formula evaluation, you can use algebraic expressions to find an unknown value, as shown in the preceding chapters on proportion (chapter 4), percentage (chapter 7), and interest (chapter 12). You will need to learn a few manipulative skills.

Solving simple algebraic equations is very much like finding the unknowns in the proportions of chapter 4. In this chapter, however, you will use algebraic notation.

Algebraic Notation	
X is the unknown variable (the unknown need not be called X).	$5 + X = 12$
The product of a number and a variable is shown without the multiplication operator.	$3 \times X = 3X$
Division is shown in fraction form.	$X \div 3 = \frac{X}{3}$

Two algebraic expressions that must have the same value form an **equation.** The equations in this chapter are true for one value of the unknown. You must find the value of the unknown which causes the equation to be true.

This equation is true when the value of the unknown X is 2.	$3X + 5 = 11$ $3 \times 2 + 5 = 11$
This equation is true when the unknown X is $\frac{3}{4}$.	$4X + 3 = 6$ $4 \times \frac{3}{4} + 3 = 3 + 3 = 6$

Think of **inverse operations** as opposite operations.

The four arithmetic operators are:

$$- \quad + \quad \div \quad \times$$

The four inverse operators are:

$$+ \quad - \quad \times \quad \div$$

To find the unknown value, apply inverse operations to both sides of the equation.

$$X + 9 = 14$$
$$X + 9 - 9 = 14 - 9$$
$$X = 5$$
$$21X = 84$$
$$\frac{21X}{21} = \frac{84}{21}$$
$$X = 4$$

Equations that contain only combinations of these operations are called linear equations. If an equation involves a multiplication or division and an addition or subtraction, the addition or subtraction is inverted first.

$$3X - 14 = 18$$
$$3X - 14 + 14 = 18 + 14$$
$$3X = 32$$
$$\frac{3X}{3} = \frac{32}{3}$$
$$X = 10\frac{2}{3}$$

To find the solutions of linear equations, you should become familiar with the following techniques:

1. Knowing the sign rule for multiplication and division
2. Reversing the sides
3. Combining like terms
4. Removing parentheses
5. Substituting an expression for a variable

1. The **sign rule for multiplication and division** If the signs of the two variables are alike, the sign of the product or quotient is plus. If the signs of the two variables are not alike, the sign of the product or quotient is minus.

a	b	ab	$\frac{a}{b}$
+	+	+	+
−	−	+	+
+	−	−	−
−	+	−	−

2. **Reversing the sides** If the unknown is on the right-hand side of the equation, it is often easiest to reverse the sides.

$$\begin{aligned} 16 - 9 &= 4X - 7 \\ 4X - 7 &= 16 - 9 \\ 4X - 7 &= 7 \\ 4X &= 14 \\ X &= \frac{14}{4} = \frac{7}{2} \\ &= 3\frac{1}{2} \end{aligned}$$

3. **Combining like terms** The parts of an expression that are separated by + or − are called **terms.** The variables or numbers that make up a product are called **factors.** Terms that contain the same nonnumeric (symbolic) factors are said to be **like terms.** (Purely numeric terms are like terms.) An equation can often be simplified by combining (adding or subtracting) like terms.

Expression: $4xy + 5w - 3w$

$4xy$ $5w$ $3w$ are terms.

Factors of $4xy$: 4 x y $4x$ $4y$ xy
Factors of $5w$: 5 w
Factors of $3w$: 3 w

Like terms: $5w$ $3w$

Combining like terms:
$4xy + 5w - 3w = 4xy + 2w$

Example 1

$$\begin{aligned} 9X - 7 + 3X + 5 &= 8 \\ 9X + 3X - 7 + 5 &= 8 \\ 12X - 2 &= 8 \\ 12X &= 10 \\ X &= \frac{10}{12} = \frac{5}{6} \end{aligned}$$

4. **Removing parentheses** Some solutions require the removal of parentheses. If a **factor** precedes the parenthesized group, multiply it by each term inside the parentheses.

$7(a + b) = 7a + 7b$

$$\begin{aligned} &3a(4b - 3c + d) \\ &= 3a4b - 3a3c + 3ad \\ &= 12ab - 9ac + 3ad \end{aligned}$$

Example 2

$$\begin{aligned} 5(X + 8) &= 3(X + 2) \\ 5X + 40 &= 3X + 6 \\ 2X + 40 &= 6 \\ 2X &= -34 \\ X &= -\frac{34}{2} = -17 \end{aligned}$$

If addition precedes the parenthesized group, remove the parentheses.

$$x + (y + 2a - 3b) = x + y + 2a - 3b$$

If subtraction precedes the parenthesized group, follow these steps:

a. If the first term in the group is positive, insert a preceding plus (y becomes $+y$).

b. As you remove the parentheses, apply the sign rule for multiplication to each term in the group.

$$\begin{aligned} x - (y + 2a - 3b) &= x - (+y + 2a - 3b) \\ &= x - y - 2a + 3b \end{aligned}$$

Example 3

$$\begin{aligned}
5X + (7X - 3) &= 8X - (4X + 5) \\
5X + 7X - 3 &= 8X - 4X - 5 \\
12X - 3 &= 4X - 5 \\
8X - 3 &= -5 \\
8X &= -2 \\
X &= -\frac{2}{8} = -\frac{1}{4}
\end{aligned}$$

5. Substituting an expression for a variable If two expressions are equivalent, one can be substituted for the other without altering the value. You substituted numbers for variables when you evaluated expressions.

$$\frac{2}{a} - \frac{b}{3}$$

for $a = 5$ and $b = 3$,

yields $\frac{2}{5} - \frac{3}{3}$

In chapters 11 through 15 involving interest, the variable T (time) was often replaced by $d/365$ or $d/360$, where d stood for days. If an equation contains two different variables, one of which can be expressed in terms of the other, the equation can be solved by substituting an expression for a variable.

$$PRT = PR\,d/365$$
$$PRT = PR\,d/360$$

$X = 2m$ if $m = n - 4$ yields $X = 2(n - 4)$

Example 4 $2W + 2L = 45$
Solve for W. $L = 3.5W$

$$\begin{aligned}
2W + 2L &= 45 \\
2W + 2(3.5W) &= 45 \\
2W + 7W &= 45 \\
9W &= 45 \\
W &= 5
\end{aligned}$$

Sometimes the variable to be replaced is within another equation and must be isolated before the equivalent expression can be substituted.

Example 5 $8C + 10 = D + 3$
Solve for C. $D - 5 = 2C$

Step 1: Solve the second equation for D.

$$D = 2C + 5$$

Step 2: Substitute the expression for D into the first equation.

$$8C + 10 = 2C + 5 + 3$$

Step 3: Solve for C.

$$\begin{aligned}
8C + 10 &= 2C + 8 \\
6C &= -2 \\
C &= -\frac{2}{6} = -\frac{1}{3}
\end{aligned}$$

PRACTICE ASSIGNMENT

Plan: Solve the equations. Show each step that leads to the answer. Leave the answers in standardized fraction form where necessary.

1. **a.** $X - 3 = 19$
$X =$

b. $3X = 21$
$X =$

c. $8X = 36$
$X =$

d. $\frac{X}{4} = 56$
$X =$

e. $6X + 5 = X + 25$
$X =$

f. $4X + 5 - 2X = 19$
$X =$

g. $\frac{3}{4}X - 6 = \frac{X}{4} + 20$
$X =$

h. $10X + 3 - 4X = 18 - X$
$X =$

2. **a.** $4(X - 1) = 3X$
$X =$

b. $3(X + 4) = X + 15$
$X =$

c. $-6(3X + 5) = 2X$
$X =$

d. $-7(4X - 4) = 3X + 5$
$X =$

e. $5(X - 2) = -3(2X - 6)$
$X =$

f. $4 + (X - 2) = 4X$
$X =$

g. $3 - (2X + 4) = 2X + 15$
$X =$

h. $-6 - (5X + 5) = 3X$
$X =$

i. $2X - 3 + 4(2X - 5) = 12$
$X =$

j. $5 - 2(7 - 2X) + (3X + 4) = 4(4X - 8)$
$X =$

3. Solve for one variable by substituting an expression for the second variable. Perform the outlined steps in detail.

a. $X = 2Y$
Solve for X. $Y = 6$
$X =$

Substitution:
Solution:

b. $2M = N - 3$
Solve for M. $N = M + 5$
$M =$

Substitution:
Solution:

c. $6p - 5 = q + 2$
Solve for p. $q = 2p + 1$
$p =$

Substitution:
Solution:

d. $3w = 5(h + 2)$
Solve for h. $w = 4$

$h =$

Substitution:
Solution:

e. $6(K - 3) = 2L - 5$
Solve for K. $L = 2K - 10$

$K =$

Substitution:
Solution:

f. $2A + 7 = B - 3$
Solve for A. $B + 2 = A - 4$
$A =$

Solve for B:
Substitution:
Solution:

g. $5s - 16 = 3t + 2$
Solve for s. $t + 4 = s - 6$
$s =$

Solve for t:
Substitution:
Solution:

h. $8P - 3 = 4C + 5$
Solve for P. $C - 7 = 3P - 2$
$P =$

Solve for C:
Substitution:
Solution:

i. $3(j - 1) = 4k + 3$
Solve for j. $\frac{k}{2} + 4 = 2j - 3$

$j =$

Solve for k:

Substitution:
Solution:

j. $4(G - 11) = 6N - 8$
Solve for G. $4(N + 6) = 2(G + 1)$
$G =$

Solve for N:

Substitution:
Solution:

Chapter 32 Employment Tests

OBJECTIVES

- Analyze and classify mathematical reasoning problems
- Resolve word problems into equations
- Recognize variations of familiar problems
- Gain confidence in solving mathematical problems
- Improve chances of obtaining employment

INTRODUCTION

When you apply for a job, do not be surprised if your prospective employer requires you to complete a test or a battery of tests. Employment tests may examine language and communication skills, mathematical reasoning, technical skills, even attitudes and personality traits. The results of these tests combined with resumes and interviews help employers to choose the most qualified employees. This chapter covers mathematical reasoning tests.

Many mathematical reasoning tests must be taken without the use of a calculator, in which case the numbers are usually small or results are easily computed. (The use of a desk calculator may be tested as a technical skill.) New tests, however, are being designed even as you study this chapter, and rules may change. When you apply for a job with a company that requires math testing, find out if the use of a calculator is permitted.

Mathematical reasoning tests deal with many problems you are now familiar with: arithmetic, fractions, insurance, interest, percentage, profit and loss, proportions, ratio, and taxes. Problems within each topic may have many variations. You will find some problems that are not usually covered in a business mathematics course as well as some that you have solved only with a calculator.

Solving Interest Problems without a Calculator

1. You can compute some problems mentally, such as: What is the simple interest on $800 at 6% for 2 years?

$$I = PRT = 800 \times 6/100 \times 2 = 48 \times 2 = 96$$

If you cannot compute an interest problem mentally, express the factors I, P, R, and T in fraction form and cancel as much as possible before you complete the paper-and-pencil computation.

2. $1,850 is invested for 50 days at a rate of 5%. What is the interest? (Assume a 365-day year.)

$$I = PRT = 1850 \times 5/100 \times 50/365$$

$$= \frac{\overset{185}{\cancel{1850}}}{1} \times \frac{5}{\underset{10}{\cancel{100}}} \times \frac{\overset{10}{\cancel{50}}}{\underset{73}{\cancel{365}}}$$

$$= \frac{185}{1} \times \frac{5}{\underset{1}{\cancel{10}}} \times \frac{\overset{1}{\cancel{10}}}{73}$$

$$= \frac{185}{1} \times \frac{5}{1} \times \frac{1}{73}$$

$$= \frac{925}{73}$$

925 ÷ 73 must be computed on paper and equals 12.671 rounded to 12.67.

3. What principal yields \$4 in one month if invested at an annual rate of 5%?

$$P = \frac{I}{RT} = \frac{4}{5/100 \times 1/12}$$
$$= \frac{4}{1} \div \left(\frac{\overset{1}{\cancel{5}}}{\underset{20}{\cancel{100}}} \times \frac{1}{12}\right)$$
$$= \frac{4}{1} \div \frac{1}{240}$$
$$= \frac{4}{1} \times \frac{240}{1}$$
$$= 960$$

Symbolic Arithmetic Problems (Also Called Literal Problems)

1. The annual salary of a machinist is K dollars more than that of an assistant. The assistant earns L dollars annually. How much does the machinist earn per month?

	Assistant	Machinist
Annual salary	L	$L + K$
Monthly salary	$\frac{L}{12}$	$\frac{L + K}{12}$

Answer $\frac{L + K}{12}$

2. During 1990, G families acquired insurance policies, representing an increase of R families over the 1988 figure. In 1989, however, the number of families buying insurance policies was N less than in 1988. There were J insurance agents in each of these three years. What was the average number of policies written per agent in 1989?

	1988	1989	1990
Number of policies:	$G - R$	$G - R - N$	G

Average number of policies in 1989

$$= \frac{\text{policies}}{\text{agents}} = \frac{G - R - N}{J}$$

Rate-Time-Distance Problems

These problems are based on the formula

$$\text{Distance} = \text{rate} \times \text{time}$$

You can solve all variations of these problems if you remember this fundamental formula.

1. Two hikers start walking from the same spot at different times. The second hiker, whose speed (rate) is 4 miles per hour, starts 3 hours after the first hiker, whose speed is 3 miles per hour. What is the elapsed time and the distance covered when the second hiker catches up with the first?

During the 3-hour headstart, the first hiker, whose speed is 3 miles/hour, has walked 9 miles. The difference between the rates of the hikers is 1 mile/hour. Every hour the second hiker comes 1 mile closer to the first hiker. Therefore, it will take the second hiker 9 hours to catch up to the first. The distance covered is rate × time = 4 × 9 = 36 miles.

2. The hikers from problem 1 start walking toward each other along a road that connects two towns 50 miles apart. How much time will elapse before the two hikers meet?

$$\text{Distance 1} + \text{distance 2} = 50$$
$$\text{Rate 1} \times \text{hours} + \text{rate 2} \times \text{hours} = 50$$
$$(\text{Distance} = \text{rate} \times \text{time})$$
$$3 \times \text{hours} + 4 \times \text{hours} = 50$$
$$7 \times \text{hours} = 50$$
$$\text{hours} = \tfrac{50}{7} = 7\tfrac{1}{7}$$

During the $7\frac{1}{7}$ hours:

The slower walker walked $3 \times 7\frac{1}{7}$ miles $= 21\frac{3}{7}$ miles

The faster walker walked $4 \times 7\frac{1}{7}$ miles $= 28\frac{4}{7}$ miles

$$21\tfrac{3}{7} + 28\tfrac{4}{7} = 50$$

Time-Work Problems

1. If worker A does a job in 6 days, and worker B does the same job in 3 days, how long will it take both of them working together to finish the job?

A: 1 job in 6 days or $\frac{1}{6}$ job in 1 day
B: 1 job in 3 days or $\frac{1}{3}$ job in 1 day
Both: $(\frac{1}{6} + \frac{1}{3})$ job in 1 day
$\frac{1}{2}$ job in 1 day
1 job in 2 days

2. Two water pipes together fill a pool in $4\frac{1}{2}$ hours. The second pipe (pipe B) alone can fill the pool in 10 hours. How long, in hours and minutes, would it take the first pipe (pipe A) alone to fill the pool?

A and B: 1 pool in $4\frac{1}{2}$ hours or $\frac{2}{9}$ of the pool in 1 hour
B: 1 pool in 10 hours or $\frac{1}{10}$ of the pool in 1 hour
A: $(\frac{2}{9} - \frac{1}{10})$ of the pool in 1 hour
$\frac{11}{90}$ of the pool in 1 hour
1 pool in $\frac{90}{11}$ of an hour
$= 8\frac{2}{11}$ hours
= approximately 8 hours 11 minutes

Dilution Problem

A marketer of fruit drinks stocks 32 gallons of pure orange juice. Pure juice sells for \$1.50 per gallon. How much water must be added to reduce the unit price to \$1.20 per gallon? What is the concentration of orange juice in this drink?

Assume that the cost of water is negligible and the gross receipts remain the same. The 32 gallons of undiluted orange juice sell for $32 \times 1.50 = 48.00$. To receive \$48 at \$1.20 per gallon, the pure orange juice must be diluted to $48 \div 1.2 = 40$ gallons. Thus, 8 gallons of water must be added to sell the drink at the unit price of \$1.20 per gallon.

The concentration of orange juice in this drink is

$$\frac{32}{40} = \frac{4}{5} = 80\%$$

Simple algebra enables you to solve word problems that at first glance may look quite puzzling. You can resolve many word problems into simple equations. With practice, you can become skilled at transforming essential information into equation form. The next four problems illustrate the technique.

Age Puzzles

1. John is 6 years older than his sister Ann. In 8 years their combined ages will be 52 years. How old are they now?

Suppose Ann's age is X. The first sentence states that John's age is $X + 6$. The second sentence states that in 8 years, when Ann's age will be $X + 8$ and John's age will be $X + 14$,

$$X + 8 + X + 14 = 52.$$

$$\begin{aligned} X + 8 + X + 14 &= 52 \\ 2X + 22 &= 52 \\ 2X &= 30 \\ X &= 15 \end{aligned}$$

Check:
Ann is 15. John is 21.
In 8 years:
Ann will be 23 and John will be 29.
$23 + 29 = 52$

2. Sidney is three times as old as Ellen. Ten years from now, Sidney will be twice as old as Ellen will be then. How old are they now? Let X be Ellen's age. Sidney's age is $3X$.

$$3X + 10 = 2(X + 10)$$
$$3X + 10 = 2X + 20$$
$$X = 10$$

Ellen is 10 and Sidney is 30.

Number Puzzle

The larger of two numbers is one less than three times the smaller. The difference between the two numbers is 9. Find the two numbers.

Let X be the smaller number. The larger number is $3X - 1$.

$$3X - 1 - X = 9$$

$$\begin{aligned} 3X - 1 - X &= 9 \\ 2X - 1 &= 9 \\ 2X &= 10 \\ X &= 5 \end{aligned}$$

Check:
The smaller number is 5.
The larger number is $3 \times 5 - 1 = 14$.
$14 - 5 = 9$

Geometry Problem

An employment test may include simple geometry problems.

A rectangle is $3\frac{1}{2}$ times as long as it is wide. The perimeter is 45 feet. Find the length and the width of the rectangle. The perimeter is the total path around the figure.

L = 3.5 × W

W

$$\begin{aligned} 2W + 2(3.5 \times W) &= 45 \\ 2W + 7W &= 45 \\ 9W &= 45 \\ W &= 5 \\ L = 3.5 \times 5 &= 17.5 \end{aligned}$$

Check: $2 \times 5 + 2 \times 17.5 = 45$

Strategies for Taking Mathematical Reasoning Tests

When faced with an actual mathematical reasoning test, you will be under time pressure and you may be nervous. The strategies that follow are designed to help you take these tests successfully.

Take a little time to look over the problems. Classify them as:

Class 1. Obvious. You have solved this type of problem many times before.

Class 2. Requires some thinking. This is a variation of a problem you have solved before.

Class 3. Difficult to do at this time under pressure. This is a type of problem you have never solved before.

First, solve all class 1 problems quickly but carefully. Then, take a little more time to solve class 2 problems. Finally, try some class 3 problems if there is time left. Even if you cannot do any class 3 problems, you may have already achieved a good score. The greater the variety of problems you solve during test preparation, the smaller the chance you will encounter class 3 problems on a test.

Some employment tests consist entirely of multiple-choice questions. Do not guess the answers; work out each problem. To discourage guessing, most of the sample test questions in this chapter are not multiple-choice.

The following pages contain two mathematical reasoning tests. Hints for each problem in the first test appear at the end of that test. Complete the first test, using the hints if necessary. Then go on to the second test, for which there are no hints. Books containing additional study material for taking employment tests are available at many of the larger bookstores. Check your school's bookstore. (If you do not find this material, speak to the store manager.)

MATHEMATICAL REASONING TEST 1

1. What is the sum of $6\frac{2}{3}$, $3\frac{1}{2}$, $5\frac{3}{4}$ and $8\frac{7}{8}$?

2. 18.6 divided by 0.31 is: ______

3. 10% of $345,000 is: __________

4. A vending machine contains $8 in quarters, $4 in dimes, and $4 in nickels. The machine contains: _____

a. Twice as many quarters as dimes

b. Five times as many nickels as quarters

c. As many dimes as nickels

d. Twice as many nickels as dimes

5. A department's sales increased by 25%. This means that the number of sales: _____

a. Increased by 25

b. Increased by $\frac{1}{4}$

c. Increased by $1\frac{1}{4}$

d. Multiplied by 4

6. Five bricklayers can complete a project in eight hours. If only four of them show up for work, how many hours will it take them to finish the project?

7. A printer is listed at $1,000. A customer can purchase it at discounts of 20% and 10%. If the seller's profit is 25% based on selling price, and operating expenses are 15% of the selling price, how much did the printer cost the seller?

8. If a family earns $2,350 per month and uses 22.5% for housing expenses, how much does the family spend each month for housing?

9. A motor home cost $29,000. After one year its value was depreciated by 40%. The next year the reduced value was depreciated by 20%. The third year the second reduced value was depreciated by 15%. How much was the motor home worth at the end of the third year?

10. An individual borrows $1,500 at 8% interest. How much interest is due after 90 days? (Consider each year as 360 days.)

11. A student borrowed $2,500 at 9% interest on September 1, 1990, and paid the loan with interest on November 15, 1991. What was the total amount paid? (Consider each year as 360 days.)

12. If a property is assessed at $85,000 and the tax rate is $3.15 per $1,000, how much tax must be paid on the property?

13. A city has experienced C plant closings during a ten-year period, causing E employees to be laid off. What is the annual rate of employees laid off due to plant closings?

14. A car trip of 800 miles is planned. If the average speed is 40 miles per hour for the first two hours, 55 miles per hour for the next six hours, and 45 miles per hour for the remaining time, how many hours and minutes will the trip take?

15. A city's waste disposal system can process 700 tons of garbage in a 24-hour day. The system is operational 18 hours each day. If the city has 300,000 residents, how many pounds per resident can be handled in an operating day?

16. Two boys can build a tree house in 6 days if they work together. The first boy would take 9 days to do it by himself. How long would it take the second boy to build it by himself?

17. If three salespersons can take care of 27 customers in three hours, how many customers can two salespersons handle in one hour?

18. There are 42,000 spectators at a baseball game. Half of them are children. There are 8,700 married couples. Of the single adults, there are $1\frac{1}{2}$ men for every woman. How many single men are there?

19. A 5-quart supply of vegetable soup stock is worth $1.20 per cup. How many cups of water must be added for the soup to be worth $0.80 per cup?

20. An appliance salesperson sells three different models of a refrigerator. Model 12F is priced at $450, model 15N at $500, and model 18J at $650. If total sales are $19,200 and the number of each model sold is the same, how many refrigerators were sold?

21. A company is planning to buy a fax machine. One dealer offers discounts of 15% and 10%, while the second one offers 20% and 5%. Of the two offers, the discount of the first dealer compared to the discount of the second dealer is: _____

a. Much larger
b. Slightly larger
c. Equal
d. Slightly smaller

22. Two types of cookies were bought for the same amount. Together, the purchase weighs 2 pounds. If one type of cookie costs $4 per pound and the other costs $3.50 per pound, how much do the $4 cookies weigh?

23. How many crates of onions worth $10 each would have to be bought to obtain 200 bags worth $1.50 each?

24. A company spent $134,550 on medical insurance last year. This represents an increase of 15% over what it spent the previous year. How much was spent on medical insurance the previous year?

25. A school received 400 boxes of chalk. Of this total, $\frac{3}{8}$ was distributed to the first grade, $\frac{1}{4}$ to the second grade, and $\frac{1}{5}$ to the third grade. How many boxes were left?

Hints for Mathematical Reasoning Test 1

1. Fraction arithmetic. Add the whole numbers and fractional parts separately.

2. Division.

3. Find the percentage mentally.

4. One way to solve is to find the number of coins of each denomination. But you can save some time by considering that the nickels amount to as much as the dimes.

5. What is the meaning of percent? No computing is required.

6. Simple variation of the time-work problem. The fewer the people who work, the longer the job takes. Assume all workers work at the same rate.

7. Find the selling price by deducting the discounts. Selling price = cost + profit + operating expense.

8. Percentage problem.

9. You could compute this problem like a multiple discount problem. But with paper and pencil it is probably easier to compute serially: Take off 40%, then 20% of what remains, and so forth.

10. Simple, ordinary interest.

11. Amount = principal + simple, ordinary interest. Find the interest for one year. Then, find the interest for Sept 1 to Nov 15 with the fraction method.

12. Real estate tax.
Tax rate per 1000 × thousands of assessed value.

13. Symbolic arithmetic. Read carefully; the problem is deceptively simple.

14. Rate-time-distance problem.
The car travels distance 1 at the first rate.
The car travels distance 2 at the second rate.
How much distance is left to travel?
How long will it take to travel that distance?
The whole trip takes time 1 + time 2 + time 3.

15. Compute the number of tons processed per operating day.
Compute the number of tons processed per resident each day.
Convert to pounds (the short ton is implied).

16. Time-work problem.

17. Extend the method used for problem 6. Fewer salespersons take care of fewer customers. The less time they spend, the fewer customers they take care of.

18. Subtract the number of children from the number of people. Then subtract the number of married people. The number of single adults remains. Next, let W be the number of single women; set up an equation that enables you to solve for that number. Then find the number of single men.

19. Dilution problem. First convert quarts to cups.

20. This might be a class 3 problem (one you have never done before).
It is, however, a very manageable problem because the number of refrigerators sold is the same for each model.
Consider the sets that consist of one of each model.
How many such sets were sold?
How many refrigerators were sold?

21. Multiple trade discounts. Assume that the two list prices are equal.
Compute the single discount equivalents.

22. This may be another class 3 problem. If, out of 25 questions, there are two class 3 problems you cannot do, you can still achieve $\frac{23}{25} = 92\%$. Study *all* problems when preparing.
When taking the test, do not spend time on class 3 problems until you have done all the other problems. This problem is solvable because the amount paid for each type of cookie is the same. (The clue is similar to the clue in problem 20.)

Amount paid on type 1 cookies = amount paid on type 2 cookies
Price of 1 pound = 3.50   Price of 1 pound = 4.00
Therefore:
Weight of type 1 cookies = $\frac{8}{7}$ × weight of type 2 cookies ($4 \div 3.5 = \frac{8}{7}$)

Now let w be the weight of the type 2 cookies; set up an equation that enables you to find that weight.

23. Simple arithmetic reasoning.

24. Find the base in an increase problem.

25. Fraction arithmetic.

MATHEMATICAL REASONING TEST 2

1. If a box of bleach weighs 1 pound 3 ounces, how much do 8 boxes weigh?

2. If the sun rises at 5:06 A.M. and sets at 7:18 P.M., what is the number of hours and minutes between the sun's rising and setting?

3. How much will $750 be worth in 9 months if it earns 8% annual interest?

4. Grade 4 has J students. Grade 5 has $1\frac{1}{2}$ times as many students as Grade 4. How many students are in Grade 5? _____

a. $\frac{3}{2J}$ **b.** $\frac{3J}{2}$ **c.** $\frac{2J}{3}$ **d.** $\frac{2}{3J}$

5. An office supply store reduced its prices for typing paper. Instead of paying $4.38 per box, a customer could purchase two boxes for $7.59. Brenda purchased four boxes of typing paper. How much did she save by buying at the reduced price?

6. If a painter can paint a house in 3 days and his helper can paint it in 5 days, how many days does it take both of them if they work together?

7. If 280 feet of fencing are required to surround a lot whose width is $\frac{3}{4}$ of its length, how many feet wide is the lot?

8. Two cars are traveling toward each other. One car is moving at 50 miles per hour, while the other is moving at 40 miles per hour. If the original distance between them is 900 miles, how many hours will it take them to meet?

9. A train can maintain an average speed of 45 miles per hour. While traveling from one city to another, it stops at 6 towns between the two cities for an average time of $11\frac{1}{2}$ minutes per stop. If the two cities are 585 miles apart, how many hours and minutes will it take for the train to reach its destination?

10. A supermarket has two coffee grinders to provide its customers with freshly ground coffee. Together, the machines can grind a day's supply of coffee in 3 hours. The larger machine alone can grind a day's supply in 5 hours. If the larger grinder is broken, how many hours will the smaller grinder take to grind the coffee?

11. The temperature yesterday was 65 degrees at noon. At noon on the day before it was 60 degrees. By what percent did the temperature rise?

12. Sam is twice as old as Joe. In 5 years Sam will be $1\frac{1}{2}$ times as old as Joe will be then. How old is Sam?

13. If 4 computers are sufficient for 7 employees, how many computers should be available for 84 employees?

14. A cash register contains $11.65 in nickels, dimes, and quarters. If there are 31 quarters and 23 dimes, how many nickels are there?

15. If a car is driven at 50 miles per hour, in how many minutes will it cover 30 miles?

16. A brand P pencil sharpener costs $17.65, a brand Q sharpener $19, and a brand R $20.50. What is the average price of the three brands?

17. Mrs. Jones bought a can of tuna for $1.55, a box of cereal for $2.29, and a bag of apples for $1.27. If she pays for her purchases with a $10 bill, how much change should she receive?

18. A drawer contains 48 loose socks that are either red, blue, or brown. If $\frac{1}{6}$ of them are red and $\frac{5}{8}$ are blue, how many pair of brown socks are there?

19. A retailer paid $2,350 for merchandise after receiving a 10% discount. What would the merchandise have cost without the discount?

20. A dealer paid $2,000 for a used car. After making repairs costing $435 and painting it at a cost of $115, he sold it at a 20% profit based on total cost. What was the selling price?

21. Rivertown has a population of T. Sea City has F less people than Rivertown and M more people than Lakeburg. If the population of Lakeburg increases by $\frac{1}{8}$, how many people live in Lakeburg? ______

a. $T - F + \dfrac{M}{8}$ **b.** $T - F - \dfrac{9M}{8}$

c. $T - \dfrac{9M + F}{8}$ **d.** $\dfrac{9(T - F - M)}{8}$

22. John took the 8 A.M. bus to Glendale, a distance of 300 miles. Charles missed the bus and started driving to Glendale at 10 A.M. If the bus's average speed is 40 miles per hour and Charles can drive 55 miles per hour, how many hours and minutes must Charles drive to overtake the bus?

23. How many months will it take $3,000 to become $3,105 if it is invested at 6%?

24. The sum of two numbers is 31. When the smaller number is multiplied by 3, the product is 8 more than the larger number multiplied by 2. What is the smaller number?

25. Three quarts of grape juice are added to 5 quarts of apple juice. If 2 quarts of this mixture are added to 18 quarts of water, what percent of the new mixture is grape juice?

Chapter 33 Fundamentals of Federal Personal Income Tax

OBJECTIVES

- Compute adjusted gross income and taxable income
- Determine tax credits
- Determine tax refund or additional payment

INTRODUCTION

The United States government, through the Internal Revenue Service (IRS), collects income taxes from businesses and individuals. The federal government uses these tax revenues to run governmental departments and services. Some states also levy an income tax. This chapter is limited to a brief introduction to the federal personal income tax.

KEY TERMS **Adjusted gross income** **Taxable income** **Itemized deductions** **Standard deduction** **Exemptions** **Tax credits**

The federal government requires individuals to pay the bulk of their income tax obligations as the income is earned. An employee uses Form W-4, the Withholding Allowance Certificate (illustrated in chapter 20), to adjust the amount of income tax to be withheld from pay. By the end of the year, the accumulated withholdings should be close to the yearly tax obligation.

By January 31, employers must report to wage earners the amount of gross earnings and the amount of taxes withheld during the past year. These figures are reported on Form W-2. By April 15, wage earners must file income tax returns. (Recently, there have been proposals to extend the deadline beyond April 15.) A tax return may result in a refund or require an additional payment.

1 Control Number		OMB No. 1545-0008	Copy C For employee's records This information is being furnished to the Internal Revenue Service	
2 Employer's Name, Address, and ZIP Code MOREHEAD STATE UNIVERSITY MOREHEAD, KY 40351 69-0610001 UNIT 070			3 Employer's Identification Number 99-6054572	4 Employer's State I.D. Number KY073492
			5 Statutory Employee ☐ Deceased ☐ Pension Plan ☒ Legal Rep. ☐	942 Emp. ☐ Subtotal ☐ Deferred Compensation ☐ Void ☐
			6 Allocated Tips	7 Advance EIC Payment
8 Employee's Social Security Number 999-64-3921	9 Federal Income Tax Withheld 2,142.39		10 Wages, Tips, Other Compensation 27,678.86	11 Social Security Tax Withheld 1,976.97
12 Employee's Name, Address, and ZIP Code ELAINE JENKINS 724 8th Street Morehead, KY 40351			13 Social Security Wages 27,650.12	14 Social Security Tips
			16 401k 1,735.18	16a Fringe Benefits Incl. in Box 10 28.74
			17 State Income Tax 1,164.67 · 18 State Wages, Tips, Etc. 27,650.12	19 Name of State KENTUCKY
			20 Local Income Tax 293.85 · 21 Local Wages, Tips, Etc. 29,385.30	22 Name of Locality MOREHEAD

Form W-2 Wage and Tax Statement 19XX

Department of the Treasury Internal Revenue Service

Source: Department of the Treasury, Internal Revenue Service.

Every year, federal income tax regulations change, and the forms may be modified slightly in response to the changes. Barring drastic revisions of the tax code, however, this chapter will help you understand basic concepts of personal income tax.

Generally, an employee's tax obligation depends on:

1. Gross earnings
2. Marital status
3. Number of withholding allowances

An objective of the Tax Reform Act of 1986 was to simplify tax return preparation for the average wage earner. With the increase in the standard deduction, more wage earners forego itemizing deductions and use the easier tax forms, 1040EZ or 1040A. On all 1040 forms, amounts can be rounded to the nearest dollar. Major advantages of the two easier forms include the time saved in completing the forms, the lesser likelihood of being audited by the Internal Revenue Service (IRS), and the money saved by individuals who formerly hired a tax service to prepare their returns.

Step	Line	Description	Sub-line	Sub-amount	Line no.	Amount
Step 4 Figure your total income	7	Wages, salaries, tips, etc. This should be shown in Box 10 of your W-2 form(s). (Attach Form(s) W-2.)			7	42,629
	8a	**Taxable** interest income (see page 24). (If over $400, also complete and attach Schedule 1, Part II.)			8a	517
Attach check or money order here.	b	**Tax-exempt** interest income (see page 24). (DO NOT include on line 8a.)	8b			
	9	Dividends. (If over $400, also complete and attach Schedule 1, Part III.)			9	420
	10	Unemployment compensation (insurance) from Form(s) 1099-G.			10	
	11	Add lines 7, 8a, 9, and 10. Enter the total. This is your **total income**.			▶ 11	43,566
Step 5 Figure your adjusted gross income	12a	Your IRA deduction from applicable worksheet. Rules for IRAs begin on page 25.	12a	800		
	b	Spouse's IRA deduction from applicable worksheet. Rules for IRAs begin on page 25.	12b	800		
	c	Add lines 12a and 12b. Enter the total. These are your **total adjustments.**			12c	1,600
	13	Subtract line 12c from line 11. Enter the result. This is your **adjusted gross income**. (If this line is less than $19,340 and a child lived with you, see "Earned Income Credit" (line 25b) on page 37 of instructions.)			▶ 13	41,966

Source: Department of the Treasury, Internal Revenue Service.

Conditions for Using Form 1040EZ	
Type of income	Wages, salaries, tips, taxable scholarships and fellowships, taxable interest of $400 or less
Amount of taxable income	Less than $50,000
Marital status	Single
Dependents	None
Deductions	Standard
Other exemptions	None
Income adjustments, tax credits	None

Conditions for Using Form 1040A	
Type of income	Wages, salaries, tips, taxable scholarships and fellowships, unemployment compensation, interest and dividends
Amount of taxable income	Less than $50,000
Marital status	Single, married, head of household, qualifying widow(er) with dependent child(ren)
Dependents	Children or other dependents
Deductions	Standard
Other exemptions	Self and/or spouse blind, over 65
Income adjustments, tax credits	Individual retirement accounts (IRAs), child and dependent care expenses, earned income credit

Adjusted Gross Income

Compute total income by adding salaries or wages (as recorded on W-2 forms), tips, business income, interest income, dividends, lottery prizes, and so forth. From total income, certain adjustments can be deducted. Adjustments include contributions to an individual retirement arrangement (IRA), penalties for early withdrawal of savings, and some others.

Example 1 Edward and Francine Mason are filing a joint tax return. Their combined salaries totaled $42,629. They also received taxable interest of $517 and dividends of $420. Each can deduct $800 for IRA contributions. What is their **adjusted gross income?**

Total income = 42629 + 517 + 420 = 43566
Total adjustments = 2 × 800 = 1600
Adjusted gross income = 41966 = $41,966

Taxable Income

Compute **taxable income** by subtracting from adjusted gross income, **itemized deductions,** including excessive medical expenses, state and local taxes, home mortgage interest, charitable contributions, and so forth, or the **standard deduction** for taxpayers who do not itemize. **Exemptions** for taxpayers and dependents are multiplied by a factor that changes each year with the rate of inflation. Total deductions and exemptions subtracted from adjusted gross income result in taxable income. After you determine your taxable income, you can find your income tax obligation from the tax table or a tax rate schedule. These tables and schedules also change each year.

Example 2 Refer to the Masons in example 1. Because their itemized deductions exceed the standard deduction, they prefer to itemize. (They must use Form 1040.) The Masons are claiming deductions of $641 for medical expenses, $1,779 for state and local taxes, $4,927 for mortgage interest, and $560 for charitable contributions. They can claim exemptions for themselves and each of their two children. The exemption factor is $2,200. What is the Masons' taxable income?

Itemized deductions = 641 + 1779 + 4927 + 560 = 7907

41966 − 7907 = 34059
Exemptions × factor = 4 × 2200 = 8800
Taxable income = 34059 − 8800 = 25259 = $25,259

PRACTICE ASSIGNMENT 1

Multiply each exemption by 2200.

1. Adrielle Schuyler earned a salary of $19,575. Forms from two banks show taxable interest of $121.34 and $32.70. She has no additional income. Her IRA contribution is $750. What is her adjusted gross income?

2. Jonathan Fitzroy's total income was $23,480. He contributed $900 to an individual retirement arrangement; he claims the standard deduction of $2,900; he has no dependents; and he claims his personal exemption.
   - **a.** What is his adjusted gross income?
   - **b.** What is his taxable income?

3. The Yamashidas, who have one dependent child, have an adjusted gross income of $39,421. If they claim deductions of $1,429 for state and local taxes, $5,017 for mortgage interest, $885 for charitable contributions, and $522 for other miscellaneous deductions, what is their taxable income?

4. Robert Karpov earned $24,150 in wages last year. His wife Sharon earned $23,618. They have two dependent children. Their taxable interest income was $142, but they were penalized $78 for early withdrawal of a certificate of deposit. They claim the standard deduction of $5,800.

   a. What is their adjusted gross income?

   b. What is their taxable income?

5. Melanie Gardner earned a salary of $26,730 last year. Her profits from a part-time business were $3,148. She also received taxable interest of $185, dividends of $62, and a lottery prize of $250. Her IRA contribution was $1,200. She can claim itemized deductions of $395 for medical expenses, $2,443 for home mortgage interest, $792 for state and local taxes, and $45 for other miscellaneous deductions. She can claim her mother as a dependent.

   a. What is her adjusted gross income?

   b. What is her taxable income?

Tax Credits

After you have determined your income tax from the tax table or tax rate schedule, you may be able to reduce your obligation by deducting certain credits. The most commonly applied **tax credits** are the child care credit and the earned income credit.

Example 3 Continue with the Masons of examples 1 and 2.

According to the tax table, the Masons' income tax is $4,316. They can claim a child care credit of $650. What is their income tax obligation?

$$4316 - 650 = 3666$$

By comparing your tax obligation with the amount withheld from your earnings, you can determine whether a refund is due or whether you must remit an additional payment.

Example 4 Refer again to the preceding examples. Edward Mason's W-2 form shows federal withholdings of $2,163 and Francine's shows withholdings of $1,691. What is their refund or additional payment?

Total withholdings = 2163 + 1691 = 3854

Refund = 3854 − 3666 = 188

PRACTICE ASSIGNMENT 2

1. Wendy and Barry Kreutzer's income tax as listed on the tax table is $2,784. They can claim a child care credit of $480. What is their tax obligation?

2. Carla Scott can claim a child care credit of $325 and an earned income credit of $530. According to the tax table, her income tax is $988. What is her tax obligation?

3. James Dubois held two jobs last year. One W-2 form records federal tax withholdings of $931; the other form reports withholdings of $573. His tax obligation is $1,559.

a. Should James remit an additional payment or file for a refund?

b. How much is the payment or refund?

4. Robert and Sylvia Seguso had a total of $3,246 withheld from their salaries last year. The tax table shows their income tax at $3,722. They claim a child care credit of $550.

a. Must they make an additional tax payment or will they receive a refund?

b. How much is the payment or refund?

5. Kevin and Toni Wells are filing a joint return. Kevin's yearly withholdings were $831, and Toni's were $859. Their income tax according to the tax table is $1,807. They can claim a child care credit of $685 and an earned income credit of $765.

a. Must they pay additional tax or can they file for a refund?

b. How much is the payment or refund?

Appendix A Equivalence Table

1 *Common Fraction*	*2* *Decimal Fraction*	*3* *Terminating or Infinite*	*4* *Rounded to Four Places*	*5* *Unrounded 8-digit Calculator Display*	*6* *Percent Form*
$\frac{1}{2}$	0.5	T	0.5000	0.5	50%
$\frac{1}{3}$	0.3 . . .	I	0.3333	0.3333333	$33\frac{1}{3}\%$
$\frac{2}{3}$	0.6 . . .	I	0.6667	0.6666666	$66\frac{2}{3}\%$
$\frac{1}{4}$	0.25	T	0.2500	0.25	25%
$\frac{2}{4} = \frac{1}{2}$					
$\frac{3}{4}$	0.75	T	0.7500	0.75	75%
$\frac{1}{5}$	0.2	T	0.2000	0.2	20%
$\frac{2}{5}$	0.4	T	0.4000	0.4	40%
$\frac{3}{5}$	0.6	T	0.6000	0.6	60%
$\frac{4}{5}$	0.8	T	0.8000	0.8	80%
$\frac{1}{6}$	0.16 . . .	I	0.1667	0.1666666	$16\frac{2}{3}\%$
$\frac{2}{6} = \frac{1}{3}$					
$\frac{3}{6} = \frac{1}{2}$					
$\frac{4}{6} = \frac{2}{3}$					
$\frac{5}{6}$	0.83 . . .	I	0.8333	0.8333333	$83\frac{1}{3}\%$
$\frac{1}{7}$	0.142857 . . .	I	0.1429	0.1428571	$14\frac{2}{7}\%$
$\frac{2}{7}$	0.285714 . . .	I	0.2857	0.2857142	$28\frac{4}{7}\%$
$\frac{3}{7}$	0.428571 . . .	I	0.4286	0.4285714	$42\frac{6}{7}\%$
$\frac{4}{7}$	0.571428 . . .	I	0.5714	0.5714285	$57\frac{1}{7}\%$
$\frac{5}{7}$	0.714285 . . .	I	0.7143	0.7142857	$71\frac{3}{7}\%$
$\frac{6}{7}$	0.857142 . . .	I	0.8571	0.8571428	$85\frac{5}{7}\%$
$\frac{1}{8}$	0.125	T	0.1250	0.125	$12\frac{1}{2}\%$
$\frac{2}{8} = \frac{1}{4}$					
$\frac{3}{8}$	0.375	T	0.3750	0.375	$37\frac{1}{2}\%$
$\frac{4}{8} = \frac{1}{2}$					
$\frac{5}{8}$	0.625	T	0.6250	0.625	$62\frac{1}{2}\%$
$\frac{6}{8} = \frac{3}{4}$					
$\frac{7}{8}$	0.875	T	0.8750	0.875	$87\frac{1}{2}\%$
$\frac{1}{9}$	0.1 . . .	I	0.1111	0.1111111	$11\frac{1}{9}\%$
$\frac{2}{9}$	0.2 . . .	I	0.2222	0.2222222	$22\frac{2}{9}\%$
$\frac{3}{9} = \frac{1}{3}$					
$\frac{4}{9}$	0.4 . . .	I	0.4444	0.4444444	$44\frac{4}{9}\%$
$\frac{5}{9}$	0.5 . . .	I	0.5556	0.5555555	$55\frac{5}{9}\%$
$\frac{6}{9} = \frac{2}{3}$					
$\frac{7}{9}$	0.7 . . .	I	0.7778	0.7777777	$77\frac{7}{9}\%$
$\frac{8}{9}$	0.8 . . .	I	0.8889	0.8888888	$88\frac{8}{9}\%$

Column 2

Dots (. . .) denote that one or more digits repeat infinitely.

For example:

$\frac{1}{3} = 0.3\ .\ .\ .$
$= 0.333333\ .\ .\ .$

$\frac{1}{6} = 0.16\ .\ .\ .$
$= 0.1666666\ .\ .\ .$

$\frac{1}{7} = 0.142857\ .\ .\ .$
$= 0.142857\ 142857\ 142857\ 142857\ .\ .\ .$

Column 3

T means terminating decimal.

I means infinite decimal.

Every fraction has a decimal representation that either terminates or theoretically repeats infinitely. However, any calculator or computer stores only a finite number of digits.

Column 4

The four-place representations were often used for paper-and-pencil computations. They are not always sufficiently accurate. Try to obtain as many digits as possible when using a calculator. If you multiply by $\frac{2}{3}$, for example, do not compute with 0.6667 or 0.6666 but let the calculator utilize as many digits as it can hold.

For example, $\frac{2}{3} \times 600$ should equal 400, but with a calculator:

.6667 [×] 600 [=] 400.02
.6666 [×] 600 [=] 399.96

However:

2 [÷] 3 [×] 600 [=] 400

or

2 [÷] 3 [×] 600 [=] 399.99996 rounded to 400.00 if representing dollars and cents.

Appendix B Chapter Test Progress Chart

Part	Chapter	Subject	Page	Date Completed	% Score
1 Fundamentals of Computing	1	Paper-and-pencil computing	______	______	______
	2	Fractions: addition and subtraction	______	______	______
	3	Fractions: multiplication and division	______	______	______
	4	Ratio and proportion, equations	______	______	______
	5	Calculators and spreadsheets	______	______	______
	6	Units of measure, metrics	______	______	______
2 Percentage	7	Percentage	______	______	______
	8	Trade discounts	______	______	______
	9	Cash discounts, invoice	______	______	______
	10	Markup based on cost or sales	______	______	______
3 Interest	11	Simple interest	______	______	______
	12	Finding the interest variables	______	______	______
	13	Promissory notes, present value	______	______	______
	14	Bank discount	______	______	______
	15	Installment plans, APR	______	______	______
4 Consumer Applications	16	Revolving credit, credit cards	______	______	______
	17	Checking accounts	______	______	______
	18	Stocks and bonds	______	______	______
	19	Life and auto insurance	______	______	______
5 Payroll & Business Expenses	20	Payroll, gross pay and net pay	______	______	______
	21	Sales and property taxes	______	______	______
	22	Property insurance	______	______	______
6 Accounting	23	Inventory methods	______	______	______
	24	Depreciation methods	______	______	______
	25	Financial statements and ratios	______	______	______
7 Mathematics of Finance	26	Compound interest	______	______	______
	27	Savings accounts	______	______	______
	28	Annuities	______	______	______
	29	Mortgages, sinking funds	______	______	______

References

Computer Graphics and Spreadsheets

Desautels, Edouard J. *Understanding and Using Computers.* Wm. C. Brown Publishers, Dubuque, Iowa, 1989.

Johnston, Randolph. *Microcomputers: Concepts and Applications.* Mitchell Publishing Inc., Santa Cruz, Calif., 1987.

Napier, Albert, and Judd, Phillip J. *Mastering and Using Lotus 1-2-3.* Release 3. Boyd & Fraser Publishing Company, Boston, Mass., 1990.

O'Leary, Timothy J. *The Student Edition of Lotus 1-2-3.* The Benjamin-Cummings Publishing Company, Menlo Park, Calif., 1987.

Shelly, G. B., Cashman, T. J., and Waggoner, G. A. *Computer Concepts with Microcomputer Applications, Lotus Version.* Boyd & Fraser Publishing Company, Boston, Mass., 1990.

Credit, Insurance, and Personal Finance

Board of Governors of the Federal Reserve System. *Consumer Handbook to Credit Protection Laws.* Washington, D.C., 1986.

Internal Revenue Service. Various publications. Washington, D.C.

Miller, Roger Leroy. *Personal Finance Today.* West Publishing Company, St. Paul, Minn., 1979.

Rejda, George E. *Principles of Insurance.* Scott, Foresman and Company, Glenview, Ill., 1982.

Wolf, Harold A. *Managing your Money.* Allyn and Bacon Inc., Boston, Mass., 1977.

Financial Mathematics

Cissell, A., Cissell, H., and Flahspoler, D. C. *Mathematics of Finance.* 7th ed. Houghton Mifflin Company, Boston, Mass., 1986.

Greynolds, Elbert B., Aronofsky, Julius S., and Frame, Robert J. *Financial Analysis Using Calculators.* McGraw-Hill Publishing Company, New York, N.Y., 1980.

Pierce, R. C., Jr., and Tebeaux, W. J. *Operational Mathematics for Business.* 2d ed. Wadsworth Publishing Company, Belmont, Calif., 1983.

Roueche, Nelda W. *Business Mathematics.* Prentice-Hall Inc., Englewood Cliffs, N.J., 1969.

Simpson, T. M., Pirenian, Z. M., and Crenshaw, P. H. *Commercial Algebra.* 3d ed. Prentice Hall Inc., New York, N.Y., 1950.

Williams, Walter E., and Reed, James H. *Fundamentals of Business Mathematics.* 2d ed. Wm. C. Brown Publishers, Dubuque, Iowa, 1981.

Index